OIL

Also by Tom Bower

OIL

MONEY, POLITICS, AND POWER IN THE 21ST CENTURY

Tom Bower

GRAND CENTRAL
PUBLISHING

NEW YORK   BOSTON

Grand Central Publishing Edition

Copyright © 2009 by Tom Bower
All rights reserved. Except as permitted under the U.S. Copyright Act of 1976, no part of this
publication may be reproduced, distributed, or transmitted in any form or by any means, or stored
in a database or retrieval system, without the prior written permission of the publisher.

Grand Central Publishing
Hachette Book Group
237 Park Avenue
New York, NY 10017
www.HachetteBookGroup.com

Printed in the United States of America

First Grand Central Publishing Edition: June 2010
10 9 8 7 6 5 4 3 2 1

Grand Central Publishing is a division of Hachette Book Group, Inc.
The Grand Central Publishing name and logo is a trademark of Hachette Book Group, Inc.

The Library of Congress Control Number: 2009936536
ISBN: 978-0-446-54798-7

To George and Sylvia Bower

Contents

◄◊►

Illustrations

———◄○►———

John Browne, Larry Fuller and Peter Sutherland. *(Press Association Images)*

The damaged Thunder Horse platform in the Gulf of Mexico. *(AP/Press Association Images)*

Browne inspects a leaking pipeline in Alaska in 2006. *(AP/Press Association Images)*

Attacks by armed gangs in Nigeria's Delta region hampered production in the area. *(Sipa Press/Rex Features)*

Oil pollution in Nigeria. *(AP/Press Association Images)*

Ken Saro-Wiwa. *(© Tim Lambon/Greenpeace)*

Marc Rich. *(Photograph by Jim Berry, Camera Press London)*

John Deuss. *(Photo by David Skinner)*

President Heydar Aliyev of Azerbaijan with the Queen at Buckingham Palace. *(Press Association Images)*

President Boris Yeltsin with leading Russian bankers. *(AP/Press Association Images)*

Mikhail Khodorkovsky and Mikhail Fridman. *(Tass/Press Association Images)*

Fridman and Vladimir Potanin. *(AP/Press Association Images)*

Lee Raymond and Lou Noto. *(AP/Press Association Images)*

Lee Raymond. *(AP/Press Association Images)*

Rex Tillerson. *(AP/Press Association Images)*

Russia's President Vladimir Putin meets leading US oilmen in Moscow. *(AP/Press Association Images)*

Raymond and Khodorkovsky at the World Economic Forum in Moscow in October 2003. *(Tass/Press Association Images)*

Nymex traders. *(AP/Press Association Images)*

Andy Hall. *(Nicholas Richer/PatrickMcMullan.com)*

Tom O'Malley. *(Roberto Arcari/contrasto/eyevine)*

Putin and Phil Watts. *(AFP/Getty Images)*

Walter van de Vijver. *(epa/Robert Vos)*

Oleg Mitvol. *(AP/Press Association Images)*

Jeroen van der Veer. *(Reuters/Itar Tass)*

Oil pollution in Russia. *(© Igor Gavrilov/Greenpeace)*

US Vice President Al Gore with President Nursultan Nazarbayev of Kazakhstan. *(Reuters/Mark Wilson)*

Venezuela's President Hugo Chávez and German Khan. *(AP/Press Association Images)*

Fridman and Putin. *(Yuri Kadobnov/AFP/Getty Images)*

Rodney Chase. *(Press Association Images)*

Igor Sechin. *(AP/Press Association Images)*

Alexei Miller and Tony Hayward. *(Reuters/STR New)*

Saudi Arabia's oil minister Ali al-Naimi. *(LANDOV/Press Association Images)*

Preface

—◄◦►—

DESPITE THE CROWDS of journalists and TV cameras jostling in the small entrance hall of OPEC's Vienna headquarters, the atmosphere was mellow. After the frenzy of oil prices soaring and then crashing during 2008, the arrival of Ali al-Naimi, the dapper Saudi oil minister, seemed undramatic. The small man was smiling after his 20-minute walk from the Grand Hotel, but his serenity was deceptive. Known as "Mr. OPEC," or the leader of the Organization of Petroleum Exporting Countries, al-Naimi had uncharacteristically sought publicity before the 11 members of the price-fixing cartel began their 153rd meeting that morning.

"We think prices will rise," he had puffed during a 6 a.m. run along the Baroque Ringstrasse the previous day. As usual, he was seeking to influence the markets in New York and London. To his satisfaction, over the next 24 hours speculators had bid up oil prices by $1 to $63 a barrel, a 100 percent increase in eight months. Over the next weeks, al-Naimi hoped, if the speculators could be persuaded, prices would rise by another $20; more truculent OPEC members, he knew, would hope that prices would eventually rise by a further $70 to $150, an all-time high. Billions of dollars had flowed from the Western countries into the oil producers' coffers in previous years, but al-Naimi was determined to defy economic law. Demand for oil had fallen since the crash in July 2008, record amounts of oil were in storage, and the world had plunged into recession. Yet he was talking up prices. Of course he understood that an excessive price hike would endanger the world's economy and annoy Saudi Arabia's allies in Washington but a limited increase would benefit his interests. If he was to succeed, he would have to persuade the other OPEC members, an exclusive but quarrelsome club, to endorse his strategy.

Unlike the exotic pageant of presidents and kings who had attended OPEC's meetings during the 1970s, the 10 ministers and their aides who followed al-Naimi into the shiny office block were colorless placemen. Journalists no longer witnessed dramas like Saddam Hussein embracing his bitter enemy the Shah of Iran in 1975 while the stylish Sheikh Yamani, al-Naimi's predecessor as the Saudi oil minister, hovered in the background. In those days, the dictators posed as brokers of the world's future. Having wrested control of their oil from the cabal of dominant Western companies known as the Seven Sisters, the OPEC countries had freed themselves of the imperialist, and occasionally racist, attitudes that had formerly dictated their fates. The American and British corporations, blamed for their willful blindness about realities in West Africa, Central America and the Arab world, had ceased to be the guardians of the common destiny.

Nevertheless, oil remained the world's biggest business. Every aspect of mankind's lives depended on the refinement of crude oil into energy, plastics, chemicals and drugs. For a century the commodity has been on a roller coaster, swinging from surplus to shortage. Cheap oil has fueled booms while high prices have plunged the world into recession. Finding a balance has been elusive. Always the target of mistrust, oil has now become a tougher, more unpredictable business than ever before.

In 1975 Anthony Sampson, the redoubtable author of *The Seven Sisters*, a groundbreaking description of the relationship between the seven major oil companies and OPEC, described the Arab–Israeli war of 1973 as the "last battle" to control the industry. "The fascination of oil history," he wrote, "lies in the ever changing form of the battle to control supply." Focused on the "collision course" between the governments of the oil producers, the oil companies and the governments in Washington and London, Sampson, like others, did not anticipate Iraq waging war against neighboring Iran and Kuwait, or that America would twice lead invasions of Iraq, characterized by some as "blood for oil." He would have been struck, as I was, by the candor of the vice president of one of America's biggest oil companies whom I asked in passing in 2007, "Was George W. Bush's invasion of Iraq about oil?" He replied, "Absolutely, yes." Some argue that the ideological Cold War has been replaced by "resource wars." In Sampson's era, the "resource war" revolved around disputes about prices between the oil companies and OPEC.

In the two years after the 1973 Arab–Israeli war, the OPEC leaders defied American forecasts that their cartel would collapse, because the consequence of oil prices quadrupling would be a recession in the West. But OPEC's defiance was rewarded, and despite the nationalization of many Western-owned oilfields, the unnerved oil companies collaborated with their expropriators. Stripped of their mystique and their arrogance, the American and British giants were transformed into paper tigers. Anxious to guarantee oil supplies and to maintain their share of markets, the companies that had discovered and developed the oilfields and refined the crude became supplicants. To many, OPEC's ascendancy appeared to be irreversible. Only a few wise oilmen mentioned the fact that cycles never changed. Permanently fixing the market was beyond any mortal, even the OPEC nations.

In the years after 1990, OPEC's lurking threat had indeed diminished. The procession of technocrats following Ali al-Naimi to the second floor of OPEC's headquarters understood that oil had become democratized: prices were set by traders in New York's and London's markets rather than by OPEC's edict. Yet, even though they were no longer brokering mankind's destiny, the ministers retained control of 40 percent of the world's oil supplies—sufficient to wield considerable influence.

Since oil is the OPEC countries' principal, and usually only, source of income, the 11 officials who attended that 153rd meeting had every incentive to seek the highest prices. Between the certainty of extracting oil from the Saudi desert for $2 a barrel or risking $100 billion to drill a speculative well four miles below the sea in the Gulf of Mexico, is the insoluble mystery of establishing the true price of a simple product. The conundrum is to identify the dividing line between reasonable businessmen and villains. Since 1990, that division has become obscured.

In that year Daniel Yergin wrote *The Prize*, a magisterial description of oil's influence on modern history. Oil, he commented, remains "central to...the very nature of civilisation." But many of the political trends of the previous century that he described were changing. The major oil companies were becoming minnows, and OPEC's power was being challenged by non-OPEC oil-producing countries, especially Russia and the states around the Caspian Sea.

The moral keynotes were also changing. Historically associated with corruption, civic corrosion and civil war, the relationship between the governments in Washington and London and the oil companies had been a dominant topic for a century. Posing as representatives of mankind's interests, but beyond mortals' control, the corporations' chairmen appeared detached from national governments. Uncertain who was using whom, many debated whether the oil companies should be supported, controlled or investigated. An important theme explored by Yergin and Sampson was the battle waged by America's federal and state governments against J. D. Rockefeller, the creator of America's oil industry. The epic legal contests against oil companies had usually ended in the governments' defeat, spurring public anger about Big Oil. "His lack of scruple and his mendacity," wrote Sampson of Rockefeller, "provoked a continuing distrust of the oil industry."

Oil provokes irreconcilable emotions. Moralizing sermons about oil have never stopped, but since Sampson's and Yergin's books, some issues have changed. Destitution in the Niger Delta, the contamination of Alaska's pristine wilderness, the destruction of Canada's forests and spreading corruption across Africa are all blamed on oil companies. "African oil did not create the system or its failings," wrote Nicholas

Shaxson in *Poisoned Wells*, accusing Shell, ExxonMobil and the French oil corporation Elf of destroying idyllic communities. Serious authors have claimed that the oil industry is "among the least stable of all business sectors," and that supplies are "utterly dependent on corrupt, despotic 'petrostates' with uncertain futures." The riddle is whether, in pursuing their priority of caring for their shareholders and their customers, the oil majors should refrain from interfering in the internal affairs of Third World countries, or accept a duty to prevent the "institutionalized pillage" of impoverished populations and to oversee the fate of their nations' oil wealth.

These issues continue to be exhaustively debated. Besides the technical and corporate histories, there are many descriptions of evil corporations exploiting the Third World and causing environmental catastrophe. In addition, a new dominant theme has arisen: "The End of Oil." Predictions laced with alarming statistics foreshadow permanent shortages, blackouts and soaring prices. "Terminal decline" is the favored phrase of those speaking in apocalyptic terms about the world imminently running out of oil. One authoritative tract is *Twilight in the Desert* (2005) by the investment banker Matthew Simmons, who claims that Saudi Arabia's oil production is "at or very near its peak sustainable volume (if it did not, in fact, peak almost 25 years ago), and is likely to go into decline in the very foreseeable future." Even Simmons's critics acknowledge the value of his polemic. If Saudi Arabia's supply of oil does indeed decline, the world's destiny is questionable. However, despite Simmons's insistence that his doom-laden prediction is "not a remote fantasy," Saudi Arabia increased its production capacity from eight million barrels a day in 2005 to 12.5 million in 2009, when the world's daily consumption was 86 million barrels. In the aftermath of prices crashing in July 2008 and surplus oil sloshing around the world, the prophets of doom disappeared from the television studios and newspapers. Since then, the wailing about the world's endangered oil supplies has reverted to blaming the oil companies for restricting their investment in the search to find new oil.

Unlike oil's first century, over the last 20 years no single nation, government, cartel or corporation has controlled its fate. Markets have determined prices and investment; but there has been a twist. Because the oil-producing countries retain up to 90 percent of the profits, the Western oil companies have the delicate task of persuading rightly self-interested governments to share their wealth and sell access to their reserves. In Africa, Asia and South America, impoverished nations may be ecstatic about the sudden promise of effortless wealth; but it is only realizable with the marketing, organization and technology invented by Western companies.

In pulling the strands of the oil industry together, this book takes no sides in the arguments among the specialists and partisans. Rather, I have recognised that many of those employed in the oil industry are remarkably intelligent individuals pursuing their ambitions with expertise and inspiration, rather than being inextricably entangled, as the alarmists suggest, in corruption, conspiracies and cover-ups.

United by a smelly, unattractive product, most of the millions of employees who work in the oil industry are strangers to each other. Unlike manufacturing cars or planning a space program, oil offers no natural bond. The gas-station attendant, the crews of the supertankers, the offshore engineers, the dedicated geologists, the excitable traders, the sober accountants, the nationalistic politicians, the rig workers in the prairies, deserts and jungles, the refinery workers and the corporate chieftains are all interdependent in their efforts to produce and convert crude oil. Yet there is no bond between them to overcome their separation and rivalry. Oil unites all their destinies, but they are professionally isolated. Since the late 1980s, however, there has been a common thread: some squeeze markets, some squeeze rocks, some squeeze crude oil through refineries, while others squeeze governments and rival corporations. Oil is not a business for fools or the faint-hearted.

To chart for the first time what has occurred over the past 20 years, I have followed the careers of some of the principal personalities who have determined oil's fate as its price rose from $7 to $147 a barrel. As

I interviewed nearly 250 people across the world, I gradually came to understand the perpetual conflict between the oil companies and the nations that control the reserves, the arguments between consumers and the proponents of the end of oil and climate change, the over-whelming influence of the oil traders, the ingenuity of the explor-ers, the ambitions and frustrations of the chieftains who manage the world's biggest corporations, and the agendas of politicians anxious to control the world's lifeblood.

The story of oil over the last two decades is fascinating, but to understand all the disparate elements—the personalities, the corpora-tions, the governments, the traders and the geologists—requires grad-ual introduction. Unlike the straightforward structure of a standard history or the description of a particular event, this book takes the reader on a journey through the lives and eyes of the major characters who have dominated the industry. As readers become familiar with the labyrinthine complexity of the subject, I hope that, like myself, they will become excited by the discovery of an epic story at the heart of all our lives.

After interviewing nearly 250 key people, I selected a handful to reflect the turbulent era. Over the past 20 years, John Browne of BP was undoubtedly the dominant personality in America and Britain. His rival chief executives, including Lee Raymond of Exxon, Phil Watts of Shell and David O'Reilly of Chevron, were similarly robust, but were begrudgingly compelled to follow his course. In a parallel universe are the oil engineers like Dave Rainey, challenging scientific boundaries to discover oil six miles beneath the seabed. Unknown to the engineers are the oil traders, churning billions of dollars every day in speculation about unpredictable prices. After speaking to dozens of traders and reporters, I chose to follow Andy Hall, an understated multimillionaire hailed by his generation as a genius. Of all the oil-producing countries, events in Russia became far more interesting than in the OPEC countries. Fortunately, as oil prices rose from $30 a barrel toward $147, I hitched myself to Mikhail Fridman, an oil oli-garch in the midst of a fierce battle with BP as President Putin and other politicians, government officials and lifelong experts all sought to influence oil's fate. Challenging their assumptions and decisions are committed environmentalists. All those personalities and interests are weaved into a narrative that makes no attempt to be encyclopedic, but simply to tell an astonishing story.

This is my eighteenth book, and I have found that a career charting the lives of politicians, tycoons, murderers and charlatans was the perfect background to grapple with the intricacies of the oil industry. Fortunately, I encountered few refusals to my requests for help. Across the world, many key players offered me their insights. What has emerged is a story that reveals how we are all simultaneously both the victims and the beneficiaries of conflicting realities in the search, production and trading of oil.

In Vienna in May 2009, Ali al-Naimi was gambling against a squeeze by those speculating that prices would fall. One week after his prediction during his Ringstrasse run, the price of oil had risen from $62 to $68 a barrel. After the summer, the prices hovered around $80. His gamble had been rewarded. Those who had speculated on falling prices had been squeezed by a counter-squeeze, talking up prices. Markets, like the subterranean rocks where oil is found, are unpredictable. Squeezes are often followed by bursts, and there are always casualties. Oil is a uniquely human story.

Tom Bower
July 2009

OIL

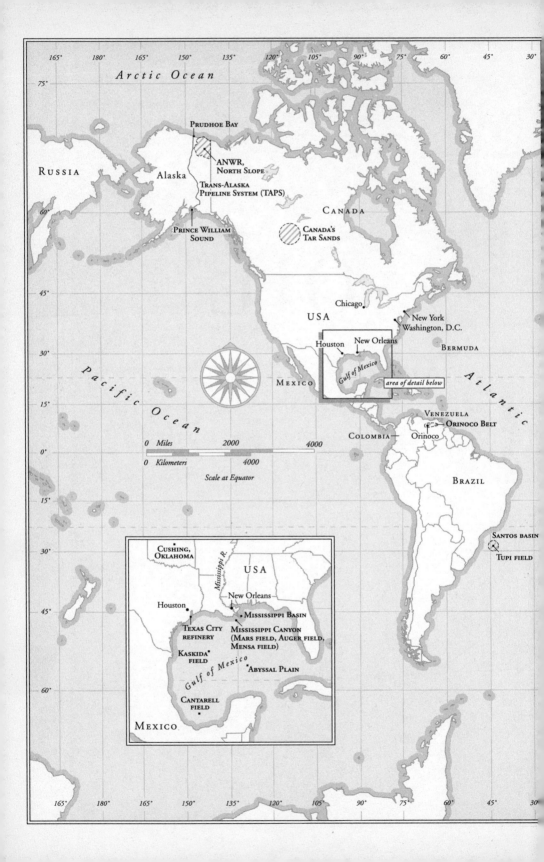

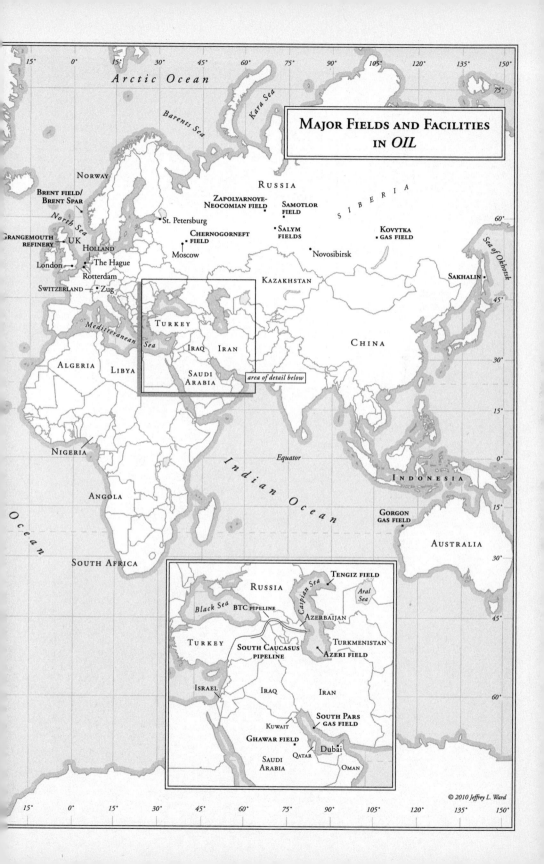

MAJOR FIELDS AND FACILITIES
IN *OIL*

Chapter One

The Emperor

———◄◊►———

LEE RAYMOND DID NOT CONCEAL his impatience. The Russian president was 30 minutes late. Speaking in muted voices, the three other men and one woman waiting with Raymond in the Waldorf Astoria suite speculated whether Vladimir Putin had abandoned the meeting. "I'm sure he'll come," suggested one. Raymond's irritation was not assuaged.

Dealing with dictators was usual for Exxon's 65-year-old chairman and chief executive. In his experience, oil was mostly controlled by feudalists, kleptocrats, zealots and fanatics. "Go to the top, do the deal and the rest follows," was Raymond's way. Over recent years the chemical engineer born in South Dakota had encountered many of the world's oil-rich despots. Renowned for his reserved, focused and analytical manner, he had run all those negotiations just the way he ran ExxonMobil itself—with clockwork efficiency. Oil, according to ExxonMobil's textbook, never surprises; principles never changed, only the circumstances. Vladimir Putin, Raymond believed, was no different from other authoritarians except that he had nuclear weapons and controlled the world's biggest oil and gas reserves. That justified the flight from Dallas and the unpleasantness of meeting another stranger.

Although outspoken and prone to steamroller those he disdained by the sheer weight of his intelligence, Raymond was awkward in the limelight. No concessions were offered to friends or opponents. Unglamorous and conscious of his harelip, he personified the arrogance that united the oil world in hatred, envy and admiration of ExxonMobil. Imbued with ExxonMobil's genes, Raymond's sense of the world was insular. Most non-Americans, in his opinion, especially those from the Third World, were disagreeable.

Today, however, was not the moment to betray his prejudices. Other heads of state had been exposed to his scorn, but, nearing the end of his 40-year career, Raymond ached to clinch this deal. If, as expected, Putin agreed in principle to ExxonMobil's $45 billion offer, the company's status as the world's biggest oil corporation would remain unchallenged. Merit and the odds, Raymond calculated, were tilted in his favor.

For 18 months a small team under Rex Tillerson, Raymond's deputy and heir apparent, had secretly vetted Yukos, a private Russian company that produced 20 percent of the country's oil. During his negotiations with Mikhail Khodorkovsky, the billionaire Jewish oligarch who controlled Yukos with a 44 percent stake, Tillerson, a well-dressed Texan who despite appearances was just as tough as his boss, reassured himself that the company would be an outstanding purchase. With oilfields stretching across western Siberia, Yukos was a gem.

During the last weeks, the proposition had improved beyond Tillerson's original imagination. Putin had approved Yukos's merger with Sibneft, another private Russian oil company. Together, they would rank fourth in the world league, controlling one third of Russia's oil production and growing at 20 percent every year, three times faster than Russia's state-owned oil industries—and beyond Putin's control. The marriage of the two companies had been blessed by Mikhail Kasyanov, the prime minister, as "a flagship for the Russian economy." Combining Yukos's production of 2.16 million barrels a day—no less than 2 percent of the world's output—with ExxonMobil's similar production would eclipse all ExxonMobil's rivals. Tillerson's main concern remained the Kremlin's reaction. In mid-2003 he had asked Khodorkovsky if the deal was politically acceptable. Khodorkovsky had replied emphatically, "Let me take care of this. I've spoken to Putin and it's okay." Nothing, Tillerson believed, had changed in the last three months. On the contrary—the meeting with Putin was intended to seal the deal.

Khodorkovsky's self-confidence was reassuring to the inflexibly direct Tillerson, who would be described by Dave Godfrey, a New York lawyer representing Yukos, as a "caricature of the top arrogant Czar giving out that it was an honor for me to negotiate with ExxonMobil." Having successfully established an oilfield on Sakhalin, a Russian island in the Pacific, as ExxonMobil's most profitable operation, Tillerson felt comfortable navigating through Russia's political and economic turbulence. Khodorkovsky, he reassured Raymond, could deliver. Naturally, Raymond did not entirely rely on Tillerson. He had met Khodorkovsky in Dallas and Moscow, and got on well with him. Money cemented their mutual respect. Raymond, like Tillerson, was inclined to accept Khodorkovsky's good faith. But while Raymond acknowledged that there were events he could not control, Tillerson lacked awareness of the limits to ExxonMobil's authority. The heir, most agreed, was nicer and more personable than the chairman, but not as wise. The less kind regarded them as more or less identical ExxonMobil models, except that Tillerson smiled.

As the chairman of the world's biggest privately owned corporation, Lee Raymond's priority was to look after the interests of ExxonMobil's shareholders. "ExxonMobil is a company, not a government," he would tell those who urged the corporation to consider global warming, social inequalities and international relations with the oil-producing countries. Those topics, and especially ExxonMobil's relationships with Third World governments, had become critical in recent years. Maintaining production had become a problem for all the major oil companies—ExxonMobil, BP, Shell, Chevron and Total. ExxonMobil in particular was engaged in contractual disputes about its operations with several governments. Like all the oil majors, the company was handicapped in its search for new reserves by Third World governments refusing to grant access on acceptable terms. In the Middle East, South America and Asia, self-interested nationalism was denying ExxonMobil commercially advantageous deals. This was partly a result of Raymond and his rivals alienating the rulers of the oil-producing nations. Unable to gain access on their terms to those countries, the oil majors held back, pleading that exploiting their reserves was "risky," "unprofitable" or "unviable." Even in Saudi Arabia, the company's trusted partner and the world's biggest oil supplier, there was animosity. At the end of some particularly acrimonious negotiations with Crown Prince Saud al Faisal, Raymond had become exasperated by the Kingdom's refusal to give Exxon a fair return. "We have better things to do with our money," he snapped. "If you don't agree to what I'm offering, I'm off to play golf."

The consequences of declining oil supplies to the global economy were incalculable, and the danger of a shortage was accelerating, although ample reserves lay under the earth. Russia's vast untapped oil reserves promised some relief from that vicious circle. Both Raymond and Tillerson recognized Russia as representing ExxonMobil's best opportunity to reverse the company's recent stagnation. Khodorkovsky was offering Raymond the ultimate prize, but more was at stake than ExxonMobil's fortunes. Access to Russia's oil would reduce OPEC's supremacy and its self-interested pursuit of higher prices. Ever since the collapse of the Soviet Union in 1989, Russia had been a magnet for oil prospectors, and over the previous decade the Western oil majors had negotiated profitable deals. Some were judged, particularly by President Putin, to be excessively profitable, exploiting Russia's vulnerability after the collapse of communism. Nevertheless, oil and gas had become the cornerstones of the country's economic growth. Winning Putin's trust, Raymond knew, was critical to completing the deal. If he failed, the Russian president's antagonism could contribute to jeopardizing the global economy. Those wider concerns had not troubled Raymond during his negotiations with Khodorkovsky. As always, ExxonMobil's interests were his sole concern.

Fortunately, the risk of Putin blocking Exxon's deal with Khodorkovsky had, in Raymond's opinion, been lifted three months earlier when the president had visited London to witness Mikhail Fridman, another oil oligarch, sign an agreement with John Browne, BP's chief executive, to sell 50 percent of TNK, Russia's third-largest oil company, for just $10 billion. Raymond did not underestimate the importance of that deal, which was the largest-ever foreign investment in Russia. It confirmed Putin's favorable predisposition toward BP and, irritatingly, Browne's decisive influence over the industry. Since 1998 Browne had not only transformed BP from an also-ran into a major challenger to Exxon, but by acquiring American oil companies he had set the pace. Raymond's ambitions were frequently compared to Browne's considerable achievements.

"He's a bandit," Raymond had said in testament to the diminutive Browne's tough negotiating skills in Alaska, where Exxon and BP shared pipelines and other facilities. Raymond disliked the Englishman's aggressive takeovers, belligerence and business model. Although both had earned their reputations cutting costs, he dismissed Browne as a generalist and a non-engineer. "We don't have hero leaders like BP," Raymond's associates would observe. "BP," Raymond would say, "will have its moment of truth." Some suspected that Raymond's unease sprang from his disapproval of Browne's homosexuality. While tolerated within Exxon, homosexuals were barred from claiming partnership benefits for expatriate expenses and services, and overt displays of their sexual preference were discouraged. Raymond refused to add sexual orientation to Exxon's non-discrimination agenda. Others believed Raymond was irritated by Browne's transformation of BP from a withered wreck to challenge Exxon's rank as number one. After two groundbreaking takeovers, a successful rebranding of BP as an environmentally friendly corporation, and big oil strikes in the Gulf of Mexico, there were rumors about Browne seeking a merger between BP and Shell to claim Exxon's crown as the world's largest oil producer. Closing the Yukos deal would terminate that menace.

"If BP can do it, anyone can do it," was the attitude among the

Exxon executives waiting for Putin in New York. "BP's deal is border-line, and Exxon can do better." Mikhail Fridman, a tough operator, had offered TNK around the industry before BP bit, but Yukos was more enticing. The irritant was Browne's success. Like Exxon, in the aftermath of the TNK purchase all BP's smaller rivals, including Total, ConocoPhillips and Chevron, felt compelled to find a similar deal in Russia.

In common with all the oligarchs, Khodorkovsky had obtained his original fortune illicitly. Smart, intelligent and arrogant, he was not flashy. Despite his wealth, estimated by *Forbes* to be $8 billion, the 40-year-old owned no yachts, was driven in a standard S-Class Mercedes and a small BMW, employed just one bodyguard, owned only one house in Zhukovka, the billionaires' enclave outside Moscow with its own clubs and restaurants, and was entranced by the latest electronic gadgets and toys. He was never seen flaunting a mistress, owned only one private jet, and, like all Russia's billionaires accustomed to the wreckage caused by vodka, he rarely drank. His ambitions nevertheless were serious. Since buying Yukos for $350 million in 1995 he had, with the help of Joe Mach, an abrasive and brilliant American oil engineer hired from Schlumberger, the world's biggest provider of services to the oil industry, transformed the company's fields by reeducating the Russian engineers. Previously large amounts of oil had been stolen by the oilfields' managers and organized criminals, and no one cared about water contaminating the wells. The oligarchs, many suspected, merely wanted to strip the assets and carelessly allow oil production to decline. Khodorkovsky had transformed Yukos into Russia's best oil company. Valued at $1 billion in 1999, it was worth $40 billion by 2003, and he retained a 44 percent stake. Overseeing 100,000 employees from his gleaming headquarters, Khodorkovsky would quietly listen to advice, never yell, and never waste time with stupid ideas. Driven by ambition, he had stopped his public relations managers propagating the message of beating Lee Raymond and turning Yukos into Exxon, instead promoting the idea of selling stakes in Yukos as a stepping stone to political power in Russia.

Khodorkovsky's ambition to topple Vladimir Putin was undisguised. During his conversations with Raymond in Moscow in early 2003, he anticipated his supporters winning 40 percent of the seats in the Duma, the Russian parliament, and himself becoming prime minister after the elections in December that year. By June 2003 he was assumed to have bribed sufficient members of the Duma, including some of Putin's supporters, to defeat the government's legislation on tax reform. If the legislation became law, Khodorkovsky was heard to predict, Putin would "get fired." Putin's irritation had not troubled Khodorkovsky, and relations between the two had deteriorated as the oligarch flaunted his ability to bribe members of the Duma in the months before the elections. Russia's internal strife did not concern Exxon's chairman, although he knew that the Kremlin was the only obstacle to a deal. "I'll look after the government," Khodorkovsky had reassured Raymond. "Don't rely on that," Raymond was advised. "We need our own approach." Other than Putin himself, only three other people could decide Yukos's fate, and those power-brokers, everyone assumed, were not sure themselves what would happen next. Igor Sechin was among those who could provide reassurance.

Igor Sechin was Putin's trusted "Mr. Oil" in the Kremlin, the gatekeeper to Russia's oil policy. The two had become friends while working together for the mayor of Saint Petersburg, Anatoly Sobchak, during the Yeltsin era. Sechin was also chairman of Rosneft, a state-owned oil company. A key member of the *siloviki*, the Saint Petersburg crowd of hard-faced former KGB and military officers surrounding Putin in the Kremlin, 43-year-old Sechin was regarded by outsiders as a xenophobe resentful of the Jewish oligarchs who controlled about 50 percent of Russia's economy. Convinced that Sechin would oppose a deal between Yukos and Exxon, Exxon's representatives in Moscow had sought an alternative route through the Kremlin's hierarchy to secure Putin's approval. Igor Shuvalov, Putin's economic adviser, was chosen as the conduit. Based at Old Square, the former Communist Party headquarters linked to the Kremlin by an 800-yard tunnel, Shuvalov had heard about Exxon's "equity investment" in Yukos soon after Khodorkovsky initially approached Tillerson. Since the size of Exxon's proposed stake was deliberately concealed, he had no reason to object. "I will inform the boss and get back to you," he had said. One week later, Shuvalov telephoned Exxon's office: "This is interesting. We are supportive."

Since then, greed had infected the negotiations. As Exxon's experts grasped Yukos's true value, caution was abandoned. A desire for a 20 percent stake was replaced by wanting everything. "We hunt for whales, not sardines," said Raymond. "We won't be a junior partner in Russia. We'll only invest in Russia when the terms are right." Khodorkovsky also became greedy. Aware that the oil majors needed new reserves and were envious of BP's deal, he wanted top dollar for his shares. The timing, he said, was perfect. The ruble had devalued, and Yukos was aggressively valued. At $35 a barrel, oil prices, he believed, were peaking. Like every oil executive, he could not imagine where oil prices were heading over the following five years.

In June 2003, Khodorkovsky anticipated success. To celebrate Yukos's record profits, he rented a luxury yacht to sail from Moscow to Saint Petersburg. Four days later, his business partner Platon Lebedev was arrested and charged with fraud. "Don't worry," Khodorkovsky told his entourage. "He'll be in jail for three months and then we'll get him out." No one was quite sure whether Khodorkovsky genuinely believed his own bravado, but he refused to flee Russia. "I'm not going to become the next insane Berezovsky," he said, referring to the oligarch who, after helping Putin to power, fled as a permanent exile to London. In the event Lebedev was found guilty of tax evasion and sentenced to eight years' imprisonment.

Raymond remained oblivious to the changing mood. Unlike John Browne, he was unversed in Russian culture and sensitivities. While Browne respected Russian history, Raymond saw a greenfield site. Exxon lacked experts who could provide genuine insight about the Kremlin's intentions, especially Putin's ambition to use the world's dependence on Russia's energy resources as a tool to reassert the nation's status as a superpower. Whether Putin regarded Khodorkovsky as a serious obstacle to that ambition in summer 2003 is uncertain. Raymond did not suffer any misgivings. Impervious to subtleties, he approached the final deal, as always, by squeezing sentiment out of the negotiations. In July 2003 he visited Putin in the Kremlin. His pitch to the president was familiar: "We can help you elevate your country by extracting your oil resources." During that visit Mikhail Kasyanov, the prime minister, assured Raymond that Exxon would be allowed to buy a stake in Yukos.

Raymond, accustomed to negotiating with kings and presidents as an equal, shared their lifestyle. Arriving in Moscow with six bodyguards, he secured motorbike outriders from Moscow's police department for his limousine's dash into the city, and the whole top floor of the Kempinsky, Moscow's most expensive hotel, was assigned to him. Putin would be intrigued to hear about two eccentricities during that visit. Since Raymond intended to leave Moscow on a Sunday, the city's Baptist church was opened on Saturday to allow him and his wife Charlene the chance to pray.

During a previous visit to Moscow, Charlene had spotted some sculptured wooden figures in a store that she wanted to inspect again. Exxon's security officers had declared that revisiting the store was excessively dangerous, so arrangements were made just prior to the Raymonds' arrival for a room on the Kempinsky's top floor to be converted into a display and filled with 60 wooden sculptures. After nearly two hours in the room, Charlene announced, "Yes, I'll take them."

Exactly 30 minutes late, Putin entered the Waldorf suite, accompanied by Igor Shuvalov, Sergei Priodka, a foreign affairs adviser, and the Russian ambassador in Washington. Putin would have been conscious that he was nearly the youngest in the room. He and Raymond shared a three-seater divan, while Rex Tillerson and Anna Kunanyansay, Exxon's Russian-speaking adviser, took chairs. Kunanyansay, a Jewish émigréé from Kiev, had made the arrangements with the Russian ambassador for the meeting. Trusted by Lee and Charlene Raymond, she was suspicious of Putin.

After a few minutes' pleasantries, Raymond cut to the chase. His tone was deferential, but not obsequious. "As you know, Mr. President, we have been in negotiations for some time with Yukos."

"Yes, I know," said Putin. "For 25 percent of the shares plus one share. A minority stake."

"Well, Mr. President," replied Raymond, "I think we must be clear. I want you to understand that we will only buy 25 percent if we can see a way to buy total control, and that's why I'm here to see you today. To check that that's okay with you. Our ultimate goal is to buy a majority stake in Yukos."

Putin did not flinch visibly, but the translators heard exasperation in his reply, "This is the first time I've heard that. Khodorkovsky didn't tell me." Raymond pursued his theme, explaining that the deal would improve Russia's relations with America. "Well, we'll see," said Putin evasively. "These details are for my ministers. You must deal with them." Raymond was not discouraged. Putin's impatience was lost in the translation, and he had not actually rejected the idea of a deal. Raymond failed to spot the significance of Putin repeating three times: "This is the first time I heard about this. Khodorkovsky never told me about this."

On September 26, after meeting leaders of the American business community including Raymond at the New York Stock Exchange, Putin flew to see President Bush at Camp David. While he was there he mentioned Exxon's bid for Yukos, and found to his surprise that Bush was unaware of the deal. Putin did not know that ever since Standard Oil was dismantled by the US government in 1911, American oilmen had been indoctrinated not to confide unnecessarily in government officials, including the president. Inevitably, the rule was broken whenever Exxon needed Washington's help. After all, despite the corporation's culture of distrusting governments, America was at the center of its universe.

Five days later, Raymond arrived in Moscow to participate in a meeting of the World Economic Forum starting on October 2. He was to share a platform with Khodorkovsky. Putin and Roman Abramovich, the oligarch and co-owner of Sibneft, would be in the audience. Before Raymond left Dallas, Khodorkovsky's demand for $50 billion for his shares had been considered. Raymond, the master of the hard bargain, declared that he would offer $45 billion. The Exxon team flying to Moscow was confident that the deal would be finalized and announced during the conference. Publicists were drafting the announcement.

On the top floor of the Kempinsky, Raymond waited for Khodor-kovsky to haggle over the $5 billion. Khodorkovsky had heard a garbled report of the meeting between Raymond and Putin in New York, which was described as "the final nail in the coffin for Khodorkovsky's rela-tionship with the Kremlin," and "the beginning of the end." Khodor-kovsky showed no concern, even after Yuri Golubev, the chain-smoking, heavy-drinking cofounder of Yukos, heard from a Kremlin official that "the meeting in New York was bad." Raymond was regarded as having been excessively blatant, and Golubev heard that Putin felt misled by Khodorkovsky, and annoyed at being placed in an "uncomfortable position" by Raymond's "inappropriate behavior." Self-interestedly, Golubev did not mention to Khodorkovsky Putin's anger at the oli-garch's failure to mention Raymond's true ambition.

In reality, the situation was worse than Golubev imagined. On his return to Moscow Putin had summoned a meeting. Poring over an "oil map" stretched across a conference table, his experts identified the existing foreign ownership of Russia's oilfields. BP's recent deal with TNK and Exxon's prospective purchase of Yukos and Sibneft would place half of the Siberian oilfields under Western control. Oil and gas made up 40 percent of Russia's exports. Putin became agitated, and rejected the arguments of the modernizers in his government that Western oil majors were more efficient than Russian producers, and their claim that Russia would retain all the profits through taxation. Suspicious that Exxon was conspiring to threaten Russia's national interest, Putin reflected the familiar mixture of Russian attitudes toward the West—simultaneously craving respect while suffering an inferiority complex. The notion of Russia's oil being under Western control sparked his insecurity, envy and resentment. His grievances were echoed by the "Gray Cardinals," his xenophobic ex-KGB cronies. Like their predecessors employed by Yeltsin, they lusted for personal wealth. Allowing ExxonMobil to move further into Russia threatened their ambitions, which were already limited by BP's deal. By September, Putin was becoming convinced that Khodorkovsky had planned for two years to fund his takeover of Russia by selling Yukos. The president feared that Khodorkovsky could even buy the prosecutor general, or at least organize his dismissal. The resurgence of Putin's national conscience had been anticipated by a handful of realists in Yukos's hierarchy: "It's all crazy to think Putin will allow a crown jewel to be sold to foreigners to benefit a group of Jewish bandits." But Khodorkovsky, they agreed, was "running high." The turbulence influenced Khodorkovsky's negotiations with Raymond.

Khodorkovsky arrived at the Kempinsky with his trusted translator Peter Laing. Over two hours he argued with Raymond about the $5 billion. Reams of paper were covered with figures as the translators interpreted the sums. "Ego" one of the translators would say, was preventing the two men splitting the difference. "They're chiseling," he concluded. Unwilling to concede, Raymond stormed out of the room and slammed the door of his bedroom without saying good night. In hindsight, he would conclude that Khodorkovsky was double-dealing, dangling an alternative deal to Chevron. "The meeting did not go well," Khodorkovsky told his staff after returning from the hotel. "We won't be able to do the deal with Exxon the way they want, but let's keep them involved to keep the pressure on Chevron." By then, Khodorkovsky's fate had been sealed.

In that familiar territory, few were surprised by the news on Saturday, October 25. Late that night Mikhail Khodorkovsky's private plane was "delayed" on the tarmac in Novosibirsk, Siberia. Suddenly, a group of masked security officers burst into the plane, shouting, "Weapons on the floor or we'll shoot!" Khodorkovsky was arrested on charges of fraud and tax evasion. His arrest prompted his close associates to flee to Israel and the USA. "This is the signal that politics has trumped even the appearance of rule of law," said Robert Amsterdam,

Khodorkovsky's American lawyer. But Khodorkovsky's arrest was popular in Russia, except among the other oligarchs, many of whom fled overseas in their private jets to watch events unfold in safety. Mikhail Fridman arrived in Mexico that night on a private jet, for a planned holiday with friends.

Rex Tillerson shed no tears. He shared Khodorkovsky's contempt for the swamp of corruption among ministers with no interest in the country; but he also had no sympathy for the oligarchs. Raymond was upset that his conversation with Putin in New York may have caused Khodorkovsky's arrest, not least because the Kremlin began pursuing Russians employed by Yukos, causing several to flee to the West. But he expressed no concern that the meeting had contributed to triggering the oil industry's renationalization, changing the atmosphere for Western business in Russia. "If they don't understand, then they'll have to learn," Raymond told an aide. "We won't be a junior partner in Russia. We'll only invest when the terms are right."

In December, Tillerson acknowledged the deal was dead. Yukos was under investigation for tax evasion and Khodorkovsky was charged with serious offenses. "Russia is closed," announced an Exxon executive. "It's impossible to put the genie back in the bottle." Western shareholders were about to lose billions of dollars. Putin had done more than terminate a deal; he had curtailed the immediate modernization of Russia's oil industry with Western technology and the chance to balance OPEC's power. Ten years after President Clinton exploited America's Cold War victory to prize the oil reserves around the Caspian Sea from Russian influence, Putin had begun to reverse the humiliation. Russia's prestige and power, he decided, depended upon high energy prices or, more potently, refusing to commit his government to satisfying all Europe's energy requirements. Security of supply rather than the price of Russia's oil and gas would determine the fate of the world's economy. Satisfying the increasing global demand for oil partly depended on increasing Russia's oil production from 10 million barrels a day to 12 million. That increase hinged on Khodorkovsky's modernization of Yukos. After Khodorkovsky's arrest, Yukos's self-improvement program gradually withered, Russia's oil production slipped and China's growing demand could not be satisfied. Unforeseen, the debacle contributed to the oil crisis in 2008 and the global recession.

By then Raymond had retired with a record $398 million payoff and pension. Looking back on his Russian experience, he could draw on the Exxon homily that there was nothing new in oil—only the players in each country were different. In the balance of risk, he had won some and lost some, but the cycle had never changed. Exxon was in better shape than it had been when he had taken over. In his Exxon-centric manner he ignored the problems he had created. The mergers and consolidation among the oil majors orchestrated by himself and John Browne had created a new arrogance and blindness toward the oil-producing countries, alienating their governments from granting the oil majors access to their reserves. Putin's reaction against Exxon was echoed by governments across the globe. In unison, they regarded Exxon, BP and Shell as selfishly unwilling to share their profits. More pertinently, Raymond and Browne, while worshipping the cycle, had misjudged their scripture.

Both had inherited similar lessons about limiting risk. Their forefathers, they had been taught, had been scorched during the 1960s by investing too much. By the late 1970s the industry was hampered by bottlenecks. Mastering the cycle, Raymond and Browne knew, was perilous, just as predicting oil prices was impossible. Their predecessors had failed to foresee the collapse of oil prices in 1986, 1993 and 1998, and none had anticipated the huge increases after 1973. Learning the lessons had proved difficult. In 2003, Raymond and Browne did not anticipate that the cycle had again turned and prices would rise. More eager to instantly satisfy their shareholders than to care for the long-term security of oil supplies for Europe and the United States, both were buying back shares rather than investing in new oilfields. They would blame oil nationalism for preventing efficient exploration and production, but Raymond's insensitivity toward Putin justified the president's suspicion.

Chapter Two

The Explorer

————◆◇◆————

GATHERING THE MASTERS of the underworld at BP's concrete campus in Houston's sprawling suburbs in early 2009 was a cruel ritual. The muted light cast a harsh sheen across the weary faces of 12 men and one woman in the "Big Brain Room," a small cinema formally known as the HIVE, the Highly Immersive Visualization Environment. In the center of the front row sat David Rainey, BP's head of exploration. Peering at the curved screen through battery-powered spectacles allowing "sight" of the whole reservoir, the audience scrutinized the computer-generated three-dimensional images of a possible oil reservoir four miles below the waves of the Gulf of Mexico. Hand-picked to assess the risks, none of the 13 was a buccaneer; they were rather proven company loyalists temporarily united by one credo: if their $100-million gamble to discover whether oil existed deep in the unknown was successful, BP could pocket $50 billion over 10 years. But they would be cursed, not least by themselves, if their calculations were wrong. For oil explorers, the license to make mistakes was limited. The humiliation of failure was permanent.

The mood in the HIVE was inevitably influenced by BP's decision to locate its headquarters within a modern concrete zone. Despite some scattered trees, the disfigured Texan landscape embodied the cliché that oil is either an old, difficult and dirty business or "new, good stuff."

"Nothing is more exciting than drilling," smiled Rainey. The Ulsterman, born in 1954, personified the oilman's permanent restlessness. Easy oil — "the low-hanging fruit" — was now history, and breaking frontiers to find new oil was "incredibly difficult." Although BP's skills in exploration were acknowledged by its rivals, the search beneath the Gulf of Mexico was particularly brutal. Excluded from most playgrounds, at best only one in three of BP's operations would strike oil.

At the end of the show the 13 headed for a hotel conference room, each clutching a personalized folder listing 50 potential sites for test holes off West Africa's coast, in Asia, South America and the Gulf of Mexico. Over the next four days they would decide where to spend more than $1 billion drilling through sand, salt, clay and rock. BP's future depended on finding new oil but there were no guarantees. Although the exploration business was dependent on science, much remained beyond their control. Even the best geologists tended to deploy just three words: "possibly," "probably" and "regretably."

Like the others in the room, David Rainey had learned his craft during four years in Alaska. In 1991 he had moved to the Gulf of Mexico. "I've been there when we've hit," he sighed, "and also when we missed. A dry hole and you feel like jumping out of the window. The emotions are indescribable." In 1999 some had been convinced that "Big Horse," a test drill in the Gulf, was a certainty, but the news from the geologist on the rig that the fossils brought up from the deep were Upper Cretaceous rather than Miocene cast a gut-wrenching gloom across the Operations Room. "We're 60 million years out," moaned the blonde team leader. Any oil would have been "overcooked." Every one of the experts in the HIVE had suffered similar agonies.

In recent years, dry holes had wrecked major oil companies. The skeletons of Gulf, Texaco, Arco and other past icons mercilessly testified that only the fittest and bravest survived. By placing enough bets to balance the odds, BP's executives calculated that what the industry uncharitably called "orphans" would not sink their company. Success depended on taking risks and limiting mishaps, not least thanks to inspired luck. BP had made a fortune in Alaska when Jim Spence, the company's chief geologist in Alaska, struck oil in 1969 after deciding to drill on the rim of a potential reservoir, because the cost of the license on the "sweet spot" was too expensive. Its rival Arco, drilling in the "sweet spot," found only non-commercial gas. Alaskan oil saved BP, but did not make the company immune to future errors. In 1983 it invested $1.6 billion to drill in the frozen waste at Mukluk in Alaska. That they would find at least a billion barrels of oil, BP's geologists told newspapers, was "certain." Instead, they hit salt water. The oil had leaked away. "We drilled in the right place," said Richard Bray, the local chief executive, "but we were simply 30 million years too late." For the next 10 years, BP became complacent and chronically risk-averse, searching for oil in the wrong places.

Rainey enjoyed the rigorous challenges during those impassioned days in the Exploration Forum. "Nothing gets through on salesman-ship and goodwill," he warned. The debate ranged between "concepts," immature proposals that were a twinkle in someone's eye; to "play," which was work in progress; and finally to "prospects," which offered a serious chance to find oil. "We've got to focus on the big stuff," Rainey reminded his experts. Like its major rivals, BP could only survive by finding huge reservoirs, or "elephants." "Little things make no differ-ence to BP," John Browne had ordained, knowing that finding a small field could take as long as finding a big one. Failure, Rainey knew, would delight his rivals. Across the globe, Shell, Chevron, Exxon and smaller adversaries were holding similar conferences. Amid ferocious competition, the challenge was to accurately assess the cost of failure. Like Exxon and Shell, BP had been accused of being averse to risk, too eager to return money to shareholders rather than to invest in finding new oil. "Volume versus risk," said Rainey, echoing an oil industry tru-ism. Reducing the 50 potential wells to 20 eliminated some risk. The holes chosen, Rainey predicted, would "glow in the dark."

His self-confidence reflected oil's changing fortunes. Twenty years earlier oil had sold at less than $10 a barrel. Without money, exploration was limited. In the late 1980s the Gulf of Mexico had been classified as an area where inadequate technology prevented new oil being found. Rising oil prices since 2003 had invigorated the search, and technological advances delayed the death certificate. With prices hovering around $25 a barrel, the public assumed that the international oil companies would continue to produce unlimited supplies. The oil chiefs knew the opposite. Finding new oil was becoming harder, and opportunities to enter oil-producing countries were diminishing, although new technology consistently embarrassed the pessimists. Within the Big Brain Room were the architects of BP's latest success, which had restored the company's credibility. In 2004 "Thunder Horse," a 59,500-ton, semi-submersible cathedral, the world's biggest platform, had been towed from Korea and positioned over an "elephant" reservoir in the Mississippi Canyon, identified by the US Department of the Interior as Block 778, 125 miles south of New Orleans. Designed to extract an astounding 250,000 barrels of oil and 200 million cubic feet of natural gas from four miles beneath the waves every day, it led to chatter among the Gulf's aficionados that BP was overtaking Shell, the pathfinder in the region.

Since 1945 oil had been extracted from the Gulf's shoreline waters, especially by Shell. For years the deep-water limit was assumed to be 1,500 feet. John Bookout, the head of Shell's exploration in the Gulf, challenged that assumption, believing that the Gulf, like Prudhoe Bay in Alaska, would minimize America's reliance on imported oil. In May 1985 the drill ship *Discoverer Seven Seas* began boring 12 exploratory wells in 3,218 feet of water. Oil was found, and a Ram-Powell platform weighing 41,000 tons was towed to the site. That project, also financed by Amoco and Exxon, confirmed that oil could be recovered from the depths and be piped 25 miles along the seabed to terminals.

Bookout next focused on the nearby Mars field, 130 miles southeast of New Orleans. In 1987 Conoco had lost millions of dollars drilling dry holes there. Unable to afford further exploration from rigs floating 3,000 feet above the seabed, the company sold the rights to Shell. Bookout was convinced that the drill should have been placed just 400 yards away. Soon after Shell's purchase, Jack Golden, BP's head of exploration in the Gulf, offered to buy a third of Shell's investment in return for sharing a proportion of the cost. Passive investment, or "farming in," by competitors was not unusual in big projects. Even the mighty oil corporations needed to mitigate their risks. Golden had regretted BP's tardiness in bidding for the US government's first round of 10-year licenses for deep-water exploration in the Gulf, and his irritation was compounded by Shell's perfunctory rebuff of his offer. Shell's executives did not want to share their potential profits, especially with BP. Over the previous decade they had enjoyed watching BP's struggle to survive, and some hoped their rival might even go out of business, allowing Shell to absorb the wreckage. But just one year later the companies' fortunes were reversed. Shell had wasted $300 million drilling a succession of dry holes in the Chukchi Sea off Alaska. In urgent need of finance, the same executives had reluctantly agreed to Golden's offer to share in the Mars field. In return for paying 66 percent of the well's costs, BP would receive one third of Mars's income. In May 1991, Shell struck oil. "Getting Mars was a bonanza in 1988," said Bob Horton, BP's chief in America. "Mars saved BP from bankruptcy." Dean Malouta, Shell's skilled Greek-Italian inventor of sub-sea technology, would bitterly agree: "We are crazy to give BP a lifebelt. They brought nothing to the table except money."

Shell's discovery, and the introduction of new engineering techniques, washed aside a whole lexicon of uncertainties and prejudices that had gripped the Gulf's explorers. Not only had Shell's engineers drilled deeper than anticipated, but the gush of oil was far greater than anyone had expected. Even before the rig for Mars was built and towed from Italy, Shell had broken another world record. In 1993, using a rig tied to the seabed by barn-size anchors in 2,860 feet of water, the company's geologists had found a giant reservoir called Auger 5,000 feet below the seabed, while in 1995 at the nearby Mensa field, abandoned in 1988 as technically too difficult, Shell's new technology and about $290 million enabled oil and gas to be extracted from 5,400 feet.

Finding those big reservoirs of oil had been coups for the geologists.

In their Houston office, John Bookout's team had plotted and recreated an area of the Gulf called the Mississippi Basin. Located just beyond the mouth of the Mississippi River, they traced where the river's sand had been deposited 25 million years earlier, and deduced the sites of potential oil reservoirs. Their findings were confirmed in 1995. Predictions that production at Mars would peak at 3,500 barrels a day were far outstripped as it hit 13,500 barrels a day, with the promise of 30,000 in the future. Dean Malouta was an equal architect of that success. At Auger's wellhead, 5,412 feet below the sea's surface, Shell installed a production system and pipeline to bring the oil onshore. The production rig was held in position by six thrusters on its hull, linked by computers to acoustic beacons on the ocean floor that transmitted signals to hydrophones on the rig. Shell's triple success reinforced the entrenched despondency in BP's offices across town.

Ever since David Rainey arrived in Houston in 1991, the gloom in BP's headquarters had been seared on his mind. After three years' work, BP had hit yet another dry well. "Sycamore" in the Gulf's KC Canyon had wasted $20 million. Jack Golden had taken the failure personally. "Every time we hit a dry hole," the wizened American explorer told Rainey, "we look back and see that we didn't have to do this." In the race for survival, Golden was as conscious as others that the oil majors' share of the world's reserves had fallen to 16 percent, and the national oil companies, driven by politics rather than economics, were less inclined to give them access to their oilfields. Five years later, BP's continuing depressing record imperiled the company's existence. At least BP could rely on its share of the profits from Shell's success at Mars—where two more reservoirs would be found at deeper levels, promising to deliver 150,000 barrels a day—and learn lessons from Shell's success, replicated in "Bongo 1," 14,700 feet below the sea off Nigeria's coast. "We're taking two years off and focusing on learning," Golden declared.

In 1996, Shell's success turned sour. The company struck a succession of dry holes in the Gulf, as did their rivals at BP, Texaco and Amoco. After the seventh dry well, everyone stopped. Exxon's explorers congratulated themselves for their refusal to risk millions of dollars just as oil prices were falling, and for waiting until others had neutralized the hazards. The failures coincided with the US government's announcement of a second auction of leases for deep-water exploration in the Gulf. Shell's breakthrough should have triggered a boom to buy new leases, but the rash of dry wells caused head-scratching across Houston.

The explorers gradually realized that a mile-thick layer of salt beneath the seabed, below the silt that had poured out of the Mississippi River and above the oil-bearing rocks, was causing scientific mayhem. Finding oil relies on plotting formations of rock created up to 60 million years ago. Based on a century's experience, geologists know which rocks are likely to contain oil. Their knowledge guarantees some predictability in the Middle Eastern deserts, the Siberian tundra and the North Sea. In those areas, the question was not whether oil would be found, but whether the quantity was sufficient to make its exploitation commercially viable.

Identifying rock formations 70,000 feet below the Gulf's surface was technically feasible. Ships dragging seismic equipment were regularly crisscrossing the Gulf, firing sound bops to the seabed and, every millisecond, recording the pattern of echoes zooming back from below. Old-timers recalled watching pallets of magnetic tape of seismic data being unloaded by forklift trucks: processing them through nine-track computers took three months. Twenty years later, all that information could be stored on an iPod and analyzed by computer within two hours. But either way, the results in the Gulf were notoriously inaccurate. As the seismic soundwaves passed through the salt, the ricocheting bops from the rock strata were grossly warped. "Recording the sound through salt," Rainey realized, "is like photographing through frosted glass. The image and the sound is distorted." Identifying the location of oil through salt was impossible. Shell's early successes had been due to nothing more than luck. "Don't worry," David Jenkins, BP's head of technology, assured John Browne. "You'll find more Mars-like oilfields once we can see through the salt."

Texaco and Amoco had developed computer programs to show two-dimensional images of rocks, slightly reducing the risk of dry holes. During the 1990s the experts predicted that 3D, and even 4D, images would further reduce the risk but only drilling produced conclusive evidence. On the grapevine, BP's executives heard Shell's boasts about its success with Chevron at the Perdito field in the Gulf, which it claimed was the result of superior seismic processing. "It's a strong indicator of our success," said Dean Malouta. Rainey was dismissive about Shell's reliance on seismic evidence rather than "human experts." In wild frontier areas, Rainey believed in geology. He could cure the salt problem, but the cost would be $100 million. BP could not commission any trials unless a rival corporation agreed to share the expenditure.

At the time, BP was a junior partner with Exxon in unsuccessfully exploring a block in the Gulf called Mickey. Faced with poor seismic images, Rainey tried to persuade BP's richer associate to finance more expensive tests. The latest computers producing three-dimensional images of the rocks were being fed seismic data recorded by ships traveling half a mile apart. BP had financed the development of software using seismic echoes recorded from cables just 12 meters apart, considerably improving the 3D image. But gathering raw data across a 300-square-mile block would be hugely expensive. "We need to go back from geophysics to geology," Rainey explained to Exxon's geologists. "We need to put everything back in its proper place." Renowned for their technical excellence, Exxon's executives are also infamous for believing that anything not invented by Exxon is certainly wrong. Ideas offered by an enfeebled, recently denationalized British operator were thus automatically suspect. Unlike BP, Exxon had focused on finding oil in West Africa, especially Angola, and with its enormous spread of interests the corporation lacked the financial imperative to find oil in the Gulf of Mexico. However, Exxon's technicians were eventually convinced to finance the experiment, and Rainey's idea was proven to be correct. Other oil companies were spurred to adopt the enhanced seismic measurements, reducing the cost for BP.

By itself, the intense mapping of rocks was worthless. Identifying the location of oil depended upon producing accurate geological maps. The oil companies raced to recruit mathematicians and geophysicists to compose computer programs based on algorithms to rectify the seismic data. Rainey's challenge was to recruit better mathematicians than his rivals, especially Amoco, the master in this field. The breakthrough coincided with BP leasing a nine-square-mile block called Mississippi Canyon 778 off Louisiana, recently abandoned by Conoco after a succession of dry wells.

The opportunity to buy the block arose after Conoco had failed to find oil at Milne Point in Alaska. As oil prices slid, the company needed to cut its losses, and BP agreed to trade Milne Point for acreage in the Gulf of Mexico. Nonchalantly, the BP negotiator said, "There's a value gap in the deal. We'll agree if you throw in Block 778." Conoco's negotiator was happy to oblige. Conoco, the BP team believed, had committed a cardinal error by misreading the geology at an unexplored depth. Concealing BP's calculations from its rivals across town, Rainey was confident of success, even though the whole Mississippi Canyon area covered 5,000 square miles.

"Everyone in the Gulf is making the same mistake," Rainey said in 1996. "The model's wrong. We're focusing on the geophysics." Rainey was convinced that his unique understanding of the Gulf would enable him to pinpoint a reservoir: "Shell and Chevron are fixated by seismic tests. They're too rigid. They're forgetting about the geology." While Alaska's rocks had taken three years to master, the complications in the Gulf took 40 years to understand. "Everyone in the Gulf is focused on 'top down,' relying only on the seismic and forgetting the rocks! It should be 'bottom up.'" Rainey insisted that BP's rivals were looking at seismic images corrected by computers, and not at the rocks themselves. In their quest to find the rocks that 10 to 20 million years ago had heated up and generated oil, they had ignored the key factor: less dense than rock, oil attempts to escape. "The deeper I go, I can see the traps, but I can't see the hydrocarbons," said Rainey. "We need to find the plumbing" — shorthand for the "migration pathway" where the oil had flowed and become trapped.

Peering at the 3D images generated by the computers in the HIVE, Rainey reminded his team: "The Gulf is the most complex area on the planet. You've got to stay humble because you can never crack the Gulf. Just as you think you have mastered it, some rocks come up and kick you in the backside. Science is helpful but in the end success depends on human understanding." The team debated whether the white columns spiraling out of the rocks on the screen were salt or sand. If they were sand, the oil would have leaked away and a $100-million test drill would be wasted. "Follow the salt," Rainey urged. The salt was an obstacle, but also an asset. The secret was to find a lump or hill rising within the rock: that would be the trap where the oil would gather, unable to leak out, sealed by the impenetrable salt. "I need people who think like a molecule of oil—where will it go into the rock?" said Rainey. In his efforts to resolve the problem he had abolished the demarcation between geologists and geophysicists. Working together, they could determine whether the rocks had ever contained oil and whether the oil was still trapped. Like the pioneers in the space race, Rainey sought innovations, but the best he could hope for was an informed guess.

Risk is the oxygen of oil companies. Success and survival depend on tilting the risk in the company's favor. In January 1996, Jack Golden told John Browne that BP's explorers had understood the lessons of Sycamore and the salt. The corporation, he urged, should make the leap. His team calculated that, rather than their rivals' estimates of 10 billion barrels of oil within the rocks below the Gulf, there were probably 40 billion barrels. In the second round of bidding for 10-year leases in the Gulf, BP should outbid Shell and Chevron. Browne agreed: the company would buy more acreage in ultra-deep water than any of its rivals.

The investment coincided with the industry's slide toward disaster. 1998 was a dog year in the oil trade. The price of oil slumped below $10 a barrel, the lowest in 50 years. There was surplus of production, and cut-price gasoline was being sold across America and western Europe. The protection enjoyed by vested interests was crumbling. Thousands of experienced engineers were fired, rigs lay unused or could be hired for 25 percent of the old rates, and bankruptcies ravaged the industry. "I can't tell you absolutely this is the bottom, but we haven't seen anything like this," admitted Wayne Allen, the chairman of Phillips Petroleum. Potentially, the only profitable activity was deep-water drilling in the Gulf of Mexico, but hiring rigs to drill to a new-record 7,625 feet below the seabed and bore down to 12,000 feet cost $200,000 a day. New rigs were being designed to moor in over 10,000 feet of water and drill nearly 30,000 feet into the rock. The 3D image of Block 778 suggested there was oil somewhere four miles below the sea bed. A test bore in Block 778 would cost $100 million. The unanswered question was, where precisely to drill a 12-inch hole four miles through the rock?

At first, the debate among the 13 explorers was sterile. Red dots from lasers darted around the screen, identifying strengths and weaknesses for the drill's path. At last the discussion became animated, and a route was chosen. The privilege of naming Block 778 was given to Cindy Yeilding, an attractive blonde geologist—an unusual sight in a male-dominated world. Having a passion for Neil Young's music, she chose "Crazy Horse," the name of his band. Protests soon arrived from the Sioux Indians, defending the memory of their chief, so the plot was renamed "Thunder Horse."

On January 1, 1999, Rainey and Yeilding sat in a bland, windowless second-floor office, dramatically named the "Operations Room," following the progress of a computer-guided drill gouging 29,000 feet through silt and salt toward the porous sandstone and shale where they believed oil had been trapped for eight million years. Only two rigs in the world were able to drill to such depths. Fortunately one of them, Discoverer 534, had already been hired by Amoco, which had just been bought by BP. The cost was $291,000 per day. Reservoir engineers had produced a computer program to steer the bit around perilous flaws, after which it was hoped that oil would gush through the metal casing to the surface. Several drill bits were broken and replaced, but the geologist on the rig reported that the rocks brought up from the depths were the right age. "We're at 13.6 million years," he told Houston, hoping that fossils 14.7 million years old, indicating the possible presence of oil reserves, would soon appear. In real time, Rainey and Yeilding scrutinized the constantly changing numbers flashing on a bank of screens for evidence of oil. One sensor attached to the drill reported whether gamma rays detected clay — a negative reading indicated oil. Another sensor measured resistance to electricity — a positive reading indicated oil and gas, because neither conducts electricity. For the next 186 days other members of the team followed the drill's progress on their laptops, at Starbucks or in their beds at night.

"Our sandbox has just got bigger," Rainey exclaimed on July 4, as the drill's sensors reported oil. Nine months later, the size of the reservoir was confirmed: one billion barrels of oil, the biggest-ever discovery in the Gulf of Mexico. "The prize was beneath the salt," said Rainey, ordering everyone to secrecy until all the neighboring acreage had been signed up by BP. After weeks of around-the-clock work, the explorers and their families discreetly celebrated their success with champagne and dinner.

Around Houston, BP's triumph was greeted with mixed emotions. In normal times, the city fathers would have been thrilled. More oil would mean a boom, but at $10 a barrel, that was not going to happen. The American public, seemingly prepared to pay more for a bottle of water than for a gallon of gasoline, were manifestly ungrateful for any Big Oil success. Unaware of the technological achievements involved, the oil industry was taken for granted by a generation of Americans who had grown up regarding cheap gasoline as their God-given birthright. Filling their gas tank did not make anyone feel good. Ever since nearly 11 million gallons of oil had spilled from the tanker the *Exxon Valdez* into Alaska's pristine waters in March 1989, the public's antagonism toward Big Oil had become entrenched. Big Oil had overtaken Big Tobacco as a focus of hatred. Within the American public's DNA was a belief that oil was a decrepit rust industry unfairly extracting tax from honest citizens. Few appreciated that Thunder Horse would fractionally reduce America's dependence on imported oil, which provided

60 percent of its daily consumption. "Guns, God and Gasoline" may have represented freedom for many Americans, yet the oil companies, apparently ambitious for ever more power while remaining unresponsive to the public, were neither understood nor trusted.

In that hostile environment, BP's achievement was acknowledged only by its rivals. The company's reputation had been soaring since 2000 because of aggressive acquisitions. Exxon, Shell and Chevron anticipated their own successes, although the timing was uncertain. While the kingdoms of the major oil companies were diminishing, BP, the largest oil producer in America, was more admired than hated. David Rainey was proud to have met the architect of that success, BP's chief executive John Browne.

As the guest of honor at a packed dinner in Houston in August 2002, Browne had been hailed as a hero. BP's dapper chief executive, regarded as an idealist and a maverick, was loudly applauded for describing the Gulf as the "central element" of BP's growth. No one in his audience underestimated BP's importance. The company had become the Gulf's largest acreage-holder, and owned a third of all the oil discovered there. In the oil business, strong personalities made the difference, and Browne, like an evangelist, was wooing his audience. "We're going to spend $15 billion here over the next decade," he promised, "drilling between four and seven wells every year." His enthusiasm was understandable. Oil that had been inaccessible in 1998 was now, he knew from Rainey, within their grasp. If the Houston team was successful, BP would outdistance its competitors. Only a handful of doubters suspected that Browne loved being treated like a rock star more than he loved rocks and their contents. Older members of his audience knew that oil had always attracted the ambitious and the larger-than-life. The same man who controlled 90,000 employees and pledged to serve mankind could also behave unaccountably. That was the nature of multinationals.

Exploration for new oil had barely increased over recent years. Since the mid-1970s, over 1,800 new wells in the Gulf of Mexico and in the Atlantic Ocean off Brazil, Angola and Nigeria had promised to deliver 47 billion barrels of oil. But, like a herd, the major oil companies assumed that prices would not rise, and feared risking their profits and their share prices. Their investment in the search for more oil was cut, and many wells had been abandoned. Yet, on reflection, Thunder Horse was recognized as marking a small revolution, and formerly abandoned areas were reconsidered. "Elephants" meant big, fast profits. Thunder Horse meant there were at least another 100 billion barrels of oil to be found under the sea in the Gulf and the Atlantic. Those who believed oil supplies would "peak" between 2011 and 2013 were challenged to reconsider their doom-mongering predictions. The only disadvantage was the cost. Convinced that oil would not rise above $30 a barrel, Browne congratulated himself that his sharp reduction of BP's costs would ensure Thunder Horse's profitability.

Positioning the Korean-built steel rig 6,050 feet above a small hole in the seabed caused jubilation among BP's beleaguered staff. "The serial number of each piece of equipment is 001," exclaimed Rainey with pride. No one on the platform expected to actually see oil. Gushers of crude soaring into the air were relics of history. Oil produced in the Gulf was diverted as it emerged from wells into the Mardi Gras system, a network of about 25,000 miles of pipelines crisscrossing the seabed from Texas to Florida. BP's task was to link Thunder Horse to the system. The obstacles were the depth and distance to the terminals: divers could not survive a mile beneath the surface. But finding elegant solutions to apparently intractable problems caused oilmen's hearts to beat faster. BP's answer was to use robotic underwater vehicles, powered by batteries and guided by sonar from the Houston control room, to find a route for the pipes to cross the furrowed, steep Sigsbee escarpment of mountains and valleys, and then to lay and weld the pipes and valves. On July 18, 2005, Thunder Horse was nearly ready. But then Hurricane Dennis hit the Gulf of Mexico, and under American regulations every engineer was compelled to abandon the rig.

The team closed the operation down, but those who gave the order from Houston forgot that the complicated procedures had never previously been executed. After the hurricane passed, the returning teams discovered the rig tilting at a dangerous angle. Defective valves in the hydraulic control system had allowed water to drain out of the ballast tanks. Oil was also leaking from equipment on the seabed that linked the well to the pipeline. BP's engineers attributed the problems to the poor quality of the manufacturers' work. None of BP's designing engineers had taken into account the fact that only valves manufactured from nickel could sustain the extraordinary pressures and temperatures on the seabed; and the welding had been faulty. The flaws were superficially simple, and exposed BP to ridicule from its rivals. Sending divers to carry out repairs a mile down was impossible, and the damage was too great to repair with robots. The equipment would have to be brought to the surface. It was not clear where the blame lay, but the sums involved were too large to reclaim from the designers and the Korean shipyard. Publicly, BP reported that the rig would be unusable until 2007, and that the repairs would cost £250 million. Such optimism caused wry smiles across Houston.

In normal times, the employees of the major oil companies cooperated to serve their common interests, but in the competitive atmosphere of the time mischievous gossip raged across Houston, and the spirit of BP's humiliated team faltered. Thunder Horse was more than just a tilting platform — it was symbolic of the company. "Poor design and supervision," smiled Shell's head of design about the calamity. "BP always shoot from the hip," said a Shell technician, characteristically dismissing the abilities of a rival. "Their technology and engineering is second rate. They're always coming to us for help." He dismissed BP as a late arrival, hanging on to Shell's coattails, copying its rivals or outsourcing. A colleague agreed that BP was a fast follower, depending on "off-the-shelf go-buys."

David Rainey was indignant at such criticism. History, he believed, undermined Shell's claims of superiority. He felt the company had rested on its laurels, and that following the success at Mars it had been closed to new ideas in the Gulf. "Deep Mensa," an $80 million well bored by Shell in 2001, had been a disaster. Technicians monitoring the data witnessed the "crash out" — the uncontrolled vibrations that smashed the drill as it struggled through fractured rock. Even the best explorers risked embarrassment on the frontiers of the industry. Mortified, Shell's engineers had taken a year to rectify their mistakes.

Shell's expensive errors had been concealed from the public. But Thunder Horse appeared to be a warning to Russia and other national oil companies not to rely on BP. The company's explanations were gleefully rebutted by a Chevron vice president: "It's defeatist to say 'Stuff happens.'" That criticism was also rebutted by Rainey. During the 1980s, he recalled, Chevron had suffered multiple drilling failures that had crippled the company. Cooperation in the Gulf with Chevron, Rainey said, had caused arguments. In 2001, BP's explorers had collaborated with Chevron to test-drill the "Poseidon" block. "They're off the structure," Rainey had complained, urging Chevron to reconsider the test location. Chevron insisted on its expertise, but missed the oil reservoir. Expressing condolences for the failure, BP negotiated to inherit the "barren" field. Rainey's team had precisely calculated the top of the reserve's "hill," hit a billion barrels of oil, and renamed the well Kodiak.

BP's engineers were, however, not protected from the reproaches of a leader of Exxon's exploration team. As the junior partner in Thunder Horse, Exxon was suffering losses caused by BP. Lee Raymond's jocular description of John Browne as a "bandit" found many echoes among Exxon's executives, especially from the technical director who recalled a fault at the BP's Schiehallion oilfield off the Shetland Islands that had compelled BP to lift equipment off the seabed not once, but twice. On two occasions the company's engineers had failed to spot valves installed upside down by the contractors. While Exxon's engineers would at worst have spotted the fault and learned the lesson, BP's management system was not equipped to evaluate the technology, neutralize risks and absorb the lessons.

Exxon, as the industry leader, proudly avoided technical disasters. Since the days of John D. Rockefeller, the nineteenth-century founder of Exxon's forerunner Standard Oil, the corporation had standardized the rigorous management of costs and processes to prevent financial or technical errors. Like God, the system and the company were infallible. Relying on a culture developed since Standard Oil's creation in 1870, Exxon was built on tested foundations. By comparison, BP in 2004 was a conglomerate including former Standard Oil companies—Sohio, Arco and Amoco—still struggling to replicate Exxon's excellence and standardization. While Raymond concealed uncomfortable truths by cultivating a mystique and keeping outsiders at a distance, Browne was constantly selling himself and his improvized company. Nevertheless, both men could justifiably claim considerable technical achievements to ameliorate oil shortages; yet their skills were spurned by oil-producing countries.

One manifestation of the mistrust of BP, Exxon and the other major oil companies lay across the Gulf, in Mexico. The country, the world's sixth-largest oil producer, owned vast quantities of unexplored oil beneath its coastal waters. To Browne's frustration, Mexico's national constitution forbade the participation of foreign companies in its oil industry, and 1938 nationalization laws had expelled American oil corporations, damaging Mexico itself. Pemex, the national oil company, mired in intrigue and patronage, had become notorious for its inefficiency, and as a slush fund for local politicians. Like so many national oil companies, Pemex was expected to provide employment—there were 27 workers on each of its wells, compared to the industry's average of 10. And those employees, lacking technical skills, relied on services provided by Schlumberger, which posed no challenge to Pemex's sovereignty.

In 2002 Mexico's president Vicente Fox sought to change that situation. The facts were alarming. Mexico's oil production was falling. The reserves in Cantarell, Mexico's biggest field in shallow water, which accounted for 60 percent of the country's production, were declining by 12 to 15 percent every year. In 2002 the government borrowed and spent $50 billion to pump more oil, but it had spent only $5 billion on exploration in four years, none of which was in deep water. Consequently, Mexico's proven reserves—the oil that was technically and economically recoverable—had been reduced within three years from

15.1 billion barrels to 11.8 billion. The country had neither the expertise nor the money to undertake deep-sea drilling, and its plight was compounded by its inability to refine sufficient crude for its domestic consumption. Instead, Pemex exported crude oil to the USA and paid mounting prices for the gasoline and other refined products imported from America. Natural gas was flared, or burned, at Cantarell because Mexico could not afford to collect and pipe it across the Gulf. Within a decade, the country would need to import oil. Fox urged the vested interests to change the 1938 constitution and allow foreign investment, with the condition that any benefits would materialize only after a decade. His exhortations were ignored. Mexico's political leaders cared even less about their introverted and protectionist neighbor than about their own plight, an attitude that weakened the oil majors and encouraged the ambitions of the Chinese and other consuming nations to make unrealistic offers to Mexico and neighboring Venezuela, which was even more beleaguered by falling production. For those governments, local politics and world prices were more important than America's energy needs.

These seemingly disparate events around the Gulf of Mexico became interlocked in the summer of 2005. In August Hurricane Katrina hit the Gulf, passing over Thunder Horse and devastating New Orleans. Winds of 220 mph destroyed old rigs, and struck the Mars rig and 11 refineries. One quarter of all America's oil production and one half of its refining capacity were paralyzed. Overnight, Americans understood the vulnerability of oil and gas production in the Gulf. Four weeks later, Hurricane Rita hit the area, damaging deep-water platforms and compounding the difficulties of repairs. Fifteen years of low fuel prices in America were over.

Although BP's oil traders in Chicago and London rank among the most aggressive, David Rainey was unaware of those who were profiting from these calamities. He had nothing in common with that breed, speculating in the darkness, welcoming the probability of oil shortages.

Chapter Three

The Master Trader

———◄◇►———

A NDY HALL WAS cheered by the reports from the Gulf of Mexico. Bad news from oilfields usually satisfied the tall, unshaven trader. Moving from his barren cubicle into the adjoining trading area, he gazed at one of the 15 screens and calculated how much he was up that day. As usual, at 5 p.m. he headed off to practice calisthenics for an hour with a ballet teacher in Norwalk, near the Connecticut coast. The rising price of oil in spring 2005 seemed to confirm Hall's bet that the world was running out of crude. "The trend is your friend," he frequently told his staff. "Ignore the trade noise. Play it long, because I've got ample time to pay." Anyone, Hall knew, could buy oil. The skill was to sell at a profit. Ever since John Browne had predicted in November 2004 that oil prices would stick at around $30 a barrel—although they had already reached $50—and had gone unchallenged by oil's aristocrats, including Lee Raymond, Hall had believed that his massive gamble on soaring oil prices was certain to pay off. Although he was coy about the exact amount, his first stakes were quantified at around $1 billion as oil hovered at about $30, and the price, Hall believed, was heading toward $100 and possibly higher.

Lauded for being "clever as sin, outgunning everyone in the brains department," and referred to as "God" by rival traders, Hall immunized himself from daily market sentiment because he was not part of the herd. An Oxford graduate and art connoisseur, soft-spoken and deceptively shy, he abided by the old adage, "Oil traders work in a whorehouse, so don't try to be an angel in this business." Originally trained by BP, he understood the mentality of Big Oil's chiefs, and believed that Lee Raymond, John Browne and the rest were in denial. Some of the smaller oil producers, like the Austrian and Italian national oil companies, had even bought hedges pricing oil at $45 to $55 a barrel, which would lead to huge losses as prices rose. In March 2005, two years after Hall had made his first bet, and oil was at $55 a barrel, Arjun Murti, a Goldman Sachs analyst, predicted that the price would reach $105 "in a few years." This was greeted by widespread skepticism, and Murti was criticized for serving the bank's interests. Unusually, Henry "Hank" Paulson, Goldman Sachs's chief executive, was required to defend him. By late spring 2008, as the oil price rose beyond $105, Hall had personally pocketed over $200 million in bonuses, and expected to make even more. Murti was being hailed in some quarters as brilliant.

Hall had traded oil for nearly 30 years. Since he had arrived in Manhattan in 1980, disenchanted by England's claustrophobic social system, he had metamorphosed into an aggressive trader. "I'm basically interested in one thing—business," he told his trusted circle. "I come in every day to make money." Whatever the oil price's wild fluctuations, and regardless of whether he was earning or losing millions of dollars, Hall coolly controlled his emotions: "This is not a zero-sum game because we've been doing it for too long to get excited. Emotionally the ups and downs get evened out." Over the years Hall had attracted both praise and loathing for perfecting the "squeeze"—causing the oil market to change, and forcing other traders to buy from him at a premium. "We're not here to help others," he said. In the old days when trading was carried out on the floor of the stock exchange, and dealers had occasionally yelled, "Am I fucking long or fucking short?" Hall had smiled about the screaming losers who always heaped blame on everyone except themselves.

Experience honed Hall's pedigree. Unlike his younger rivals, he had started his career in BP's supply department in the midst of the first oil crisis in 1973. Until then, BP and the other oil majors—Exxon, Mobil, Shell, Chevron, Gulf and Texaco, together known as the Seven Sisters—who controlled 85 percent of the world's oil reserves, had perfected a cozy arrangement to fix the world price. Their representatives met regularly to discuss their costs and calculate their required profits. Blessed by a near-monopoly and a surplus of oil, the seven chairmen would travel as statesmen to the Middle East and inform the Arab producers the price the cartel would pay for their oil the following year, usually around $25.25 per ton, or $3.60 a barrel. The chairmen acknowledged each other's "turf" and, acting like governments, used their intelligence agencies and military supremacy to impose one-sided agreements. The Arab producers meekly signed fixed-price contracts, Exxon formally announced the price, and the crude continued to flow from the Middle East to refineries in Europe and America, although the USA could rely on its own plentiful supplies, supplemented by additional oil from Venezuela and Mexico. Before 1939, Europe imported 90 percent of its oil from America, but after 1945 it switched to Middle Eastern oil, which cost 20 cents a barrel to produce compared to 90 cents for oil from Texas. Even American oil companies increased their imports. To placate small US producers, who were protesting about competition from Arab oil, in 1956 President Eisenhower limited imports, thus increasing the glut in the Middle East. Four years later, without consultation, Exxon and the other Sisters unilaterally cut prices for oil producers. Resentful of the cartel, Saudi Arabia and four other leading Middle Eastern oil producers met in Baghdad in 1960 to form OPEC, to challenge the Seven Sisters' ownership of their reserves.

The new, unfocused group confronting the Western cartel remained ineffectual until the Six-Day War in 1967. Resentment against America and Britain sparked the declaration by Saudi Arabia of an oil embargo, but this show of bravado descended into farce when the Seven Sisters efficiently organized increased supplies from Iran and Venezuela, and Saudi Arabia's income plummeted. The fiasco emboldened Muammar Gaddafi after his coup in Libya in 1969. "My country has survived

5,000 years without oil," he told Peter Walters, BP's managing director, during their first tense meeting in 1970, "and unless we get more money we will stop supplies." A huge spurt in demand had prompted Exxon to forecast for the first time a world shortage of oil, and the fear of scarcity, plus America's increase in imports to 28 percent of its consumption, served the interests of OPEC. The Seven Sisters, OPEC knew, could only control prices so long as there was a surplus of oil. Armand Hammer, the chairman of Occidental, was the first to capitulate, reducing production and increasing his payments to Gaddafi in May 1970. Gaddafi's success encouraged the Shah of Iran, and then the governments of Venezuela and Saudi Arabia, to demand price hikes. The oil companies feared losing their power to threaten the producers with a boycott if they rejected the prices they stipulated. Meeting in New York on January 11, 1971, 23 oil companies agreed, with the American government's permission, to breach the antitrust laws, and confront Libya and OPEC. Their unity was short-lived. During negotiations in Tehran and Tripoli in March 1971, the companies' agreement disintegrated, and prices were increased beyond their limits. "We'll never recover," Walters lamented. "There is no doubt that the buyer's market for oil is over," admitted David Barran, Shell's chairman. The Arabs, he noted, felt betrayed by the West. Sensing weakness, the Libyan and Iraqi governments began partial nationalization of Western oil interests in 1972. The United States, said Gaddafi, deserved "a good hard slap on its cool and insolent face." The Shah agreed. He nationalized 51 percent of the oil majors' Iranian interests and increased prices again. Peter Walters was meeting OPEC representatives in Vienna on October 6, 1973, when he heard that Egypt and Syria had invaded Israel during Yom Kippur, the most holy day in the Jewish calendar. The relationship between the OPEC producers and the Seven Sisters had changed unalterably. The public and the politicians blamed the oil companies for creating chaos and making excessive profits. In the vacuum of considered energy policies, Western governments were accused of perpetuating a "fool's paradise" by relying on arrogant oil executives to supply civilization's lifeblood. Eric Drake, BP's chairman, admitted to Andy

Hall and other graduates recruited during that epic year that oil would probably rise from $2.90 to the unprecedented price of $10 a barrel. Prices actually rose to $12, provoking the Seven Sisters' disintegration and the industry's transformation. Oil was no longer a concession or a product for refining, but became a tradable commodity attractive to cowboys.

Until 1973, oil traders hardly existed except for a fringe group who, to the irritation of BP and Shell, shipped crude from Russia to Rotterdam to supply West Germany and Switzerland. After the Seven Sisters were disabled, BP and Shell no longer felt obliged to protect the Arabs' monopoly. Whenever the corporations had a surplus of crude, they traded it for instant delivery in Rotterdam, one of the world's largest oil-storage areas. Anro, a subsidiary of BP managed by Yorkshireman Chris Houseman, began speculating in oil and refined products based on "spot" prices quoted in Rotterdam, and Shell established Petra, a rival trader in the port. Gradually the two companies replaced the fixed-price contracts agreed upon with OPEC with contracts based on prices quoted among traders on the day of delivery in Rotterdam. Oil became a traded commodity in an unregulated market, subject only to finance from banks and counterparty risk.

The treatment of oil as a commodity akin to sugar, rice, coal and particularly metal ores caught the attention of Marc Rich, a secretive trader employed by Philipp Brothers, the world's largest supplier of raw materials, based in New York. Ambitious for wealth, Rich would achieve notoriety in 2000 when, in the last moments of the Clinton administration, the president granted him a pardon on charges of tax evasion. Rich's journey had begun in the late 1960s. Accustomed to playing both sides in order to control the market for any mineral buried in the earth, he and his partner Pincus "Pinky" Green had realized that the Seven Sisters' control of the oil surplus would eventually be challenged and replaced by the producers' governments. Like the handful of rival traders in London, Rich understood both the complications and the simplicity of oil. After sophisticated technology had found a reservoir, basic project management would efficiently pipe the crude to a tanker for delivery to a refinery. To earn a real fortune from trading oil, Rich knew, required understanding of refining—heating crude oil to boiling point and separating the parts: naphtha for chemicals and the distillates to make gasoline, jet fuel, heating oil, kerosene and diesel. Making a profit from the manufacture of those fuels depended on understanding the constraints of the 600 refineries in the world, each calibrated to process a particular crude from roughly 120 different types. If a refinery calibrated for Iranian crude was denied supplies, the adjustment to process the alternative heavy, "sour" sulphur crude from Saudi Arabia or the lighter "sweet" crude from Iraq was expensive and time-consuming. Profiting from oil, Rich knew, depended on anticipating the circumstances that could cause a disruption of the market or spotting a potential shortage, and securing alternative supplies.

The biggest profits were earned by breaking embargoes, of which none was more high-profile than that against the apartheid regime in South Africa. A company called Sigmoil, loosely connected to Philipp Brothers, dispatched laden tankers from New York to South Africa. In the middle of the Atlantic, the ships' names were changed by rapid repainting, successfully confusing the hostile intelligence services in South Africa. In that atmosphere, Rich was looking for his own niche.

In early 1973, Rich heard rumors about a forthcoming Arab invasion of Israel. That war, he believed, would lead to an oil embargo and soaring prices. Rich was focused on Iranian oil, which in the event of war would be withheld. If he could accumulate and store Iranian oil, its value would rocket after the crisis erupted. Rich was able to find Iranian officials close to the Shah, the pro-Western dictator imposed on the country after a CIA coup in 1953, who were prepared to break their government's agreement to supply oil exclusively to the Seven Sisters. Working in the shadows, Rich flew to inhospitable locations to supervise the loading of crude onto tankers destined for refineries in Spain and Israel and, more importantly, storage in Rotterdam. In exchange for selling the oil below the world price to Philipp Bros., but unbeknownst to the company's directors, the Iranian officials, it is alleged, received "chocolates" in their Swiss bank accounts. Even the corrupt, Rich always acknowledged, were clever. In New York, however, Philipp's directors disbelieved Rich's information about an imminent war. Fearful of the financial risks of purchasing and storing Iranian crude, they ordered the stocks to be sold. Philipp Bros. position has always been that it had no idea what Rich was up to.

After the October invasion, as Israel fought for survival, the oil producers met and agreed to increase prices; to prevent any supplies of weapons reaching Israel, they also imposed an embargo on Holland and the USA. In the face of queues and rationing of gasoline, there was fear throughout the West of economic devastation. Richard Nixon, fighting to retain his presidency in the midst of the Watergate scandal, supported Israel against what Henry Kissinger, his secretary of state, called OPEC's "political blackmail." In retaliation after Israel's victory, the Shah, hosting a conference of OPEC producers in Tehran in December 1973, urged even higher prices than $12 a barrel. Privately, Nixon protested about the potential "catastrophic problems" that would be caused by the "destabilizing impact" of the price increase. Iran, the Shah replied, needed to realize the maximum from its resources, which "might be finished in 30 years." Whether the Shah believed his prophecy was uncertain, but OPEC's new power was indisputable.

By then, Marc Rich and Pinky Green had quit Philipp Bros. in fury to create a rival organization. Registered in Zug, Switzerland, Rich's new company used Philipp's secrets and key staff to establish a network that spanned the globe, although the paper trail ended either in a shredder in his New York headquarters or in Zug, beyond the jurisdiction of America's police and regulators. There was good reason for destroying the evidence. Rich's growing empire was profiting by exploiting regulations introduced by President Nixon in 1973 to mitigate increasing oil prices and to encourage American companies to search for new oil. The regulations priced "old" oil higher than "new" oil. In common with many American oil traders, Rich relabeled "old" oil as "new." Unscrupulous traders, it was officially estimated, made about $2 billion from such practices between 1974 and 1978. Rich would claim that he, like his rivals, had exploited a loophole in badly drafted regulations. However, he had set himself apart from other traders by ostensibly operating from Switzerland, in order to evade American taxes. That might have been ignored if he had not planned to profit by exploiting a crisis in Iran, where oil workers were striking to topple the Shah, disrupting supplies. Oil prices in Rotterdam rose by 150 percent, the harbinger of what would be called the second oil shock. Anticipating the shortage, Rich had again purchased oil for storage from corrupt Iranian officials. Among his customers was BP, the former owner of the Iranian oilfields, which was anxious to keep its refineries operating. BP's reliance on Rich increased after the Shah was ousted from Tehran in January 1979 and replaced by the Islamic fundamentalist Ayatollah Khomeini. Fears of an oil embargo pushed prices further up.

On BP's trading floor in London, Andy Hall watched Chris Moorhouse, the lead trader, regularly run up a flight of stairs to ask Bryan Sanderson, the director responsible for the supply department, to approve contracts to buy oil at increasingly higher prices. Over those weeks Rich resold oil that had cost between $1 and $2 a barrel for around $30. Resentful traders haphazardly tried to compete, and enviously asserted that Rich had paid for the oil with weapons. More seriously, Rich's oil was occasionally exposed as substandard.

Refineries across the world relied on Iranian inspectors to certify the quality of the oil. Few realized how easy it was for Rich to disguise a tanker of low-quality crude. One tanker dispatched by Rich's company to supply Uganda's solitary power station carried, despite the inspector's certificate, unusable "layered" oil. After a day's use the power station broke down, and the country's electricity supply was cut off until another tanker arrived. Rich was aware that he was breaking the US embargo, but his profits were soaring. His good fortune was not welcomed by those queuing for gasoline across America and Europe. Big Oil was accused of profiteering from rationing supplies, and Rich was in the firing line after the seizure on November 4, 1979, of 52 American diplomats in Tehran. His profiteering from America's humiliation sparked a federal investigation into suspected tax evasion.

Rich's success also aroused the interest of two independent oil traders: Oscar Wyatt, an American famous for running over anyone who
got in his way, and John Deuss, alias "the Alligator," a scarred buccaneer based in Bermuda, born 200 years too late. The son of a Ford
plant manager in Amsterdam, Deuss's early career as a car dealer had
ended in bankruptcy. His next occupation was bartering oil between
opportunistic producers and South Africa and Israel, both of which
were excluded from normal trade by embargoes. From the profits he
bought a refinery and 1,000 gasoline stations on America's East Coast.
Compared to Marc Rich, Deuss and Wyatt were minnows. Rich's skill,
as they both appreciated, was obtaining oil by any means possible, brilliantly mastering the markets and insuring himself against losses by
asking Andy Hall to legitimately hedge his daily trade against price
fluctuations.

In 1980, Hall arrived in New York to run BP's nascent trading
operation. After BP's expulsion from Iran and from Nigeria in 1979 for
illegally trading with apartheid South Africa (exposed, according to
BP's executives, by Shell, which was eager to remove a rival), the company was seeking new sources of income. BP's directors had noticed
that as OPEC's control over prices crumbled, BP could trade just for
profit—buying and selling oil from other suppliers, and not just for its
own use. After the discovery of oil in Nigeria in the mid-1950s and in
the North Sea in 1969, the governments in London and Washington
encouraged the oil companies to flood the market in order to undermine OPEC's cartel. Hall, a novice trader, was given a short lesson on
the art by Jeremy Brennan, the trader whom he was replacing. "To find
out market prices," explained Brennan, "just tell them you want to buy
when you want to sell, and that you want to sell when you want to buy.
Keep good relations with the other majors and don't squeeze." Hall
decided to ignore the advice.

Conditions in America had changed. Although the country was
the world's largest energy producer if its oil, gas and coal were combined, the regulations introduced by Nixon in 1971 to encourage more
exploration and keep oil prices down had proven unsuccessful. The

fall of the Shah had prompted a new search for more oil and other energy sources, including nuclear power and natural gas, and energy efficiency. President Jimmy Carter encouraged the purchase of fuel-efficient cars, especially diesel engines, which used 25 percent less gasoline, and greater energy conservation. His initiative was floundering when, on September 22, 1980, Iraq invaded Iran, starting an eight-year war. Overnight, both countries ceased supplying oil, and in anticipation of shortages, inflation and a recession, oil prices soared. The government in Saudi Arabia increased oil production to stem the emergency, and the crisis was short-lived. In 1981 Ronald Reagan, the new president, abolished price controls, and America was promised as much cheap oil as it needed. No one anticipated the turmoil this would cause. America's oil industry was booming, and the supply gap from Iraq and Iran was filled from the North Sea and Alaska. Then, just as Saudi Arabia increased production, oil demand in the West fell. Prices tumbled, and OPEC members cheated on quotas to earn sufficient income. In retaliation against its OPEC partners Saudi Arabia flooded the market, and prices fell to $10 a barrel, undercutting oil produced in America. To save jobs in Texas, Vice President George Bush toured the Middle East, urging producers to cut production. His task was hopeless. Oil was no longer a state utility but was becoming a private business. Speculators and traders, not least Andy Hall and BP, rather than politicians and the OPEC cartel, were gradually determining prices.

The major oil companies had lost their way. The nationalization of their assets in Iran, Saudi Arabia, Libya and Nigeria had shaken their self-confidence. Relying for supplies from dictatorships, Peter Walters of BP decided, had proven to be a mistake. Irate shareholders were demanding better profits. The oil companies began searching in the shallows of the Gulf of Mexico and in the North Sea, but refused to stray into the unknown. An offer to Walters in 1974 from the Soviet ambassador of exclusive rights to explore for oil in western Siberia had been rejected as too risky. Without experience in exploration, Walters did not understand the limitations of his strategy. The new world was unstable, and the future was unpredictable. Oil had become a cyclical

business. Fearful of a financial squeeze, the American majors diversi-
fied into non-petroleum industries that would eventually include coal
mining, mobile phones, high-street retailers, nuclear power, chemicals,
button manufacturing and minerals. Exxon invested in Reliance Electric; Occidental bought Iowa Beef Processors; Gulf considered buying
Barnum & Bailey circus; BP bought a dog-food factory. Astute trading was another solution to compensate for low prices and the loss of
oilfields.

To exploit the political uncertainty, Andy Hall was urged to trade
aggressively. In the era before computers and screens, the market was
inefficient. Traders were constantly scrambling to identify the last trade
in the market and the latest price paid by rivals. In 1981, ascertaining
future prices was difficult. At the beginning of the Iran crisis, experts
had predicted that oil would rise beyond $40 a barrel, but instead it had
remained at around $30, and sometimes lower. Politicians and OPEC's
leaders blamed London's traders and the Rotterdam spot market. The
oil companies, having bought massive quantities of oil to cover every
eventuality, were dumping their stocks. The volatility of prices caused
OPEC and most of the major oil companies concern, but BP seemed
well placed to profit from the new uncertainty. Unlike other traders,
Hall noticed that besides the increasing amounts of oil being imported
by the USA and the simplicity of trading tankers of crude oil on the
daily Rotterdam spot market, there was an opportunity to speculate
about future prices by using schemes devised in the financial markets.
The rapid changes in prices made those profits potentially lucrative.
The second oil shock had hastened the development of speculation.

The impetus for the change was BP's discovery of oil in the North
Sea. Before the discovery of the Forties field in 1970, few experts had
believed that any riches would be found under the gray water. The
surprise breakthrough fired a stampede, akin to a gold rush. Among
the biggest reservoirs was "Brent," discovered in 1971 beneath 460 feet
of water, which would provide 13 percent of Britain's oil and 10 percent of its gas. Developed by Shell across 10 fields and 13 platforms,
the reservoirs were 9,400 feet below the seabed, and the oil was piped

92 miles to Sullom Voe, a terminal in the Shetlands, using unique technology. In 1976 Shell's experts estimated that production would end in the mid-1980s, and on that basis the oil companies were allowed to take the oil cheaply, without paying special taxes. But as the North Sea reserves' true size became apparent and their productivity was extended for at least a further 35 years, the British and Norwegian governments imposed punishing taxes just as on other national oil companies, and reaped the same consequences of the oil majors refusing to search for new oil.

Initially the North Sea produced about 24 tankers of oil every month. As production increased, a few American refineries switched to the "light and sweet" North Sea crude and abandoned Saudi Arabia's heavy "sour." Although the quantities of this oil were small, their effect on the market was significant. After 1976, North Sea production was controlled by the British and Norwegian governments. To avoid oil shortages in Britain and to thwart profiteering, the government agency BNOC (British National Oil Corporation) intervened at the taxpayers' expense to undercut OPEC prices, and directed that crude should be sold only to refineries. In the early 1980s these restrictions were breaking down, and North Sea oil was leaking onto the "spot market," attracting dealers in London and New York. Although the quantities traded were small, the free market of Brent oil became the price-setter or benchmark for oil produced in North Africa, West Africa and the Middle East. The Saudis complained of chaos, but the traders loved the opportunities for speculation. BP and Shell fixed Brent prices, and using BP's oil and information, Andy Hall began trading Brent oil aggressively. Both oil companies had to accept that the market had become opaque.

To introduce transparency into the "forward" market while controlling prices, Peter Ward, Shell's senior trader and the self-appointed guardian of the Brent market, formalized in 1984 the idea of "15-day Brent." On the 15th of every month the oil majors were assigned a cargo of 600,000 barrels of Brent crude at Sullom Voe for delivery the following month. At that point, once the oil major named the day for delivery, the Dated Brent could be traded, and speculation started. "Forward" meant a standard paper contract to physically deliver oil at an agreed destination but in an unregulated market beyond the exchanges in London, New York and Chicago. Tankers carrying 600,000 barrels of oil were sold and resold 100 times before reaching a refinery. Ward believed he had created an orderly market at fixed prices. He had not anticipated that Hall and others would profit by legitimately squeezing rival traders. As the oil traveled across the North Sea, it was bought and sold by traders playing a dangerous game—buying more Dated Brent than had been sold, knowing that others had sold more than they had bought, in the expectation of eventually balancing their books. Since the quantity of Brent oil available every month was limited, Hall could profit by buying large quantities for future delivery, hoping that rival traders would eventually be compelled to buy from him at a premium price. Squeezing the market—compelling rival traders needing the oil to fulfill their own contracts to buy at his price—added uncertainty and volatility to prices. As the Dated Brent was sold to refiners, the price of the 15-day Brent rose because there was less on the market, rewarding the squeezer. In that topsy-turvy world, Hall perfected the squeeze, attracting charges of price manipulation. The squeeze, Hall knew, was not illegal. On the contrary, the British system invited speculators to buy large quantities of Brent for future delivery, despite the fact that Hall's tactics precipitated a 15-year battle to draw a line between aggressive dealing and manipulation of the annual $30 billion trade.

As the spot market grew and prices moved depending on disruption of supplies, Hall became a substantial participant in the "futures market" for the sale of oil—taking advantage of regulated trade on an exchange. His advantage over other traders was BP's own information. Only BP knew how much oil would be piped from its Forties field through its own pipeline to the terminals at Hound Point. Working with Urs Rieder, a Swiss national at BP's headquarters in London, and under the supervision of Robin Barclay, BP was not only anticipating how prices would vary, but was actually causing the market to change.

That power transformed the company's image. Buoyed by BP's constant participation in the physical market, Hall traded uncompromisingly against smaller competitors. Leveraging the market to the hilt was not illegal, but entrepreneurial. Rieder's move from BP to Marc Rich strengthened the relationship between Hall and the American trader. To outsiders, BP had become the elite of traders. BP's traders were a special breed, stamped by pedigree and lifelong friendships. Not only were they numerically astute, they were also internationalist, aware of historical, religious and cultural tensions dictating the price of oil. Among them, Hall shone as the head boy of a new school.

Hall's casualties included Tom O'Malley, Marc Rich's successor at Philipp Bros. Shrewd, intriguing and charismatic, O'Malley possessed an instinctive understanding of oil trading, bending rules but, unlike Rich, not breaking them. Profiting from the oil industry's inefficiency and the market's ignorance, he occasionally exported cargoes of oil from America's West Coast to the East Coast merely to boost prices on the West Coast, but he was occasionally stung by Hall's squeeze when he was contracted to supply Brent oil in New York. To enhance his business and remove the competitor treading on his toes, O'Malley offered Hall a job. Simultaneously, Hall also received an offer from Marc Rich. At the climax of his negotiations with O'Malley, Hall asked for the terms and conditions of his employment and a company car. "Terms and conditions," snapped O'Malley, "is BP bullshit. You come to Philipps to become rich." Hall's resignation from BP in summer 1982 was regarded as a bombshell in London. Rising stars and potential board members never left the family.

Combined, Hall and O'Malley were feared as "crocodiles in the water," and became notorious for analyzing markets, buying large, long positions in Brent oil, and holding out if there was insufficient volume until rivals screamed for mercy. In the Big Boys' game, a rival trader's scream was an invitation to squeeze harder. Philipp Bros., or Phibro, was good at squeezing, because there were large numbers of small traders—at least 50 in the US alone. To outsiders, Phibro personified the separate world inhabited by oil traders. "You're ignoring the rule, 'Don't steal from thy brethren,'" London trader Peter Gignoux complained. The British government's remaining control over North Sea oil prices crumbled as Phibro aggressively traded primitive derivatives and futures against rival traders. The "plain-vanilla swap" compelled the customer either to take physical delivery of the oil or pay to cover the loss.

For the first time, global oil prices were influenced by traders speculating as proprietors, regardless of the producers or the customers. The OPEC countries, especially Saudi Arabia, hated their game, and even Shell was displeased that their precious commodity created profiteers and casualties. In 1983 the market became murkier when Marc Rich remained in Switzerland and escaped facing criminal charges including tax evasion. Despite the scandal, Phibro and others continued to trade with him and Glencore, his corporate reincarnation in Zug. Phibro's aggression invited retaliation. During that year, Shell took exception to Phibro squeezing Gatoil, a Lebanese oil trader based in Switzerland. Gatoil had speculated by short-selling Brent oil without owning the crude. Subsequently unable to obtain the oil to fulfill its contracts because Phibro had bought all the consignments, it defaulted on contracts worth $75 million. Refusing to bow out quietly, Gatoil reneged on the contracts and sent telexes to all its customers blaming Hall's squeeze. Shell's displeasure was made clear at the annual Institute of Petroleum conference in London, where every trader was warned not to attend Phibro's party featuring Diana Ross. "A puerile idea to boycott our party," scoffed Hall, furious that the "clubby clique of traders around Gatoil and Shell obeyed and we were on the other side." Shell levied a $2-million charge, and Phibro paid.

Michael Marks, the chairman of New York's Mercantile Exchange, Nymex or the Merc, attempted to put an end to the chaos in 1983. Dairy products had been traded on Nymex since the market was established in 1872; Maine potatoes were added in 1941; and later traders could speculate on soya beans, known as "the crush." Marks introduced trading of heating oil, an important fuel in America, and crude oil futures, dubbing the price spread "the crack." The reference for prices was the future delivery of West Texas Intermediate (WTI,

America's light sweet crude oil) to Cushing, a small town of 8,500 people including prison inmates in the Oklahoma prairies. Several oil companies were building nine square miles of pipelines and steel container tanks in Cushing as a junction linked to ports and refineries in the Gulf of Mexico, New York and Chicago. Prices quoted on Nymex, based on those at the Cushing crossroads, rivaled those at London's International Petroleum Exchange, trading futures in Brent and natural gas delivered in Europe. Instantly, the last vestiges of Saudi Arabia's stranglehold over world prices were removed. With the formalization of a futures market, OPEC's attempt to micromanage fixed prices was replaced by market forces. The fragmented market became more efficient, but also murkier. Dictators producing oil were unwilling to succumb to regulators in New York, Washington and London. Instead of sanitizing oil trading, Nymex lured reputable institutions to join a freebooting paradise trading oil across frontiers without rules. "I wish we were regulated," one trader lamented. "Why?" he was asked by Peter Gignoux. "So I could tweak the rules."

In 1982, Phibro had faced an unusual problem. The profitable commodities business was handicapped by a lack of finance. Its solution was to buy Salomon Brothers, the Wall Street bank, and begin issuing oil warranties. Manhattan was shocked at a commodities trader owning an investment bank. Overnight, Hall and O'Malley were established as super-league players among oil traders, yet Hall was upset. "Traders and asset managers don't mix," he announced. "I don't want to be part of a bank." Phibro moved to Greenwich, Connecticut, to be as far from Salomon as possible, operating as a hedge fund before hedge funds became widespread.

Across Manhattan, Neal Shear, a pugnacious gold trader at Morgan Stanley, had watched Hall's success with interest. Recruited in 1982 from J. Aron & Co., a commodities trader owned by Goldman Sachs, to start a metal-trading business to compete with his former employer, Shear envied the easy profits Hall and Rich were making. Compared to gold, he realized, oil trading was much more sophisticated and profitable. Without transaction costs or retail customers, and blessed by general ignorance about differing prices in Cushing and elsewhere in America, traders could pocket huge profits. In economists' jargon, oil trading was "an inefficient market." Shear's business plan was original: "Our concept is not to be long or short but flat, to profit from transport, location, timing and quality specifications." Initially he wanted Morgan Stanley to copy and compete with Hall and Rich, but Louis Bernard, one of the bank's senior partners, understood that the rapid changes in oil prices guaranteed better profits than speculating in foreign exchange. On Morgan Stanley's model, the volatility of oil prices could be 30 percent, while in the same period foreign exchange could move just 8 percent. Investment bankers who had traditionally offered their clients the chance to manage risk in foreign currencies could make much more by offering them the chance to manage, protect and hedge crude prices against the risk of price changes. In 1984 Bernard hired John Shapiro, a trader at Conoco, and Nancy Kropp, a trader employed by Sun Oil, to trade crude. To ensure a constant stream of information about the market's movements ahead of its rivals, the bank leased a few oil storage containers from Arco in Cushing. Hour by hour the traders in New York would be aware of whether there was a surplus or a shortage of WTI in Oklahoma, which determined prices on Nymex. Shapiro invented oil options, explaining the new idea to the oil industry at its annual conference in London in 1985. "We're not taking speculative positions," he explained. "This is defensive, as a hedge, leaving Morgan Stanley to manage the residual risk. We've no desire to do an Andy Hall." Andy Hall had also "invented" oil options, offering to the public the chance to invest in the oil trade. In the same year, by a different route, Goldman Sachs established another group of oil traders.

As the gold market deteriorated in 1981, J. Aron & Co., a conservatively managed precious metals dealer, had been sold to Goldman Sachs for $30 million, although the rumored price was $100 million. Goldman Sachs's partners had only agreed to buy what one called a "risk-averse pig in a poke" because they assumed that Phibro's purchase of Salomon's must be clever, and Aron would give them additional international experience to earn a slice of the commodities trade.

Three years later, 30 Aron metal traders were ordered to start trading oil. Under the leadership of Steve Hendel, Charlie Tuke and Steve Semlitz, they were to rival Morgan Stanley. Among their new ventures was speculating in heating oil contracts. By offsetting any order to buy or sell heating oil for future delivery, the bank earned its profit on the arbitrage regardless of future prices. "Arbing on the difference in price" depended on whether the speculator took a bearish or bullish view, but the risk was taken by the customer. The bank's books were nearly always balanced. Whenever an order to buy was booked, the bank's traders made sure that the order for the future was fulfilled by finding a supplier. In those early days, neither Goldman Sachs nor Morgan Stanley was a proprietary trader betting on the price, and they were blessed that British banks were either too sleepy or too small to compete.

In 1985, to profit from the "cash and carry possibilities" of heating oil and crude, Goldman Sachs's traders also acquired storage containers in Cushing and New York. The two American investment banks had become players in physical and paper oil. Oil prices, they realized, were determined not only by demand, but also by supply and international events. In that jigsaw, they traded only if they had the edge. Recognizing that accurate prediction of prices was impossible, the traders did not bet on prices going in a particular direction, but traded on the volatility itself as Brent fell from $30 a barrel in December 1985 to $9 in August 1986. Fast and furious, dealers traded huge volumes even to earn just half a cent on a barrel. The watershed in their trading—before computer models had eradicated the club atmosphere—was the formal introduction of derivatives ("Contracts for Difference"), allowing traders to own huge "paper" positions to influence the market. Bankers, oil traders, the oil companies and the OPEC producers were plotting against each other to master and manipulate the market. The trade in futures, or "paper barrels," was as much a banking business as an oil trader's speciality.

1986 was the beginning of oil's Goldilocks years. Survivors of the crash were destined to earn fortunes because of the volatility of prices. Regardless of whether these went up or down, the traders could profit.

During the boom in the 1970s, oil prices had soared fivefold, and pundits had predicted $100 a barrel. In the mid-1980s, Sheikh Ahmed Zaki Yamani, the Saudi oil minister, became worried that high prices would encourage the West to search for alternative sources of energy. In that event, he anticipated, the floor for oil prices would be $18. Others including Matt Simmons, a Houston banker, predicted a crash. "Stay alive till '85" became the mantra of groups characterized by Simmons as "insular and unreliable" for failing to understand the effect of the growing excess capacity. Contrary to their expectation, oil prices had fallen despite the Iran–Iraq war. Falling prices appealed to President Reagan. According to rumors, in 1985 he urged King Fahd of Saudi Arabia to flood the world with oil in order to destroy the Soviet economy; at the same time, Margaret Thatcher ended BNOC's monopoly in the North Sea, deregulating prices of Brent oil. During December 1985, Simmons's pessimistic forecast began to materialize. Prices were falling from $36 as Saudi Arabia flooded the world with oil, and they fell further as unexpected surpluses of oil from Alaska, the North Sea and Nigeria were dumped on the market. Traders in the speculative Brent market played for huge profits as prices seesawed. The value of 44 to 50 tankers carrying 600,000 barrels of crude oil every month from the North Sea terminals to refineries assumed global importance among the 50 players—oil companies, banks and traders. All crude oil in the world beyond America was priced in relation to Dated Brent, the benchmark of oil prices. Two traders fixing a future price for oil produced in Nigeria would base their contract on the price of Brent on the material day in the future. By squeezing the price of Dated Brent, traders could directly influence the price of crude sold by Nigeria, or Russia, or Algeria. Fortunes could also be made by manipulating the market prices of other oils across the globe based on Brent.

In that hectic atmosphere, a group of traders regularly met at the Maharajah curry house off Shaftesbury Avenue in central London to agree to joint ventures to reduce risk and decide what price they would bid for Brent. In the era before the computerization of the markets, the traders, unable to know at an instant the price of oil elsewhere in the world, relied on gossip and trust, knowing that rivals would pick their pockets whenever possible. The atmosphere in the curry restaurant was akin to a club where "everyone was prepared to screw but not kill." Over 50 percent of the trading market was governed by self-interest rather than laws. "Can you break a law when laws don't exist?" asked one club member rhetorically. Unscrupulous traders seeking to achieve the desired price on a Dubai contract would try to squeeze the price of Brent oil on that day. Shrewd traders noticing a rival taking up a perilous position would step aside to avoid a crash. The unfortunates who screamed "help" could expect assistance, but at a price. The hostility was not tarnished by malice. Those meeting in the restaurant were deal junkies playing for pennies on each barrel, and at the end of the day they jumped into their Porsches to party and celebrate all night with Charlie Tuke before starting to trade at 6 o'clock the following morning. Among the reasons to celebrate was the crash of Japanese trading companies in London. In previous years, their traders had been paid commission on turnover and not profits, and were thus keen to accept any contract. Those traders were known as "Japanese condoms" because they would be left holding all the contracts. As oil prices fell in the mid-1980s, the Japanese traders had been forced to pay huge sums to the London traders before their companies closed. "Hara-kiri all round," toasted the profiteers, enjoying in that helter-skelter era the fierce competition by rival markets in New York and Chicago to attract their trade.

Falling oil prices in early 1986 terrified the Saudi rulers. President Reagan had lifted all controls, allowing supply and demand to determine oil prices. Many predicted huge rises, but instead prices began falling from $26 a barrel in January. By April they were at $11. America's high-cost producers could not compete in the new markets. Domestic production in Texas and California collapsed. Across the country oil wells were mothballed, dismantled and closed. Laden by huge debts, property prices across the oil regions fell by 30 percent, followed by bankruptcies and a smashed economy. Hardened oilmen grieved about "the dark days." In spring 1986, US vice president Bush flew to Saudi Arabia to plead with King Fahd to stop flooding the market.

OPEC's first attempt to stabilize prices by cutting production in early May 1986 by two million barrels a day temporarily restored prices to $15. Any OPEC country that broke the rules, Yamani warned, would be punished. OPEC reduced output by another million barrels to 15.8 million barrels a day. Traditionalists believed that Saudi Arabia's bid to control prices would succeed. Prices rose to $17 on May 19, but secret sales of Saudi crude to sustain the country's expenditure exposed Yamani's political weakness as prices tumbled again to $12. OPEC had lost control. The industry was in chaos. In July prices fell below $10. British prime minister Margaret Thatcher refused a Saudi request to cut production in the North Sea. Her reasons were not political but were intended purely to raise taxes, regardless of the fact that the price collapse was causing havoc in Texas's oil industry and the American economy. Bush's plea had been too late. By August that year the customarily riotous margarita lunches in Houston had dried up, sales of Rolex watches ceased, 500,000 jobs disappeared, bankruptcies proliferated and Texas was devastated. In the darkest days, oil was $7 a barrel. In October Yamani was fired by the king, and Saudi Arabia cut production from nine million barrels a day to about 4.8 million.

Fearful of continuing low prices, the oil majors' enthusiasm for exploration and improved production evaporated. Less glamorous, but nevertheless critical to the future, the profits from refining oil began a long, permanent decline. Convinced that low prices would last for years, the major oil companies sharply reduced their investments. "It's the end of the party," said Peter Gignoux, noting that the world could no longer rely on the Seven Sisters as guaranteed oil suppliers. Liberated from that responsibility, the major oil companies resorted to skulduggery to reduce their taxes. By churning trades of oil to reduce Brent prices to absurdly low rates, they could reap lucrative tax advantages from the 15-day market. In 1986 Transnor, a Bermudan company, claimed to be a victim of a squeeze over Brent oil orchestrated by Exxon, BP and other oil majors. To seek relief, its directors litigated against the companies in America.

The oil companies became alarmed. The Brent trade was an unregulated international business, not subject to American or British laws. Squeezing Transnor was part of the game to manipulate prices and secure tax advantages. The companies' initial ploy in the American court was to persuade the judge that 15-day Brent was similar to "a forward contract" used by farmers to secure guaranteed prices for their crops, and was therefore not subject to the Commodity Futures Trading Commission (CFTC), the American regulator.

Created by Congress in 1974 "to protect market users and the public from fraud, manipulation and abusive practices" in the commodities trade, the CFTC initially supervised 13 commodity exchanges with staff recruited from Congress, especially the agricultural committees. Political favorites, some with limited experience, were appointed commissioners to supervise those monitoring the markets. Relying on the traders' reports submitted to Nymex as its primary tool to identify suspicious price movements, the agency was deprived of adequate funding by Congress, undermining its prestige from the outset. After 1984, as the trade of contracts tripled and the trade in options multiplied tenfold, the 600 staff struggled with an inadequate computer system and a falling budget to identify market manipulation and excessive speculation in 25 commodities, the value of which was growing toward $5.4 trillion a month. That bureaucracy was anathema to Exxon and BP. The oil majors adamantly denied the agency's authority over their business.

Transnor argued the opposite. Its agreement to buy Brent oil, the company argued, was a speculative or hedging "futures contract," which was subject to American law and the CFTC. In 1990 Judge William Conner found in favor of Transnor, ruling that trading Brent was illegal in America. The oil majors were dismayed. Oil traders, they argued, were big enough to look after themselves without a regulator's protection. To persuade the US government of their cause, they stopped trading with American companies and lobbied the director of the CFTC in Washington to reverse the judge's ruling. The CFTC, a lackadaisical regulator caring primarily for farmers and agricultural contracts, had never experienced the pressure of oil lobbyists. Within days the companies declared victory. Fifteen-day Brent was declared to be a "forward contract" and beyond regulation. The oil companies could administer their own "justice," especially when they fell victim to a squeeze of Brent oil orchestrated by John Deuss, the sole owner of Transworld.

Transworld was based in Bermuda, with trading offices in London and Houston, and Deuss's micromanagement stimulated the sentiment among his traders that the only compensation for suffering his obnoxious manner was the unique lessons in oil trading he could provide. Oil spikes, Deuss believed, occurred once every decade, and in the intervening years traders should tread water, manipulating the market with squeezes. The best squeezes, he boasted, passed unnoticed.

During 1986, Deuss decided to execute a monster squeeze on the Brent market. Mike Loya, Transworld's manager in London, was delegated to mastermind the purchase of more oil than was actually produced in the North Sea. In that speculative market, the cargo of a tanker carrying 600,000 barrels of North Sea oil was normally sold and resold a hundred times before it reached a refinery. If prices were falling, traders who bought at higher prices were exposed to losses, while those selling short would expect to profit. Starting in a small way, Deuss and his traders in London bought increasing amounts of 15-day Brent every month. Seeing that by tightening the market they were pushing prices upward and earning extra dollars, they became bolder. Summer 1987 was the self-styled "Eureka Moment." To allow maintenance work, monthly oil production had been reduced to 32 cargoes. Traders at Shell, Exxon and BP had as usual sold 15-day cargoes, expecting to buy back at the end of the period any oil they needed for their refineries. Now, however, their offers were ignored. Mike Loya, the traders noticed, had bought over 40 cargoes, so owned more oil than the fields produced. And having bought everything, Loya was not selling. Transworld's squeeze was felt in London and New York. Prices rose and the protests grew. The oil majors needed Brent to produce specific lubricants that were unobtainable from other North Sea crudes. Without that oil, the refineries could not operate. Contractually bound to supply Brent, they were compelled to pay Transworld an extra $2 a barrel, earning Deuss $10-million profit for one month's work. Unexpectedly, the majors then suffered a second blow. Because of the complexities of oil trading, while 15-day Brent prices increased, Dated Brent prices fell. That fall directly cut the prices of oil produced in West Africa and the Gulf, so the producers lost money on supplying it to other refineries. Deuss's squeeze had caused chaos. "Very painful," admitted BP's senior trader, suspecting that the squeeze had been profitably shadowed by Goldman Sachs and Marc Rich.

After Loya's summer coup, Transworld's traders earned more profits from smaller squeezes until, in December 1987, Deuss believed he had the information to strike a spectacular bonanza. Focused on an audacious coup against the oil companies, he was convinced by the golden fable that no regulator, stock exchange or even country could control the oil market. Like every trader, Deuss nurtured his OPEC contacts, and few were more important than Mana Said al Otaiba, the oil minister of the UAE, who was also a co-owner with Deuss of a refinery in Pennsylvania. Al Otaiba convinced Deuss that in order to force up oil prices, OPEC would agree at its meeting in January 1988 to significantly cut production. If OPEC's production fell, Brent prices would rise.

"Buy Brent," Deuss ordered. Transworld's traders in London bought 41 out of 42 Brent cargoes for $425 million, but prices barely moved. No other traders appeared to believe that OPEC would cut production. Then prices began to fall. "Buy more," Deuss ordered, to shore up his position. To achieve a squeeze, he simultaneously also bought Brent oil from rival traders for delivery in the same period. Those traders, unaware of Deuss's plot, had expected to buy those cargoes from BP and Shell once they were produced. In the common usage, the traders were "short"—selling oil without owning it. As the moment of delivery approached, Deuss demanded delivery of the oil. The unsuspecting traders discovered that no oil was available. Deuss expected to hear

screams appealing for mercy. The traders faced two options: either pay
Deuss a penalty for defaulting on their contracts, or buy their cargoes from Deuss in order to resell them to him, inevitably suffering
a hefty loss. But instead of hearing screams, Deuss became perplexed
by the "shorts" silence. Unknown to him, Peter Ward, Shell's trader,
had agreed with Exxon to sabotage the squeeze by producing extra
oil. "Deuss is a buccaneer," Ward declared. "Let's teach him a lesson."
There was, he decided, a fine line between combat trading and corrupt
trading.

To embarrass Deuss, a Shell trader gave the details of the failing
squeeze to the *London Oil Report* and BBC television. "Everyone's
ganging up against him," noticed Axel Busch, the *Oil Report*'s editor.
"It's become a free-for-all." Deuss calculated the cost. Not only could
he not afford to pay for the 41 cargoes he had bought, but the storage
costs if he did take delivery would be crippling. Urged by his staff to
continue buying up to 60 cargoes, Deuss blinked. Unable to bear the
risk, he retreated. Summoning his London manager out of an Italian
restaurant, he ordered, "Sell everything." Within minutes, the first six
cargoes were sold. Competitors smelled Deuss's panic. With 35 cargoes remaining, prices collapsed. Transworld lost $600 million. Deuss
could pay his debts only by selling his oil refinery. "He's been bagged,"
laughed Peter Gignoux. It was the end of an era. In 1988 the International Petroleum Exchange in London opened a regulated market
to trade Brent futures. Refiners could hedge their exposure to prices.
Some believed that the squeeze and manipulation had finally been curtailed. But the traders and the oil majors knew that humiliating one
buccaneer had not legitimized the trade. The odds, Andy Hall knew,
and the potential profits, had only increased.

Chapter Four

The Casualty

———◄◦►———

SHELL'S DIRECTORS CONGRATULATED themselves on scoring a hit against those disrupting the Brent oil trade. There was a shared pride among the company's longtime employees about their company's probity and purpose. Built by Dutch engineers and Scottish accountants, nothing was decided in haste. Decisions were taken only after all the circumstances and consequences had been considered and the benefit to the value chain was irrefutable. Although BP might produce more oil, Shell earned higher profits.

Reared on that tradition, Chris Fay was bullish. With 23 years' experience in Nigeria, Malaysia and Scandinavia, Fay had become the chairman of Shell's operations in Britain. Shell's 10 oil-producing fields in the North Sea and others under development were his responsibility. Among the problems he inherited in 1993 was the fate of Brent Spar, a platform in the North Sea used to load crude onto tankers. Erected in 1976, the 65,000-ton, 462-foot-high structure had been decommissioned in 1991, and by 1994 was no longer safe. Dismantling it was a problem. There was no suitable British inshore site, while dismantling at sea would cost $69 million. Shell's engineers had considered 13 options offered by different organizations, and Fay had discussed the alternatives with Tim Eggar, the Conservative minister for energy.

With the government's public approval, Fay confirmed on February 27, 1995, that the platform would be towed 150 miles into the Atlantic and, using explosives to detonate the ballast tanks, would be sunk in 6,600 feet of water. The cost would be $18 million. The only downside of the apparently uncomplicated process was that the metal, alongside innumerable shipwrecks on the seabed, would take 4,000 years to disintegrate. Neither Fay nor Eggar was concerned. Over a hundred similar structures had been dumped by American oil companies in the sea without protest, creating artificial reefs off Texas and Louisiana. "This is a good example of deep-sea disposal," claimed Eggar, anticipating that the Brent Spar's disposal would be followed by that of 400 other North Sea structures.

Two months later, at lunchtime on April 30, 1995, four Greenpeace activists jumped from the Greenpeace ship *Moby Dick* and occupied the derelict Brent Spar. The rig, announced Greenpeace, was filled with 5,500 tons of toxic oil, which would escape and contaminate the sea and kill marine life if it was sunk. Media organizations around the world were offered film of the occupation, with close-ups of Shell's staff aiming high-pressure water hoses at the protestors. Any viewer who doubted that Shell was the aggressor was reminded by Greenpeace about the company's poor environmental record. In March 1978 the *Amoco Cadiz*, a tanker carrying a cargo of 220,000 tons of oil, broke up in the English Channel, contaminating the French coastline. Shell owned the oil and was blamed for the disaster, a tenuous link motivated by anger at Shell's refusal to boycott South Africa during the apartheid era and by its supply of oil to Rhodesia's rebellious white settlers despite international sanctions after they declared independence in 1965. The accumulated anger against Shell took Fay and his co-directors in London and The Hague by surprise, especially the accusation that Shell was untrustworthy. Taking the lead from Lo van Wachem, the former chairman of Shell's committee of managing directors, who remained on the board of directors, Shell had already declared its ambition to lead the industry in the protection of the environment. In advertisements and meetings, directors mentioned the possibility of withdrawing from some activities to avoid gambling with the company's reputation. This commitment had been disparaged by Greenpeace. To gain sympathizers, the environmental movement was intent on entrenching its disagreements with the oil companies.

Fay and his executives knew that Greenpeace's allegations were untrue: the platform contained no more than 50 tons of harmless sludge and sand. Greenpeace, they were convinced, had invented the toxic danger as part of its long campaign that mankind should stop using fossil fuels. The battle lines had been drawn after Shell's spokesmen, in common with Exxon's and BP's, had dismissed any link between fossil fuel and damage to the environment. Convinced that the truth would neutralize the Brent Spar protest, Fay appeared on television. But, unprepared for Greenpeace's counter-allegation that Shell was deliberately concealing internal reports describing the toxic inventory, he visibly reeled, fatally damaging Shell's image. His personal misfortune reflected Shell's inherent weaknesses, especially its governance.

The historic division of the Anglo-Dutch company had never been resolved. In 1907 Henry Deterding, a mercurial Dutchman who had gambled with oilfields, investing in Russia, Mexico, Venezuela and California, had negotiated the merger between his own company, Royal Dutch, and Shell Transport, a British company, on advantageous terms, giving the Dutch 60 percent of Royal Dutch Shell. The company's management, however, had remained divided. Two boards of directors — one Dutch and the other British — met once a month for a day "in conference." Each meeting was meticulously prepared, but serious discussions among the 30 people in the room — 20 directors and 10 officials — were rare. Each director could normally speak only once during these meetings, which, remarkably, lacked any formal status. After the "conference" the two national boards separated and made decisions based on the conference's discussion. Aware that the company had become renowned during the 1970s as a vast colossus employing eccentric people enjoying a unique culture, van Wachem, a self-righteous, abrasive chairman, had imposed some reforms while acknowledging that Shell's dismaying history had inflamed Greenpeace's protest.

Henry Deterding, infatuated with Hitler, had negotiated without consulting his directors to guarantee oil supplies to Nazi Germany, and in 1936 he retired to live in Germany. After the war, to remove the concentration of authority in one man, the company had created a committee of managing directors with limited powers to influence Shell's directors. That barely affected the inscrutable aura of an aloof international group of interlinked but autonomous companies immersed in engineering, trade and diplomacy.

As the friends of presidents and kings, Shell's chairmen did not merely control oilfields, but sought influence over governments. Supported by a planning department to project the corporation's power, Shell's country chairmen in Brunei, Qatar, Nigeria and across the Middle East wielded authority akin to that of a sovereign. Yet beyond public view, Shell's employees worked in a non-hierarchical, teamlike atmosphere, exalting technology and engineers who, in the interests of the industry and Shell's reputation, occasionally donated their patents and expertise for the industry's common good. That collaborative attitude was proudly contrasted with Exxon's. Unlike the American directors, whose principal task was to earn profits for their shareholders, Shell had proudly enjoyed its status during the 1980s as a defensive stock—shares that remained a safe investment even in the worst economic recession. Shareholders were tolerated as a necessary evil, and modern management techniques were disdained, emphasizing the company's increasing dysfunctionality. "I'm not saying we enjoyed it," said van Wachem about the 1986 collapse in oil prices, "but there was no panic." With more than $9 billion in cash on the balance sheet, van Wachem's strategic task appeared uncontroversial. Shell owned Europe's biggest and most profitable refining and marketing operation, and Shell Oil was the most successful discoverer of new oil in the USA. Nevertheless, van Wachem's poor investment decisions, combined with a fatal explosion at a refinery at Norco, Louisiana, in 1988, had hit Shell's profits. In 1990 they fell by 48 percent in the US, and net income in 1991 collapsed by 98 percent, from $1.04 billion to $20 million, far worse than its rivals. Shell's poor finances had compelled the sale of oilfields to Tullow and Cairn, two independent companies, and dismissing 15 percent of the American workforce.

Lo van Wachem's ragged bequest was inherited in 1993 by Cor Herkströter. The very qualities of Herkströter that attracted praise in Holland led to criticism of him in London and New York as a socially inept, cumbersome introvert whose disdain for financial markets was matched by a conviction that he was God. Content that Shell produced more oil than Exxon and enjoyed a bigger turnover, Herkströter did not initially feel impelled to close a more important gap. By limiting the influence of accountants and advocates of commercial calculations, Shell earned less per barrel of oil than Exxon. Although Shell's capitalization was $30 billion more than Exxon's, the world's biggest oil company had earned lower profits than its rival since 1981. Complications and compromises had reduced the company's competitiveness and increased costs. For nearly 20 years, to avoid making unpleasant business-related decisions, Shell had chosen to follow a path of consensus. Emollience was particularly favored by the Dutch. Although Dutch shareholders owned only 10 percent of the stock, and over 50 percent was owned by shareholders in the US and Britain, the Dutch directors disproportionately dominated the company, encouraging its fragmentation between different cultures — Dutch, British and American — and also between the different departments — upstream, downstream and chemicals. To his credit, by May 1994 Herkströter, unlike his more conservative Dutch directors, recognized Shell's sickness. Calling together 50 executives to review the company's financial performance, Herkströter concluded that Shell had become "bureaucratic, inward-looking, complacent, self-satisfied, arrogant...technocentric and insufficiently entrepreneurial," all of which was stifling efficiency and the search for new oil. The same sclerosis undermined his authority when Greenpeace boarded the Brent Spar.

"People realize this is wrong," explained Peter Melchett, Greenpeace's executive director. "It is immoral. It is treating the sea as a dustbin." Greenpeace's accusation had aroused public antagonism against Big Oil. All the oil majors were linked with Shell as untrustworthy,

environmental spoilers. Across Germany, Shell's gas stations were boy-
cotted. In Holland, managers reported that a similar boycott was crippling their operation. The decentralized company had never anticipated
that a decision in one country could trigger violent protests in another.
Even though the directors knew that Melchett lacked any evidence
to undermine Fay's honest explanation that the platform's tanks had
been cleaned in 1991, the oil executive's humiliation on BBC television
had echoed across Europe. Like his fellow directors, Herkströter was
destabilized by accusations of Shell's dishonesty and by angry disagreements between the company's managers. In particular, Herkströter was
stunned when Shell staff in Germany leaked material to the media to
embarrass the company's senior executives in Holland.

In the House of Commons, British prime minister John Major,
unaware of Shell's internal warfare, solidly defended the corporation.
As he spoke, Herkströter and his fellow directors, shaken by the boycott and the demand by European politicians, especially Helmut Kohl,
Germany's chancellor, that Shell abandon its plans, collapsed. Just
after Major's public justification of the disposal of Brent Spar, Shell's
board in The Hague capitulated. "They caved in under pressure," complained Michael Heseltine, the secretary of state for trade and industry,
outraged after Fay telephoned and ordered the British government to
cease interfering in his company's business.

The platform was towed to Erfjord, near Stavanger, and dismantling started in July 1995. Melchett was invited to inspect the contents
of the tanks, and was shown to have been mistaken. "I apologize to you
and your colleagues over this," he said publicly after negotiations. "It
was an honest mistake," said Paul Horsman, the leader of Greenpeace's
campaign. Although Shell was vindicated, Herkströter did not recover
from the stumble. Shell's directors were exposed as weak—one even
said, "Greenpeace did a wonderful job"—while Greenpeace, refusing to concede the high ground, invented a more serious campaign to
recover its credibility.

In 1995, the jewel in Shell's crown was Nigeria. Signed in 1958,
Shell's original deal with the country was hugely profitable. The

corporation paid the Nigerian government $2 for each barrel, regardless of the world price, until it reached $100. Thereafter, the royalty was $2.50. Beyond that minimal amount, Shell pocketed the remainder. War and corruption had eroded that windfall over the years. Historically there was no reason why 240 ethnic groups, Christian and Muslim, could exist within a single nation of 140 million people. Oil underpinned the artificial unity that had been constructed by the British colonial government, and keeping that fragile coalition together was the central government's priority. Any threat of secession was unacceptable, especially that declared in 1967 by General Ojukwu, the leader of the oil-rich eastern region of Biafra. Knowing that the country would disintegrate without oil, the government in Lagos launched a war to crush the rebels. Over three years, Biafra and Shell's operation were devastated. The recovery after 1970 had been sporadic. The new income created a mirage of universal wealth. If oil sold at $25 a barrel, each Nigerian citizen would benefit, although by only 50 cents per week at most. But even those profits were wasted by the government on white-elephant projects, including an outdated steel mill purchased from Russia. Simultaneously, the new wealth sucked in imports and destroyed local jobs. To alleviate the social upheaval, Shell built hospitals, schools and social centers. Contrary to advice, Shell's local country chairmen refused to consult aid agencies and nongovernmental organizations about these projects. Rashly, Shell's executives assumed that the government would provide teachers, doctors and nurses.

By 1992 Shell's 5,000 Nigerian staff, 20,000 contractors and 270 expatriate staff had rebuilt most of the wells, replaced equipment destroyed during the war and sought to compensate for losses. But, in the rush for oil, Shell applied standards that would have been unacceptable in Europe or America. Toxic gas was flared from the wells, and oil spills, seeping across farmland and rivers, remained untreated. Nevertheless, only one million barrels of oil a day, half of Nigeria's capacity, was produced, and even that was affected by corruption. Every year the company's auditors arrived from Europe to unearth endemic corruption among the company's local employees. Systematically, some of Shell's

Nigeria managers gave contracts to friends and received backhanders, or paid inflated invoices and pocketed the cash. The auditors found hefty sums paid for "travel expenses" to politicians and government officials and their families. Usually the same expenses were also paid by the government, and the officials kept the difference. At the top level, vast sums of money received from Shell in royalties and taxes were diverted by Nigeria's politicians and officials to private offshore bank accounts. Brian Lavers, Shell's country chairman until 1991, had been under pressure to pay bribes to government officials and local chiefs. To avoid participating in any illegal activity, Shell's board agreed to pay middlemen, farmers and tribal chiefs as "consultants" and for "services" to build social amenities including schools, roads and cinemas. Beyond Lavers's control, these were constructed for inflated prices, allowing Shell's local managers and their friends to steal considerable sums of money. Despite his equally fierce opposition to the Nigerian government's corruption, Philip Watts, Lavers's successor, had no alternative but to reluctantly agree under pressure in 1991 to expand Shell's operation in the country. The company increased the number of rigs searching for oil from seven to 22, agreed to pay higher royalties and, critically, agreed in return for a bonus to increase the country's officially registered oil reserves from 16 billion barrels to 25 billion barrels. "I arrived in this job," said Watts, "absolutely determined to make a difference on issues I felt strongly about. You're talking to someone who was in the eye of the storm."

Bureaucracy, inflation, aging equipment, pollution and soaring taxes amid general lawlessness were just part of Watts's inheritance. Watts, a seismologist, had worked in Borneo, the Gulf of Mexico, the North Sea and Holland before arriving in Nigeria. Intelligent and opinionated, he was intolerant of those he disdained, not least the local criminals. Oil had turned Nigeria into a magnet for villainy. In the Niger Delta, 40,000 square miles of swamps and creeks where the Niger flows into the Atlantic, gangs of Ogoni tribesmen were systematically drilling into Shell's pipelines to divert up to 80,000 barrels of oil every day into barges moored on the creeks. The cargoes were sold to untraceable tankers, chartered by European traders, anchored in the delta or offshore and resold to uninquisitive refineries, especially in nearby Ghana. The European traders could also be the victims. Lured by a succession of telephone calls, a Glencore representative arrived in Nigeria carrying a suitcase filled with several million dollars in cash to buy oil. After the suitcase was handed over, the "sellers" disappeared. If that misfortune gave Watts wry amusement, the Ogoni gangs' activities caused headaches. Explosions while siphoning oil caused numerous deaths, and the thefts from pipelines caused spillage across farmland and in rivers. The environmental damage placed Shell under pressure to pay compensation to farmers, which in turn encouraged some of them to sabotage pipes in order to claim compensation. Attempts by Watts to crack down on corruption, theft and sabotage endangered Shell's employees. Increasingly, they could only work if protected by armed militias. Continued civil unrest forced many oil wells to close down. There were 2,470 security officers employed to protect the operational staff. Although Shell's directors in Holland condemned the use of guns as "intolerable," the nature of the corruption in Nigeria left no alternative.

"There's a staggering skimming of government funds paid straight into Swiss bank accounts," Watts exploded. Since each Shell "country" was self-financing for expansion, Watts's ambitions were frustrated by the government's refusal to pay Nigeria's share of the bill. Most of the $7 billion received every year by the government in taxes and royalties simply disappeared. Hundreds of millions of dollars that the government was contractually obliged to contribute to develop new reserves had been deposited in Swiss bank accounts by corrupt officials. A succession of ministers, Watts discovered, had "not only stolen the eggs but refused to even feed the goose." In the face of wholesale corruption, even Nigeria's banks refused Shell's requests for loans and overdrafts. Watts's predicament was complicated in December 1993 by a military coup led by General Sani Abacha and the slump of Nigerian oil prices to $12. The new dictator repressed striking protestors and arrested the trade unions' leaders in the oilfields, but failed to address Shell's complaints. Exasperated, Watts threatened the minister of finance and the governor of the central bank. "If we can't pay wages or finance our development," he warned, "I'll make sure it'll be in the press. Even my driver will be protesting in the street." Talking tough appealed to Watts, although the corporation's conflict of interests—eagerness for more oil, collaboration with corrupt rulers, disregard for tribal sensitivities and discounting the social damage caused by the oil spillages—could not be disguised during an international protest.

In 1990, Ken Saro-Wiwa, a 49-year-old writer and poet, launched a campaign outside Nigeria on behalf of the Ogoni tribe, who inhabited 1.3 percent of the oil-rich delta and produced 1.5 percent of Nigeria's oil. After 30 years of oil production the Ogonis' farmland, water and air were polluted by oil spillages and the "acid rain" produced from the gas flaring above their crops and villages. In compensation, they received little income from the oil royalties. With Shell's knowledge, central government ministers refused to remit even the agreed 1 percent of the revenue to the locality, and national politicians never visited the region. Until he extended his campaign against the Nigerian government to America and Europe, Saro-Wiwa's efforts had been fruitless. But the crusade and his encouragement of an armed uprising in the delta altered Shell's relationship with the government. The Biafran experience had taught the company that any interference with oil revenues, or any demand for secession, would be squashed.

To protect Shell's oilfields, Watts felt justified in appealing to the government for protection from constant vandalism. "We're not a bottomless pit of money," he explained. Although the uprising was wrecking Shell's operations in Ogoniland, only 3 percent of the company's worldwide production was threatened. During 1993, as the disturbances increased, Watts requested the support of 1,400 armed policemen, in return for which he would provide logistics and welfare. At the company's expense, "mobile police" armed with AK-47 rifles, some of whose uniforms bore Shell's insignia, were dispatched as an "oilfield protection force" to the delta. As reports of death and destruction in Ogoni villages reached Europe and America, Shell was accused of financing "kill-and-go mobs" to brutally suppress the uprising. Amid chaotic scenes, Shell withdrew its staff and stopped pumping oil. The Ogonis would claim that on December 1, 1993, Watts thanked the inspector general of the police for his cooperation "in helping to preserve the security of our operation." His gratitude was premature.

Beyond Nigeria, Saro-Wiwa's description of the delta's desolation and the Nigerian government's oppression of the Ogonis aroused fierce protests. Shell was urged to exploit its financial influence and persuade the government to cease the violence and grant the Ogonis independence. Herkströter, supported by younger directors including Mark Moody-Stuart, the British heir apparent, resisted those demands. "We have to work with the government," the directors agreed. "We don't have a mandate to interfere." Recalling the outrage during the 1960s about American multinationals including ITT and United Fruit directly interfering in South American affairs, Shell's directors declared, "We don't get involved in politics." In arguments with representatives of the relief agencies, Moody-Stuart insisted, "Even if the government steals money, we cannot do anything about it. We are guests in the country and cannot intervene."

In May 1994 Saro-Wiwa was arrested for inciting the murder of four chiefs and government officials who had been attacked by a crowd of Ogoni youths in a meeting hall and hacked to death. The price of Nigeria's oil, said protestors in the electrified atmosphere, was blood. The promise by Abacha in July 1994 that the death penalty would be imposed on "anyone who interferes with the government's efforts to revitalize the oil industry" chilled Saro-Wiwa's supporters, especially the striking oil workers.

Brian Anderson, who replaced Watts as the local Shell chairman in 1994, visited General Abacha. Like many Europeans, he had assumed that Saro-Wiwa would receive a short sentence. Nurtured by Shell's straitjacketed culture, Anderson was immune to the nuances of the dictatorship, and his report to The Hague after his first conversation with the general did not raise any alarm. One year later, after a prejudiced trial, Saro-Wiwa was condemned to death. Only after the verdict and another visit to the general did Anderson realize his mistake. By then it was too late to influence Shell's directors. Like a supertanker, they were impervious to shocks that required an immediate change of course. By then, Saro-Wiwa's fate had become an international issue. Across America and Europe he was portrayed as the victim of Shell's conduct, and the company was accused of polluting the Ogoni farmlands and of failing to protest against the rigged trial while financing the government's destruction of the delta. President Clinton, Nelson Mandela and other international leaders protested to Abacha. The World Bank, church leaders, Greenpeace, Amnesty International, PEN, the International Writers' Association and even members of the Royal Geographical Society demanded that Shell abandon its operations in Nigeria. The opprobrium spread across all of Big Oil. Accused of exploitation, corruption, environmental damage and murder, Shell was urged to intercede and prevent Saro-Wiwa's execution.

In The Hague, Cor Herkströter and his board maintained their composure. Shell men never flapped. Shell's "Business Principles," a set of guidelines committing the company to an apolitical role, had been adopted in 1976 and subsequently updated five times. According to those principles, Shell's duty was to be decent but not evangelical. Multinationals should not interfere in sovereign states. "We must be part of the furniture," everyone agreed. "It's ridiculous that we should intervene against a military dictatorship," said one director, to approval. "If we left," said another, "we would cut off Nigeria's nose and our own. The French would replace us in a flash." The company's huge investment needed to be protected, not least because after years of frustration there was still hope that the Nigerian government would agree to build a plant to liquefy and ship the country's vast deposits of natural gas in tankers as LNG (liquefied natural gas). Shell was the master of the complicated technology necessary to freeze natural gas to $-160°C$, at which temperature it became liquid gas, which could be shipped around the world. Six hundred cubic meters of natural gas could be condensed into one cubic meter of LNG. The profits would be huge. Only a minority of British directors understood that Shell's investment in Nigeria was becoming disproportionate to the profits. The capital, they believed, could have been better spent elsewhere. That British minority believed that standing aside from Nigeria's political battles had been a mistake, and that Shell should have taken more interest in the delta's environment years earlier. Yet in the midst of the storm, changing course had become too difficult. There was no alternative but to support the wrong decision. "I never doubted that Shell would stay the course," said Watts. "We resiled from protest," observed a Dutch director. "Shell should not be blamed for an unjust government." "Shell doesn't get involved in politics," announced a spokesman. Questions were referred to the British, Dutch and American governments, which equally failed to make any forceful protest and opposed sanctions, although Nigeria exported 40 percent of its oil to the US. At the very last moment Shell and the three governments did protest to General Abacha, but on November 10, 1995, Ken Saro-Wiwa and his eight fellow defendants were executed. Greenpeace blamed Shell's silence for the deaths. "It is not for commercial organisations like Shell," replied a company spokesman, "to interfere in the legal process of a sovereign state such as Nigeria."

Stigmatized as international pariahs, Shell's directors realized on reflection that earlier intervention by the corporation might have stopped Saro-Wiwa's execution. Nevertheless, amid appeals for international sanctions, Anderson and Shell's directors met in London to decide whether to push ahead with the plan to build an LNG plant for Nigeria's natural gas. One month later, on December 15, 1995, 30 years after suggesting the idea, Dick van den Broek of Shell signed an agreement with the Nigerian government to build a $3.8 billion LNG plant. He had threatened that failure to sign would terminate any future agreements with the company. Shell's directors were ecstatic. Ninety percent of the LNG output had been presold, and Shell was a 25.6 percent shareholder. Shortly after, Shell agreed to build a giant platform offshore, in an area called Bonga. These deals aggravated suspicions about Shell's conduct during the Saro-Wiwa affair and its promise to return to the Ogoni region if its workers' safety was guaranteed by local communities. Many critics believed that Shell's managers in Nigeria had refused to protest against Saro-Wiwa's execution because of collaboration with the regime. Those censuring Shell included the World Council of Churches, whose report accused the company of polluting the Ogoni area by dumping oil into waterways and of showing "inertia in the face of the government's brutality," which included intimidation, rape, arrests, torture, shooting and looting. God, said the Council, damned Shell in Nigeria. Shell denied all the charges. Exonerating itself of any responsibility because it had withdrawn from Ogoniland in 1993, Shell derided the report for regurgitating old and previously discredited allegations, 99 percent of which, it declared, were fabrications. But the company could not win. The criticism nevertheless prompted Herkströter to admit that Shell's culture had "become inward-looking, isolated and consequently some have seen us as a 'state within a state.'" Mark Moody-Stuart was among the few who became openly disturbed that the company had misjudged the situation. "We should have been more patient," he admitted, "and less angry and offered more. There are lessons to be learned." "Nigeria," lamented John Jennings, a Shell director, "is like a house falling down. All we can do is patch it up so it leans but doesn't collapse." Watts was philosophical. "In oil, mistakes get buried in the mists of time." In June 2009, Shell would pay $15.5 million in compensation to settle a lawsuit with Saro-Wiwa's family, while admitting no wrongdoing.

Few Nigerians had attended Watts's farewell party from Nigeria in 1994, but Shell's directors were relieved that the company's investments in the country were secure. General Abacha had been persuaded that without Western expertise Nigeria's oil production and income would diminish. Unlike Venezuela and Indonesia, Nigeria had no intention of expelling the oil majors. Both sides agreed they needed stability. In view of the continuing violence targeted against the president, Brian Anderson accepted the permanent protection of Nigerian soldiers for Shell's employees. The corporation's archives for 1995, Shell's *annus horribilis*, were sealed. Reviving the company had become critical to its future prosperity.

Shareholders were demanding improved profits. Years of cautious underinvestment, Herkströter realized, were no longer sustainable. The company had been bruised like the other oil majors by the fall of oil prices, and its poor financial performance had been undermined by choosing only ultrasafe investments and its failure, other than in the Gulf of Mexico, to find "elephants." To improve value per share, Herkströter decided to stop the company befriending presidents and kings, and to focus on reform of its financial controls. Localness, previously Shell's strength, was to be curbed. Fiefdoms were abolished. One third of the headquarters staff were dismissed, and the power of the resident chairman in each country was reduced in favor of Exxon's method of governance through central control. The survivors were ordered to stop playing politics and start earning money. But Herkströter's headlines did not translate into action: little happened other than a costly joint venture in America with Texaco and Saudi Aramco (the Arabian-American Oil Company), which would prove disastrous. To prevent the balance of power tilting toward the British directors, Herkströter marshaled the Dutch directors to reject Mark Moody-Stuart's proposed purchase of British Gas (BG), a substantial oil exploration and production company, for £4 billion. Moody-Stuart was "very upset," observed Phil Watts. In 2008 BG would be worth about £35 billion.

Herkströter was equally inept in his attempts to restore Shell's reputation. "We are now being asked to solve political crises in developing countries," he said in October 1996, "to export Western ethics to those countries and attend to a multitude of other problems. The fact is we simply do not have the authority to carry out these tasks. And I am not sure we should have that authority." That opinion was opposed by Mark Moody-Stuart and Phil Watts.

Primed by his experiences with Brent Spar and Nigeria, Watts put together a list of tasks under the heading "Reputation Management." For Watts, Brent Spar had been "a life-changing experience . . . We had done a technically excellent job but we had all missed the big trick. A time bomb was ticking—we missed it and we all thought we were doing our best . . . We never dreamt we would get that much attention."

But if Brent Spar was Watts's "big wake-up call," he found that Nigeria "keeps us awake all the time." By April 1996 he had compiled a list of initiatives, including "Ethics, Human Rights, Political Involvement, and the key items for the review of the Business Principles." The "stewardship over Shell's reputation" was Watts's priority.

Greenpeace's campaign against the oil companies had focused on Shell's exploration in the West Shetland islands. Ignoring the environmental lobby, Herkströter realized, was pointless. The initiative, he noted, had been seized by BP's John Browne. Spotting the tide of opinion, Browne had, amid fanfare, delivered a speech at Stanford University urging the world to "begin to take precautionary action now" to protect the environment. Shell's directors agreed to embrace the same ideology. The corporation crafted public statements promoting its intention to be more open, to acknowledge human rights and to protect the environment by including renewable energy projects in its core business plan. In the future, said Herkströter, Shell would engage with Greenpeace to discuss the reduction of greenhouse gases in coal gasification and biofuels. Satisfied that he had fulfilled the public relations requirements, Herkströter approved the purchase of one fifth of Canada's Athabasca tar sands for C\$27 million, a relative pittance. The total estimated reserves were 1,701 billion barrels of oil. Shell anticipated extracting 179 billion barrels. Exploitation of the tar sands was uneconomic while oil was at \$15 a barrel, but would be profitable once the price hit \$40, although the process offended Shell's newfound commitment to protect the environment. The tar's extraction would require the felling of 54,000 square miles of forest, an area the size of New York State, and as a consequence wildlife would be killed and water polluted. Huge amounts of power would be required to create the steam or hot water needed to separate the bitumen from the clay, and more power and chemicals were required to separate the light petroleum from the bitumen. The whole process created three times more carbon than conventional oil operations. In The Hague, the purchase was mentioned as manifesting Shell's ability to play both sides of the argument.

At the end of 1997, Herkströter retired. Mark Moody-Stuart, his successor, was dissatisfied with his inheritance. Appointed as "Mr. Continuity," Moody-Stuart, a Cambridge geologist and a Quaker who loved sailing, regarded his predecessor's changes as timely but ineffectual. Few of the reforms had materialized. "Shell needs drastic remedial measures," he said, while fearing that the majority of Dutch directors would resist even the appointment of senior directors from outside the corporation. Shell had already missed out on two important investments. Approached by the governments of Angola and Azerbaijan to develop their oil, the company had refused requests for preliminary cash bonuses, and the opportunities were seized by BP and Exxon. Under Herkströter, Moody-Stuart lamented, Shell had even ignored the middle way. Adrift and unacclimatized to the new world, Shell had allowed its long-nurtured relationships with the governments of Oman, Nigeria and Brunei to deteriorate, and earnings were falling. In 1998 the company's profits were $5.146 billion, compared to $8.031 billion in 1997. "There will be a coming crisis if we don't change," warned Moody-Stuart. "Change is a pearl beyond price." The obstacles were Shell's fragmented culture, divided management and entrenched country barons who had successfully frustrated Herkströter's reforms. To many British employees, the Dutch engineers' arrogance was stultifying. Convinced of their superiority, they regarded their rivals at Exxon, Chevron and especially BP with measured contempt. Yet some refused appointments in unpleasant oilfields, preferring to remain in the comfort of European and American offices, focused on investment and process rather than practical work on the ground. Convinced of the righteousness of science and engineering, the LNG department had seriously advocated building a terminal near the Bay Bridge in San Francisco.

"I'm clearing out the cupboard," Moody-Stuart announced, planning instant surgery. Offices around the world were closed and country chairmen demoted, 4,000 staff were dismissed, 40 percent of the chemical plants sold, $4.5 billion of bad investments written off, capital spending cut by one third and, most dramatically, American Shell lost its independence. Appallingly managed and beyond financial control, US Shell represented 22 percent of the company's assets, yet contributed only 2.6 percent of its earnings. Walter van de Vijver, a 42-year-old engineer, was dispatched to integrate the American company with its European owner. The cost of Moody-Stuart's surgery was huge. Shell's net income fell by 95 percent, from $7.7 billion in 1997 to $350 million in 1998. There was little optimism that things would improve. The oil price in 1998, Moody-Stuart believed, was "likely to stay at $10," and the likelihood of it going above $15 was "low." At those prices, Shell's profits, like BP's and Exxon's, were certain to fall further.

Moody-Stuart's parallel agenda was to reform Shell's "Business Principles." A team had been working since September 1997 to develop a five-year strategy to resolve dilemmas involving human rights, global climate change and environmental problems. A larger question was whether any of these activities made sense in a "world of $10 oil." Moody-Stuart was emphatic that his strategy was to generate profits "while contributing to the well-being of the planet and its people." By then Watts had completed his study to alter Shell's reputation. To boost employees' self-esteem and to celebrate the "transformation process," Moody-Stuart agreed that Watts, the new head of exploration and production, should stage a stunt. At a conference of 600 Shell executives in Maastricht in June 1998, Watts was propelled onto the stage in a spaceship, dressed in a spacesuit. "I have seen the future and it was great," he yelled to his audience, all of whom were wearing yellow T-shirts emblazoned with the slogan 15 PER CENT GROWTH. The onlookers were, remarked one eyewitness, "gobsmacked" by Watts's attempt to remake his "dour, pedantic image." Everyone understood his agenda, however: Shell's reserves were falling, and targets needed to be stretched. Managers were formally urged to "improve our effectiveness." The message was "improve the scorecard." At the end of his presentation, Watts urged his flock to sing Beethoven's "Ode to Joy": "Somewhat over the top," Moody-Stuart admitted. "We all do foolish things occasionally." Galvanizing morale had been important. The oil majors were facing a torrid time. Those that failed, Moody-Stuart knew, would be buried alive. Executives from four American oil companies—Mobil, Amoco, Arco and Texaco—had approached Shell seeking mergers or to be bought. Shell's split structure made that impossible. The company, Moody-Stuart knew, needed a counterplot to resist the unexpected challenge posed by BP.

Chapter Five

The Star

————◁◦▷————

JOHN BROWNE UNDERSTOOD oil better than most. Shell's Mark Moody-Stuart, Chevron's David O'Reilly and Exxon's Lee Raymond could not match Browne's intellect and bravado, but none had as much to prove. Employed by BP since leaving Cambridge University, the son of a BP executive who had met his Romanian mother, a survivor of Auschwitz, in post-war Germany, Browne understood that trouble and taboos had been inherent within BP since its creation. During his youth he had lived with his parents in Iran and had witnessed the company's arrogance and subsequent humiliation. The industry's roller-coastering battles ever since encouraged his taste for audacious gambles to rebuild a conglomerate lacking geographical logic and natural roots.

BP was founded on disobedience and survived by maverick deeds. The original sinner was William Knox D'Arcy, a wealthy Australian who arrived in Persia in 1901 on a hunch that oil could be discovered there. D'Arcy negotiated a 60-year concession over 480,000 square miles of desert. For seven years his team drilled unsuccessfully across an area twice the size of Texas, until in 1908 he was ordered by Burmah Oil, a Scottish investor, to stop. Having started yet another test bore, D'Arcy's team ignored the message and, detecting a strong smell of gas, struck oil. There was no natural reason why that fortuitous discovery should have evolved into the formation of a famous company. Culturally, the directors of the new Anglo-Persian Oil Company based in Glasgow were embarrassingly ignorant about their faraway asset. In contrast to the American oil companies that had spawned an integrated market built on discoveries in Texas and across the prairies, Anglo-Persian, which became BP, was a colonial concession sponsored by the British government. Managed by retired military officers recruited particularly from the Indian army, its staff clung to their suzerainty. Amateurs in marketing and untrained to supervise refineries and chemical industries, they aspired to be gentlemen, and were generally indifferent to indigenous politicians, especially Arabs and Iranians, whom they regarded as inferior. Unlike Shell's country chairmen, soaked in local cultures and enjoying rapport with host governments, BP's managers carelessly alienated their hosts, offhandedly oblivious of Iraq's and Iran's vast oil wealth.

Little changed before the nationalization of BP's oilfields in Iraq in 1951. Sir Eric Drake, the corporation's conceited chairman, assumed that the confiscation would be compensated by increasing oil prices and the discovery of new reserves in Libya, Nigeria and Abu Dhabi, or by expanding into petrochemicals and shipping. Over the next 20 years, BP balanced the escalating demands of the Shah of Iran, the bellicosity of OPEC and Arab nationalism, especially in Libya, by finding new oil in Alaska in 1968 and the North Sea in 1970. The problem was the directors' lack of commitment to exploration. The discovery of a new field, noted the exploration department in 1971, evoked the reaction, "What on earth are we going to do with all this oil?" Terry Adams, BP's director in Abu Dhabi, was expected to embody that casual attitude. To finance a pipeline in Alaska, Adams was ordered in early 1973 to sell half of BP's share in Abu Dhabi's offshore interests to a Japanese company for $736 million. "This is top secret, none of the locals need to know," BP's manager Roger Bexon told him, referring to Sheikh Zaid, the leader of the state. In his anger after the sale was announced, Sheikh Zaid nationalized half of the Anglo-Japanese investment. The

Japanese never believed that BP was unaware of the impending confiscation, and the Abu Dhabians griped about BP's lack of respect. Insouciantly, the British pleaded ignorance, underestimating the profoundly negative consequence of their arrogance.

Arab irritation compounded BP's problems in the region after the 1973 war. In succession, the company's oilfields in Kuwait and Libya were nationalized. Overnight, BP's plight was dire: the company had become entirely dependent on the discovery of oil in Alaska and imminent production in the North Sea, and it had fallen in rank from membership of the Big Three to seventh among the Seven Sisters. Morale was flagging, and there were even fears that BP faced extinction. Unlike the precise management processes at Chevron, Mobil and Exxon, which ran in harmony regardless of the identity of the individual chief executive, BP's direction depended upon the chairman's vision. "There are no sacred cows," declared Peter Walters, appointed chairman in 1981, who advocated retrenchment. BP's focus was to be entirely oil. Following Exxon and Shell, Walters slowly reversed the diversification into non-oil businesses and ordered a $6 billion sale of all the nutrition manufacturers and mineral interests. He seemed unable to do much more to salvage the company from the morass. Impaired by the British government's nonchalance, BP was crippled by debts, aggravated by the government's order to repurchase about 10 percent of the company's shares from the Kuwaiti government that had been bought during a disastrous flotation. In an industry dominated by Exxon and Shell, BP had stalled, destabilized by debt. Walters never recovered his self-confidence.

Two BP directors in America regarded Walters's cuts and style as merely scratching the surface rather than offering a revolution. In 1983, Bob Horton, a brash 46-year-old fellow of the Massachusetts Institute of Technology, and his 35-year-old deputy John Browne had arrived at BP's American headquarters in Cleveland, Ohio, to supervise BP's 54 percent investment in Sohio, the successor to the Standard Oil Company of Ohio, the original John D. Rockefeller corporation. The purchase had given BP an entrée into Alaska, but London had failed to prevent the American directors buying a copper-mining company, wasting $6 billion of Alaskan profits. "Sohio's completely out of control," exclaimed Horton. "They're losing $1 billion a year." Originally acquired in 1970, Sohio was Horton's platform to prove his credentials as Walters's successor. As head of BP chemicals in 1980, he had closed 20 plants and fired two thirds of the workforce. The cure at Sohio in May 1987 was to buy total ownership for $7.9 billion (£2.5 billion) and dismiss swaths of staff. Sohio, Horton and Browne proudly announced, would earn profits of $560 million within two years. Renamed BP America, it represented 53 percent of BP's total assets. From Ohio, the warts of BP's culture in London were glaring. Deprived of courage, hope and energy, BP could only be resuscitated if the employees' historical aversion to risk was replaced by American entrepreneurship. Their successful remedy in Cleveland, Horton and Browne decided, should be applied to the whole company after they returned to London in 1989.

Like most oilmen, Horton and Browne believed in 1989 that "demand had peaked," and oil would remain cheap because high prices stunted demand. Exxon, Mobil, Chevron and other more powerful competitors argued that prices were unpredictable, and survival depended upon cutting costs. Horton encouraged Walters to follow the herd. "BP cannot survive with this culture," he told Walters after listing 11 layers of management. "It's sclerotic. Get rid of the brigadier belt. Too many have a vested interest to sabotage change." Starting from scratch, said Horton, BP needed to be repositioned and to duplicate Shell's "wonderful worldwide brand." Browne, as the new chief executive of exploration, echoed that criticism. In June 1989 he commissioned a presentation for investors in London and at the Rockefeller Center in New York. "This is dreadful," he said after previewing the slides. "We're declining." BP's access to 70 billion barrels of reserves had dropped to four billion, and were not being replaced. Production was falling from 1.5 million barrels a day to below one million. While its rival Shell had successfully retained profitable oil and gas fields in Nigeria, Oman, Malaysia, Brunei and Holland, BP would go out of business unless it found new, big prospects. Tom Hamilton, the

American chief for international exploration, was told by Browne to present a scenario for a new strategy. "I'm going away with my family on holiday," explained Hamilton. "Take the company plane and come back early," ordered Browne. "I'll need 90 days to do it," replied Hamilton. "You've got three days to calculate the best odds to discover more oil," replied Browne. In September 1989, Browne commissioned new exploration operations in Yemen, Ethiopia, Vietnam, Angola, Gabon, Congo, South Korea and the Gulf of Mexico.

Few doubted the need for brutal surgery. Peter Walters's retirement in early 1990 provided the opportunity for change. Persuaded by Bob Horton's presentation about his achievements and by his argument in favor of a cultural revolution, the board unanimously picked "Horton the Hatchet" as BP's new chairman and chief executive. "Project 1990," said Horton, "is my personal crusade to revolutionize the company." Twelve thousand employees would be dismissed and $7 billion of assets sold. Horton espoused drama as a resolution to the crisis.

Eighty-two committees at BP's London headquarters in Finsbury Square were axed, leaving just four. The 11 layers of management were also reduced to four. To inspire enthusiasm and to reincarnate BP's 120,000 staff as open-minded and freethinking, Horton participated in "cultural change workshops" with 40 senior staff to discuss the "new vision and values." His propagandists praised "the terrific buzz which motivated us to get the change moving," but others carped that the balance between pain and progress was wrong. Horton had chosen Jack Welch's operation at General Electric as his model for a centralized, focused corporation. In the oil business, no one could ignore Lawrence Rawl, the chairman of Exxon. Although Exxon was, in Horton's opinion, "wildly overmanned and too engineer- and lawyer-led," Rawl consistently produced successful results. Horton's public predictions, accompanying jerky attempts to build solid corporate foundations, compared poorly with Rawl's rare but pertinent statements about Exxon's unflustered deliberations. As oil prices gyrated in late 1990 from $40 down to $31, Rawl cautioned that uncertainty made investment decisions difficult: "This is a long-term business. We cannot turn the money off and on every time someone clears his throat in the Middle East or elsewhere as the price goes up and down."

The "cough" was Iraq's invasion of Kuwait in August 1990. America's oil industry was still struggling. Oil production had fallen every year since 1986 by between 2.5 and 6.5 percent. Banks remained reluctant to lend because of the continuing uncertainties. The oil business, it was said, was as safe as rolling dice in Las Vegas. Even Exxon lacked sufficient money and personnel to instantly boost production. The US government offered no leadership to fashion a new energy policy. In 1988 America had believed that George Bush Sr. was the oil industry's dream candidate, although as Ronald Reagan's vice president he had offered it no help, and he had in fact campaigned for the presidency as an environmentalist. During his single term Bush would dilute an energy bill giving the industry minor tax relief, would not limit imports, and would cancel the sale of eight offshore leases. Texans, surrounded by abandoned derricks, were angry that the president sent the army to Kuwait out of fear of losing 1.5 million barrels of oil a day, but that no one appeared to care about Texas's similar losses since 1986. Their anger spread to contempt for East Coast liberals and Californian environmentalists who nevertheless still harbored a sense of entitlement that energy should be abundant and cheap. Hoping for a cautious recovery from that economic devastation, Horton concluded that "the fundamental realities point to higher oil prices." BP, he decided, needed to change fast.

The hyperactive Horton lacked Rawl's gravitas. He misunderstood Exxon's foundations, created in around 1865, and built on vast untapped reserves of oil. Ever since John D. Rockefeller's retirement in 1897, the corporation had been led by domineering personalities molded by Exxon's character and caution. Unlike that prototype, Horton was not fashioning himself as a conservative, sober, confident chieftain, but was duplicating the caricature of a brash American chief executive. After four years in Cleveland, he had forgotten that BP was a British boys' club, uniting in a collegiate atmosphere people who had lived, worked and played together for 25 years. Running too fast, he was failing to implement his own plans. Instead of focusing on the cuts, he ordered BP to expand despite the continued recession. At a time when the price of oil was about $16 a barrel and slipping, he expected that it would rise to $21 or even $25. Convinced of his own genius, he welcomed personal publicity. Impulsive and careless with his language, he told the first journalist invited into his office: "I'm afraid because I am blessed by my good brain which is in advance of my colleagues', I tend to get to the right answer rather quicker and more often than most people." (He would forever regret this remark: "It came out wrong, and I have had it hung round my neck ever since — never ever did I think I was a genius, far from it.") The cover of *Management Today* featured Horton holding a hatchet, while *Forbes* magazine photographed him sitting on a throne. There was gossip within BP's headquarters about Horton asking his secretary, "Should I go to a charm school?" His insensitivity bewildered his colleagues. Newspapers began reporting Horton's unpopularity, one asking: "When Robert Horton and his wife return from their holiday in Turkey, many BP staff will hope that their plane will crash." David Simon, a managing director, was told that Horton concealed such criticisms from his mother. "Good God," exclaimed Simon. "Horton has a mother!" Another executive told Horton to his face, "Why don't you bugger off to Chessington Zoo and watch the gorillas and monkeys?" "Why?" asked Horton. "Because you might learn a lot."

Relations between Horton and his fellow directors were not improved after they arrived at Heathrow Airport on June 23, 1992, to fly to Alaska for a board meeting. Horton was overheard having an unseemly argument with the BA employee at the check-in desk. The atmosphere at the board meeting was fractious. BP would record its first quarterly net loss of £650 million ($1.24 billion) after its income fell by 82 percent, compared to a £415 million profit in 1991. The debt had increased to $16 billion and the share price had slid from 332 pence in June 1990 to 209 in June 1992. "We're bleeding cash like crazy," said one director, querying why the proposed cuts had not materialized, especially at the refineries. "You can count on BP's DNA to find an inspired route out of the trouble," countered a Horton sympathizer, only to be crushed by another director: "Exxon and Chevron don't get into trouble." Oblivious to the storm, Horton insisted during the board meeting that BP should pay a normal dividend to please investors. "Could you wait outside?" he was asked by the banker Lord Ashburton. Beyond his hearing, the reckoning was swift. "He's spent too much time with ambassadors and playing politics in Washington," said one voice. "And he's spent too little time on the details of the business," added another. "Bob is ambitious, abrasive and arrogant," concluded a third. "We need a change." The mood was summarized by Ashburton: "There's been a build-up of small flakes which has become quite a lot of snow on the ground." Three weeks later the nonexecutive directors, including Ashburton and Peter Sutherland, met at Barings bank in the City, London's financial district, on a Saturday morning to decide Horton's fate.

The unsuspecting chief executive was summoned the following Wednesday. "Robert," said Ashburton, "the board has decided to ask for your resignation." "My God," exclaimed Horton, shocked that his fate was even being discussed. "I was brought down as laughable," he reflected. "I got a head of steam. My mistake was believing change could be done so fast. I should have shown more tenderness." The public announcement was stripped of any charitable sentiment. "Hatchet Horton's" decapitation matched the cultural change he had championed, except that his dismissal was interpreted by outsiders as the final collapse of a stodgy giant. BP, rival oil companies believed, would shortly be receiving the last rites.

Horton was replaced by David Simon, a trusted team player with expertise in refining and marketing. "This is about the style of running the company at the top," Simon said about his predecessor. "It's not that I don't have an ego. It's just that it's not terribly important to me." Simon, a cerebral linguist, acknowledged his limitations. "Look, chaps," he frequently smiled during meetings, "you know I'm not very bright, so could you explain this in simple language?" Six weeks after Horton's dismissal, BP halved its dividend. Horton's intention to copy Exxon and centralize BP was reversed. Power was devolved to trusted subordinates who would be accountable to business units, an innovation introduced by McKinsey & Company, the management consultants. That suited John Browne, the head of exploration and production and the heir apparent. Although Browne's admirers described an occasionally soft and lonely character, fond of ballet and opera and not inclined to socialize, he espoused confrontation to resolve problems. BP's style, he believed, should not attempt to mimic Exxon's. Hierarchies and conformity were to be destroyed, and to encourage initiative there would be informal lunches, no lofty titles, and meetings between forklift drivers and accountants. Outsiders were greeted by charm, but employees understood the ground rules of a self-styled alpha male: "One mistake and you're out." His lesson from Sohio was the importance of consolidation and cuts.

"I'm astute enough to know what I'm doing," Browne told Tom Hamilton. In 1991, after working with him for six years, Hamilton admired Browne's negotiating skills and passion to reduce costs, but questioned his limited experience. In his early career Browne had chopped and changed between jobs, spending just nine months at the Forties field in the North Sea and the same amount of time in Prudhoe Bay, never staying long enough to see his mistakes emerge. Not only was his knowledge about operating in the mud and sand of oilfields superficial, but he lacked any taste for solving engineering problems. Working in an office filled with monitors displaying information to feed his appetite for facts, he concealed his limitations by obtaining detailed dossiers on every face and every issue in order to brief himself before meetings. Browne's impressive ability to absorb information, Hamilton feared, produced blindness about the whole picture and an inability to anticipate what could go wrong.

That weakness, Hamilton believed, stemmed from Browne's addiction to the wisdom handed down by McKinsey. Persuaded during his studies at Stanford in California that BP's experts could be replaced by consultants, he appeared to become a financial executive surrounded by accountants focused on balance sheets to satisfy the shareholders, rather than harnessing engineering skills to manage a project. "To save money," Browne had argued, "we can buy in what we need." In Browne's opinion, Hamilton did not understand the skill required to direct BP's limited cash toward prospective windfalls. BP's technicians, he felt, needed to be business-oriented. Making profits was his only criterion, whether by improved technology, lower costs, reduced interest payments or higher volumes. "The engineers in Aberdeen gold-plate everything," he complained. "They're inefficient and wasteful." BP's engineering headquarters at Sunbury, infamous for pioneering "space grease" and constantly reinventing the wheel, was to be closed. Browne saw no incongruity in an oil and chemical corporation relying on hired freelance engineers. "If this goes wrong, John," Hamilton warned, "there'll be no place in the world to run and hide."

Browne's conception of himself as a different kind of oil executive leading a different kind of oil company did not appeal to Hamilton. The final straw was an argument about cutting costs during an 18-hour flight to inspect an oilfield in Papua New Guinea. Browne's antagonism toward BP's traditional embrace of engineers irritated Hamilton. "We may have to turn back, John," he cautioned halfway through the helicopter flight across the jungle. "Cloud could prevent us landing." Just before they arrived, sunlight burst through the clouds. "So why so many problems?" chided Browne. Hamilton resigned soon after, avoiding the profound change Browne demanded in exploration. Profits, said Browne, depended on cutting costs, especially exploration costs, by 50 percent, from $10 to $5 a barrel, while at the same time finding enough new oil to start replacing BP's depleting reserves in 1994.

Accurate forecasts of oil prices had become impossible after 1986. For the first time prices were varying during a cycle of boom and bust. Conscious that the oil majors had invested too much during the 1960s, Browne pondered the revolutionization of the industry's finances. The new challenge was to balance the cost of exploration and production with the potential price of oil five years later. Oil companies, Browne knew, could only prosper if the cost of exploration and production matched market prices once the crude was transferred from the rocks to a pipeline. The yardstick for BP, the measure of future success, would be to equal Exxon, the industry's most efficient operator. Exxon's net income per barrel—the income divided by production—was about one third of BP's. In estimating the cost of all new projects, Browne ordered that regardless of whether oil prices were low or high, BP would only invest if profits were certain. With losses of £458 million in 1992, the new wisdom reflected BP's plight. The corporation could not risk losing more money. If his plan was obeyed, Browne predicted, BP's annual profits by 1996 would be $3 billion.

Predictions were also offered by McKinsey, which in 1992 forecast the atomization of the major oil companies into small, nimble operators. The consultants foresaw excessive costs burdening the oil majors, restricting their operations. Too big and too expensive to run, they would give way to small private companies and the growing power of the national oil companies. By the end of the century, according to McKinsey, the Seven Sisters would shrink and their shares would no longer dominate the stock markets. Browne rejected that scenario, believing that only the majors could finance the exploration and production necessary to increase reserves. He would be proved partly wrong. Although the oil majors' capitalization in 2000 was 70 percent of all quoted oil companies (McKinsey's had predicted that their value would fall below 35 percent in the stock markets), Browne was underestimating—albeit less than his rivals—the resurgence of nationalism. The national oil companies were increasingly relying on Schlumberger, Halliburton and other service companies and not the majors to extract their oil. But, fearful of excessive costs, he was attracted by McKinsey's formula to replace BP's conventional management structure. To a man interested in the dynamics of the industry but not in the minute detail of "what you had to do after you bought your latest toy," the idea of establishing competing business units answerable to a chief executive was appealing. By contrast, Exxon had neutralized individual emotions and relationships to standardize the response to every problem and solution. Depersonalizing employees to serve BP's common purpose, Browne believed, would be self-destructive. BP, he knew, was too raw and too fragile to emulate Exxon's self-confidence. The company's staff would be encouraged to use their own initiative in the field. Taking risks was necessary for BP to survive and grow, but those risks would be subject to Exxon's style of ruthless control of costs from headquarters.

"We're stamp-collecting in exploration," Browne told Richard Hubbard, the company's senior geologist. "We either make money or walk away." He reduced the number of countries where BP was exploring from 30 to 10, and sacked 7,000 employees. "We must focus only on elephants," he ordered. "It's the New Geography," acknowledged David Jenkins, the head of technology. BP was heading for unexplored areas previously barred by physical and political barriers.

The new ventures included offshore sites in the Shetlands, the Gulf of Mexico, the Philippines and Vietnam. The most important risk was a 50 percent stake in the search for oil under 200 meters of water at the Dostlug field in Azerbaijan, and a $200 million search at Cusiana, 16,000 feet up in the Colombian jungle. Colombia, Browne told analysts in New York during a slick presentation in 1993, was to be the hub of BP's growth: "We estimate that the field contains up to five billion barrels of oil." His optimism was conditioned by self-interest, but would yield an unexpected benefit. Oil prices, David Simon predicted in 1992, would remain at $14 a barrel until 2000, half the 1983 price accounting for inflation. The Arab countries, Simon was convinced, would welcome BP back, "and we'll get our hands on cheap oil." While OPEC complained to the British government about North Sea production undercutting the Gulf's prices, some OPEC countries, suffering reduced income, were reversing their hostility toward foreign investment. Production in Venezuela had fallen since the nationalization of its oilfields in 1976. BP was invited to bid to return to over 10 fields, including the Pedernales field, abandoned in 1985. Browne's excitement, compared to Shell's cagey hesitation, gave BP the image of a well-oiled machine. Other decisions by Browne suggested the contrary. During his "good news" speech in New York he declared that the tar sands had no future, investing in Russia was too risky, and BP would not invest in natural gas in Qatar because "the project will not provide a good return."

Browne's self-confidence was fed by the inexorable monthly rise of

BP's share price. Helped by cuts in the cost of refining and marketing, and in exploration from $4 billion in 1990 to $2.7 billion in 1994, and by the sale of $4.3 billion in assets including 158 service stations in California, profits were rising—in one quarter by 92 percent. The transformation of BP's operation in Aberdeen from loss into profit sealed Browne's reputation. Oil production had expanded in the North Sea, especially at the Leven field, and the company was certain to extract more oil from Alaskan fields newly acquired from Conoco and Chevron. Since the US preferred Alaska's light sweet oil to Saudi Arabia's sour oil, OPEC would suffer. "One swallow doesn't make a summer," David Simon cautioned, conscious that oil prices were low and that BP still relied for its entire reserves on Alaska and the North Sea, both of which were nearing the peak of production. Nevertheless, it seemed that the struggle to recover was succeeding. Browne's admirers spoke of his magic restoring a dog to its place as one of the world's oil majors. In 1995 BP became the industry's darling, overtaking Chevron, Mobil and Texaco with profits of $3 billion. Debt had been halved from $15.2 billion to $8.4 billion. "We've clawed our way back," cheered Simon, who in July 1995 became chairman, with Browne as chief executive. "We've put them through painful changes." Browne's ambition to promote himself as a different kind of oil executive and BP as a changed company had triumphed beyond expectations.

Browne's skill was to highlight his achievements and bury his failures. Several of his ambitious hunts for elephant oil reserves had produced "orphans." In Colombia, the earlier focus of euphoria, the company had become embroiled in a public relations battle with a left-wing pressure group over BP's involvement in a civil war, the narcotics business and a regime of terror waged by paramilitaries employed to protect BP's 450-mile oil pipeline. The alleged victims were native farmers whose land had been portrayed in an orchestrated campaign as confiscated, their water reserves depleted and their livestock slaughtered. Worst of all, the oil wells were producing less than half what Browne had anticipated. After substantial criticism, BP would eventually compensate the farmers. BP's rivals were suffering similar disappointments.

On the basis of promising geology, Mobil had invested heavily in Peru. "I mean, this was classic," said Lou Noto, the company's president. "This is the classic way of how to do it. Yet we came up with a dry well—$35 million later." Exxon had similar failures in Somalia, Mali, Tanzania, Mozambique, Nigeria, Chad and Morocco. Shell wasted money in Madagascar and Guatemala. Arco had wasted $163 million drilling 13 orphans in Alaska. Over the previous decade, about $14 billion had been dissipated in unsuccessful attempts to repeat the last big finds in the North Sea and Alaska. Those discoveries had cut OPEC's share of the world's oil production from 50 percent in the 1970s to 30 percent in 1985. In 1994, OPEC's share rebounded to 43 percent, while it retained 77 percent of the world's reserves. Shell fired 11,000 of its 106,000 worldwide workforce. In the same year, American production fell to 6.9 million barrels a day, the lowest since 1958, and the country became a permanent net importer of oil. With demand for oil rising, OPEC's influence appeared certain to increase. Those statistics encouraged Browne in 1995, despite his earlier reservations, to seek opportunities in Russia.

Russia's oil could replenish the oil majors' reserves and counter OPEC's influence. Despite the bribes and the gangsters, none of the oil chiefs jetting into Russia on their private jets from Texas and California hesitated to assert their indispensability in saving Russia from destitution, and US vice president Al Gore did not pause to consider the consequences of flying to Kazakhstan in December 1993 to encourage the country's split from Russia, spiting the nationalists in Moscow and Saint Petersburg. On the contrary, causing anger among the Russians excited President Clinton and others in Washington. Russia's debt crisis, declining oil production and political instability, they believed, presented an unmissable opportunity. With the US importing half its oil consumption, Clinton made the diversification of supplies a priority, and the Caspian could offer at least 200 billion barrels. To win the gamble, the politicians combined with BP's John Browne, Exxon's Lee Raymond and Ken Derr of Chevron to display utter indifference to Russia's gradual collapse.

Chapter Six

The Booty Hunters

————◄○►————

THE INTRODUCTION OF DEMOCRACY wrecked Russia's oil industry. To secure political popularity in 1989 for "Glasnost" and "Perestroika"—openness and reform—Mikhail Gorbachev had diverted investment from industry to food and consumer goods. Blessed by reopened borders, free discussion in the media and the waning of the KGB, few in Moscow noticed the crumbling wreckage spreading across the oilfields in western Siberia, an area of 550,000 square miles, nearly the size of Alaska.

Finding oil in that region after the Second World War had been effortless. Gennady Bogomyakov, the first secretary of the Communist Party in Tyumen province, was famous during the 1950s for increasing production from the easiest and best fields "at any price," regardless of the environmental cost or human welfare. In that plentiful region, Russia's oilmen were blessed with outstanding science, but cursed by problems they themselves caused—poor drilling, damaged reservoirs, neglected equipment and reckless oil spills. Instead of cleaning up the mess, wells were abandoned and the engineers moved on to new fields. Rather than halting the destruction, Gorbachev's encouragement of a consumer revolution inflamed it. Overnight the flow of money from Moscow to pay for repairs and salaries and to drill new wells stopped.

Angered by Moscow's indifference to their deteriorating working conditions, poor housing and food shortages, the oil workers in 1990 began to produce less oil, the first decline since 1945. The relationships between companies in different regions also began to fracture. Oil companies in Siberia found difficulty in persuading factories in Azerbaijan to supply equipment, especially pumps; and some oilfields in Azerbaijan, the Caspian and western Siberia refused to supply crude oil to refineries.

After the disintegration of Soviet control over Eastern Europe, Gorbachev was confronted by national governments in Azerbaijan and Kazakhstan, both oil-rich states, agitating for independence. He remained blithely unaware of the potential problems until the country was struck by shortages of fuel. Gas stations closed in Moscow, and airlines stopped flying. Beyond the major cities, towns were dark, visitors wore overcoats in their hotel rooms and the harvest in Ukraine was jeopardized. Living standards were falling, and there were threats of strikes. Reports from Siberia warned Gorbachev: "The situation is very serious. It is creating an explosive atmosphere." The ruble's value began sliding, Russia's international debt rose, and the country's oil companies began bartering oil for equipment, or even demanding dollars for domestic sales. Russia's daily oil production fell during 1989 from 12 million barrels a day to 11 million. "The atmosphere is exceedingly tense despite government promises," a trade union leader told the Kremlin. Gorbachev's indecision, complained L. D. Churilov, president of the government oil company Rosneft, was causing the crisis.

As oil production in 1990 declined toward 10 million barrels a day, Gorbachev was urged that only foreign investment and Western technology could rescue Russia's economy from collapse. There were precedents for similar appeals. Ever since the first gusher of oil had burst through a well in Baku in Azerbaijan in June 1873, Russia had allowed foreign companies to produce oil on its territory when times were bad. After the Bolshevik revolution in 1917, and again after the Allied victory in 1945, foreign oil companies had been lured into Russia, only to be expelled as production and prices improved. In 1990, admitting that

Russia's plight was "catastrophic," Gorbachev appealed to Germany for help. His choice was odd: Germany was almost the only Western country without any expertise in oil production. After his invitation was extended to all Western oil companies, many seized the opportunity as an alternative source of oil following Iraq's invasion of Kuwait. In contrast to the turbulence in the Middle East, Gorbachev appeared to be offering Western oil companies safe investment opportunities in 12 vast areas, totaling the size of the United States, with more oil and gas than the whole of the Middle East. Only a fraction of the oil under the Siberian plains and the Arctic had been extracted.

Despite the lack of any formal agreements, the oil companies could not resist the opportunity. Loïk Le Floch-Prigent, the chairman of Elf, the corrupt French national oil company, led the way. "I'm the boss," Le Floch-Prigent insisted, refusing to work with any Russian partner. The French were followed by ENI of Italy, another corporation tinged by corruption whose former chairman, Gabriele Cagliari, would later "commit suicide" in prison, suffocated by a plastic bag. Then came the Anglo-American majors. Exxon and Mobil focused on western Siberia, Chevron sent a team to Kazakhstan, BP and Amoco competed in Azerbaijan, Marathon Oil, a second-division oil corporation based in Houston, snooped around Sakhalin on the Pacific coast, all jostled by experts representing smaller companies. The Western prospectors had suspected that Russia's oil industry was, like its military services, "Upper Volta with missiles," an image conjured in the 1970s by a Western intelligence agency, comparing the impoverished West African country with Soviet Russia. None appreciated that the best of Russia's geologists and engineers were as talented as their Western counterparts; but nor had anyone imagined the chaos of Russia's oil production. Mediocrity had suffocated the flair.

The detritus was staggering. Thousands of wells had been damaged or abandoned. By 1989, isolated from the West, Russia's proud oil engineers had been unaware of technological developments in the outside world. Unable to drill beyond 10,000 feet and ignorant about horizontal drilling, the Russians had constantly pumped water into the rocks to maintain the volume of oil, leaving 80 percent of the wells contaminated. Poor engineering, bad cement, imprecise drills, failing compressors and mechanical breakdowns had caused a gigantic stain to spread across the landscape. During 1989, thousands of corroded pipes in western Siberia had broken, spilling about 51 million barrels of oil onto the ground and into rivers. Most of them remained unrepaired. The catastrophe was reflected in a single report presented to Gorbachev. In 1980, new wells had produced about 2.85 million barrels a day, but a decade later the rate had fallen to 1.28 million. Only Western expertise could reverse Russia's predicament. The benefits would be mutual. The oil majors needed new sources of crude oil, and Russia offered enormous potential.

A handful of oil executives moved around carefully "to smell the coffee and get to know the relevant people," but they encountered deep-rooted suspicion. Russia's oilmen questioned the motives of those who, after decades of NATO's embargo preventing Russia's purchase of Western technology, demanded access on a grand scale on their own terms. "Seventy years of mutual misinformation and mistrust must be set aside," said Tom Hamilton, newly appointed as president of Pennzoil, a medium-size American oil corporation. The distrust was partly a legacy of Cold War enmities, particularly doubts about America's motives after the publication of a CIA prediction in 1977 that poor conditions in Russia's oilfields would compel the country to import oil by 1985. The forecast was mistaken, but Russia's plight was, in the Russians' opinion, linked to a 1985 visit to Washington by Saudi Arabia's King Fahd. President Reagan had urged the king to increase oil production in order to cripple Russia's earnings from oil exports, which amounted to about 40 percent of its foreign income. Oil prices had in fact fallen from a peak of $50 a barrel in 1985 to around $25 in 1990, increasing Gorbachev's panic and the Russian oilmen's suspicions. Veterans who knew their history were aware that in 1917, Western oilmen had rushed into Russia hoping to pick up bargains and prevent the Bolsheviks undercutting their cartel by flooding the world with cheap oil.

Andy Hall of Phibro, among the first Western visitors to western Siberia, was undeterred by such misgivings. The region was being promoted by Houston entrepreneurs as an opportunity to acquire oil reserves for pennies a barrel, and Hall was persuaded that although it had been exploited over the previous 50 years, new technology could produce huge windfalls of oil and profits. The uncertainty created by the Gulf War encouraged his confidence, shared by most Western oil-men and governments, that Russia would provide a secure supply of oil, free of OPEC's interference. The lure to invest was made more tempting by Phibro's trading losses. Hall had overestimated the poten-tial volatility of prices caused by the war and the early stages of the 1991–92 recession, resulting in losses at Phibro's refineries at St. Rose, Louisiana, and in Texas. He had also failed to balance the increas-ing demand for diesel and the decreasing demand for gasoline, which required different crude oils. In the first nine months of 1992, Phibro lost $34 million. Calculating the odds as a trader without the advice of independent specialists, Hall assumed like others that the Krem-lin's invitation was genuine, and that profitable oil from western Siberia would compensate for the refining losses. His company White Nights promised to invest $100 million and to hire the best expertise.

Hall's investment was exceptional. The oil majors were uninterested in providing Russia with technical advice or investing in old oilfields. Their aim was to find new Russian oilfields and book the reserves. Mobil was focused on Yakutia, 1.25 million square miles of virgin ter-ritory, five times the size of Texas, with only a few wells but guaran-tees of vast reserves. Amoco's team headed for Novy Port, 1,400 miles northeast of Moscow, on the Yamal Peninsula, committed to spending tens of millions searching for oil and gas while surrounded by people surviving among leaking pipes and polluted soil and water, with high levels of cancer and without adequate heating in a region where the temperature fell to –27°C in winter. Texaco, led by Peter Bijur, began prospecting in Sakhalin, an oil- and gas-rich island on Siberia's Pacific coast. Conoco excitedly signed deals to develop oilfields in the Arctic Circle and at Shtokman, a giant discovery in the Barents Sea. Chevron offered to invest in Kazakhstan. The temptations for local politicians were overwhelming.

In the barren Kazak desert—a harsh, unexplored, landlocked region of nearly 200,000 square miles—the Russians had found large flows of "very high quality" oil in the early 1980s. With proven reserves of 39.6 billion barrels of oil and 105.9 trillion cubic feet of gas—3.3 percent and 1.7 percent of the world's proven reserves—and huge deposits of minerals, no one doubted that Kazakhstan could become one of the world's top 10 energy producers. A thousand wells had been drilled, but by 1990, with less than 20 percent of the oil extracted, most had been abandoned. Russian failure had been worse along the shallow waters of the Caspian Sea. According to folklore, the villagers had dug wells for water in the mid-1970s and found oil. In the mid-1980s Russian engineers realized that the Tengiz deposits, on the northeast shore of the Caspian, were among the world's biggest. The light, honey-colored crude was perfect for refining into gasoline. But the Russian engineers were unable to erect rigs in 400 feet of water; a pipeline 282 feet below the surface fractured because of poor-quality welding, and an offshore platform was blighted by fires. Exploring beyond the shallow waters had been impossible because the Russians had never mastered horizontal or air drilling, which would mean that the oil, mixed with poisonous gas and under hydrostatic pressure, would have risked exploding. Two billion rubles spent since 1979 had been wasted. Conceding that their performance would not improve and fearing environmental damage, the Russians had acknowledged the obvious. Only Western technology could reach Tengiz's 16 to 32 billion barrels of oil, trapped under half a mile of salt, 5,400 meters beneath the seabed. Existing technology could recover between six and nine billion barrels from one of the world's largest and deepest fields. Future technology could reach the remainder.

Despite the political, financial and engineering problems, Ken Derr, the chairman of Chevron, decided to gamble the corporation's fortunes on the prospect. Chevron's foundation, in its previous incarnation as Standard Oil of California (Socal), had been rooted in the discovery of oil just north of Los Angeles in 1879. The company's glory years had taken place in Saudi Arabia. In 1938, Socal's employees had found the first oil in the desert, and over the following 35 years the company, cooperating with Texaco in Aramco, had sporadically flourished, not least after it identified the giant Saudi oilfield Ghawar in the 1950s. But the corporation, one of the Seven Sisters, wilted after the Saudi government progressively nationalized Aramco's oil wells. The 1980s were Chevron's nadir, as inferior technology yielded a string of dry holes. Fearing that its future was at risk without new oil, in 1984 Chevron merged with Gulf, an independent oil company created by the Mellon family, operating in the Middle East. Ken Derr would admit that the merger had been a "cataclysmic event," "just messy" and "a paper nightmare." The cumbersome sale of 1,800 oil wells owned by Gulf—one well was sold for $12—spawned an expensive and stodgy bureaucracy.

Struggling with unprofitable oil and gas processing plants in America, and trying to reinvent Chevron's image, Ken Derr decided to copy John Browne. Like all American oil companies, Chevron had been compelled by Congress's restrictions to search for new oil overseas. The corporation's experience had been unhappy. $1 billion had been lost in the Sudan, and millions of dollars had been wasted in unsuccessfully searching for oil in China. After Chevron abandoned production off the Californian coast, its earnings were the lowest of all the oil majors—10 percent compared to 23 percent among the leaders, largely because it cost Chevron $6.18 to extract a barrel of oil, compared to Arco's $3.65. The company decided to sell off its American assets and, by acquiring foreign oilfields, to redefine itself as a global company. In 1985 it had owned 3,400 oilfields in America. By 1992 only about 400 remained, but Chevron's suffering had not ceased. The corporation's fate was balanced on a knife's edge. Despite optimistic pledges, its oil reserves would slump in 1994 to 6.9 billion barrels, and production was also falling, by as much as 15 percent a year. To save the corporation, Derr placed less importance on improving the quality of Chevron's engineering than on emphasizing "return on capital" and "fixing the finances." Like Browne, he understood the value of a considered gamble, and he chose to bet $2 billion on Kazakhstan initially.

Kazakhstan offered to reverse Chevron's slow demise, although there remained the unresolved question of finding a route for a pipeline to transport the oil to a harbor. Desperate to secure new reserves, Derr decided to ignore that problem. In June 1990, after negotiating between rival factions in Moscow and Kazakhstan, Chevron signed an agreement with the Russian government to explore and produce in the Caspian region. The estimated cost over 40 years was $20 billion. Payments would be staggered, depending upon the success of the operation.

The Western oilmen traveled noisily. The local workers in the Russian oilfields felt patronized by American prospectors seeking a Klondike bonanza. The knowledge that production in Russia had fallen by 9 percent in January 1991 gave the Americans a discomforting brazenness. "We know what to do," one Western oil executive told a Russian minister. "We're taking risks with our money, so don't interfere with us." The explicit threat was that those employed by Amoco, Texaco, Chevron and other corporations would leave if the Russians caused them to be dissatisfied or made their risk excessive. But none of the foreign oilmen understood the attachment Russians felt toward their "natural riches," or the psychology of people emerging from 70 years of communist dictatorship. Instead of sympathizing with their plight and satisfying Russian hunger for technology with an "option value" agreement, the American oil majors insisted that any investment would need to meet American standards of due diligence. Irritation rankled among Russians already dismayed by the introduction of the market economy. Gorbachev's supporters were criticized for succumbing to the capitalists' greed for Russia's raw materials. Chevron's concession in Tengiz especially inflamed Russian fears about a "dirty deal" by which "Russia will be plundered and sold for a mere song," while Chevron pocketed a $100 billion windfall. The news of Chevron producing oil at Tengiz's well No. 8 in June 1991 gave Russia's media the excuse to attack capitalists for exploiting Soviet resources under the guise of perestroika. Old nationalists spoke about the sale of the family silver. Regardless of Russia's desperation, they urged, foreigners should be forbidden to profit from its wealth. Buffeted by the opposition and fighting for survival, Gorbachev capitulated. Instead of maintaining the slow conversion from communism to a market economy, he switched back to secure the hardliners' support. On March 22, 1991, Russia announced a 40 percent tax on all oil exports.

Andy Hall was staggered. His $100 million investment was threatened by this unexpected turn. Unlike the oil majors, he could not easily bluster about leaving. To his relief, Gorbachev bowed to threats from the American administration on Chevron's behalf and replaced the 40 percent with a 3 percent levy. Two weeks later, on August 19, Gorbachev was arrested during an attempted coup. Released after three days, the president was too feeble to resolve the worsening oil crisis. Kazakhstan had declared independence, and in November 1991, as Chevron was planning to start exploration, President Nazarbayev arrived in London to meet BP experts to review the Chevron agreement. As Russia's oil production fell to eight million barrels a day, the Kremlin feared that the country's oil supply would shortly become crippled. Fearing chaos and unable to prevent his support splintering, Gorbachev suspended some oil exports on November 15. He was too late. On December 25 former mayor of Moscow Boris Yeltsin, a corrupt, alcoholic populist, became the new president of Russia.

Oil compounded the political pandemonium of Yeltsin's inheritance. Laws were drafted to privatize state-controlled industries and property, but the first stage of dismantling the Soviet command economy was the overnight abolition of import controls on January 2, 1992, by Yegor Gaidar, the acting prime minister. Russia's oil production fell to 7.5 million barrels a day, inflation rose to 740 percent, and Russia's bureaucracy was fragmenting as Gaidar, anticipating a counterattack by the communists entrenched in the bureaucracy, decided to privatize Russia's industries by giving stocks and shares to the managers and workers. In the oilfields, the managers, ignoring orders and laws, lost any incentive to maintain production, and seized their opportunity to grab the spoils. Yeltsin's dilemma was profound. Russia's economy was based on cheap oil, up to 47 percent of which was regularly wasted during generation and heating. Although the government had increased the price paid for oil from 2 cents to 48 cents a barrel, the same oil was being resold in New York for $19 a barrel. In an unruly economy, Yeltsin's officials were powerless to order local bosses to pay the oil workers, or to direct that oil be supplied to the refineries, or to command the oilfields to hand over the dollars earned from exports. While gasoline was being sold in Moscow in vodka bottles and refineries were limiting production, Yeltsin floundered, issuing ineffectual decrees asserting state control over oil and gas production and exports. On the brink of complete breakdown, the state was even short of sufficient dollars to hire American specialists to seal a huge blowout of a well in Mingbulak in Uzbekistan. Almost 150,000 barrels of oil had been burning every day since March 2, but news of the inferno only reached Moscow at the end of April. Alarmed by the chaos, in April 1992 Loïk Le Floch-Prigent commissioned a newspaper campaign in Moscow to persuade Russians to pull back from the brink of disaster and trust Elf. His appeal was ignored. On May 18, 1992, to dissuade oil workers from striking, the Kremlin shipped trainloads of rubles to western Siberia to pay wages and raised the price paid to the producers to $3 a barrel.

In late December 1991, the Soviet Union split into 15 independent states. The rulers of Kazakhstan, Azerbaijan and Turkmenistan, determined to keep all the profits from their oil and gas, voiced their historic antagonism toward Russia. In Russia itself the managers of the oilfields also dug in, withholding supplies. Caught in the middle of the political battle, Western oil executives became perturbed. Despite their enthusiasm for developing Russia's riches, the country's officials were uncertain about their own authority and were powerless to remedy the absence of laws, valuations, taxes and balance sheets as understood in the West. The oil executives' requests for enforceable contracts and proper accounting to safeguard their investment were met by blank stares, while their intention to earn profits aroused resentment. In that atmosphere, the oil majors reconsidered the risk of investment. "When we make multi-billion-dollar decisions," said John O'Connor of Mobil, pondering whether to develop oilfields in northern Siberia, "you have to have confidence that the system is predictable and stable. All these things are absent."

By September 1992, paralysis gripped the Russian oil industry. Twenty-five thousand out of 90,000 oil wells had been closed. In the Kremlin, Viktor Orlov, a former natural resources minister and chairman of the government's oil committee, warned Yeltsin about "very irrational and wasteful" management of the Siberian oilfields. Russia's production, he suggested, could fall to six million barrels a day, only just over half the rate in 1988. At the beginning of 1993 the forecast worsened. During a meeting in the Kremlin in April, Vladimir Medvedev, the president of the union of oilmen, told Yeltsin, "The crisis is deteriorating into a catastrophe." With 20 percent of Russia's oil wells idle, production could fall in 1995 to four million barrels a day, a million less than the country consumed. Yeltsin's dilemma appeared insoluble. The country was beset by inflation, unpaid bills, unpaid workers, an unstable ruble and crumbling infrastructure in the oilfields. With Moscow and the provinces disputing each other's authority, Siberian oil companies were illegally exporting their production, and Russian customers were not paying for their supplies. To increase output by just one million barrels a day, Yeltsin was told, would cost $15 billion. Restoring production to 11 million barrels a day to sustain the country's foreign earnings would cost over $50 billion and would require Western expertise. Not only was that amount unaffordable, Yeltsin knew, but the country was divided over whether to admit foreign investment. Even partial privatization required the removal of political and legal uncertainties over the ownership of resources. Yeltsin was incapable of resolving that conundrum. "Russia has been a big disappointment for many people," said Elf's spokesman in Volgograd. "At first people thought it might be the new Middle East." In the five years since Russia had opened its doors, the foreign rush had become bogged down by the Duma's indecision, changing laws and taxes. Oilmen were accustomed to problems, and could console themselves with the truism, "This is where the oil is," but none had anticipated being stymied by three straightforward deals that unexpectedly antagonized Russian sentiment.

In 1992, Marathon Oil signed an initial agreement with government officials in Moscow to exploit the oil and gas on a territory known as Sakhalin 2, a frozen island in the Pacific Ocean, 6,472 miles and seven time zones from Moscow. In the tsarist era, criminals were sent into exile on Sakhalin, and in 1983 MiG fighters flew from the island to shoot down KAL 007, a Korean passenger plane. Across that bleak 28,000-square-mile shelf in the Sea of Okhotsk, oil production was possible only during the summer. Between October and June, storms, strong currents and seven-foot-thick ice packs prevented work.

Oil had been produced onshore since 1923. In 1975, assured by the Russian government that there were between 28 and 36 billion barrels of oil under the sea, a Japanese company drilled some wells, but lacking expertise and money, its quest was soon terminated. A second agreement with another Japanese exploration company in 1976 ended in the early 1980s after the Japanese concluded that the venture was uneconomic. In November 1991, anxious to exploit the reserves, the Russian government issued an invitation to major Western oil companies to tender for a license. Most were interested, but nearly all became deterred by Russian politics. Valentin Fyodorov, the governor of Sakhalin, demanded that the successful company lend his region $15 billion to develop the infrastructure for a new republic. Reluctantly, some companies agreed, only to become involved in an intense debate in Moscow about whether foreign exploitation of Russian energy should be permitted. Some Russian politicians rejected outright any sale, some opposed catering to Fyodorov's audacious demands, while others argued that, in the midst of its current financial crisis, Russia had no alternative but to sell its mineral wealth for hard currency.

Eventually, in April 1992, Marathon, in partnership with the Japanese company Mitsui, was allowed to start a feasibility study. Soon after, Mobil and Shell were inserted into the consortium as junior partners. To protect its investment from Russia's punitive tax regime, Marathon negotiated over the following two years a Production Sharing Agreement (PSA) with the Russian government, giving the Americans a majority stake in the $4 billion venture and excluding any Russian participation. The PSA fixed unchangeable terms for the taxes and royalties payable by Marathon throughout the project's life. After the agreement was signed in 1994, Russian legislators, officials and ministers, realizing that Russia would receive little income from the sale of its own oil until all the costs incurred by Marathon to develop the project had been repaid, began arguing with officials in Moscow's oil, geology and finance ministries. Marathon anticipated making tax-free profits for 25 years before Russia earned anything. To protect its hugely favorable agreement, Marathon had successfully insisted that any dispute was subject to international arbitration rather than the Russian courts. The Russian negotiators, failing to hire Western bankers and lawyers as advisers, had unquestioningly accepted Marathon's terms, and were ridiculed for gullibly falling into an American trap to profit from Russian oil. The critics ignored reality. Russia was technically incapable of producing oil in Sakhalin, and without a PSA agreement no Western oil company could risk developing the island's reserves. Those arguments eventually prevailed, and Marathon's deal was approved by the Duma.

Despite the souring mood, Exxon's Rex Tillerson persuaded the Russian government to sign a second PSA for Sakhalin 1, a neighboring area, albeit on less favorable terms than Marathon's. Exxon was allowed only a 30 percent share of the project, with Rosneft holding 40 percent. Exxon would receive 85 percent of the profits, while the Russian government took 15 percent. Tillerson was pleased. Like Marathon, Exxon had successfully exploited Russia's misfortunes, with little regard for the consequences. With the support of President Clinton, Exxon had sought to make profits for shareholders rather than to win the Russian government's trust and thereby secure a lasting balance to OPEC. In Tillerson's opinion, Exxon's commercial priorities were paramount. Success in Sakhalin, he hoped, would tempt the Kremlin to allow Exxon's exploration in the Barents Sea and the Kara Sea, an Arctic zone potentially containing eight trillion cubic meters of natural gas, the world's biggest reservoir. If developed, the natural gas could be piped through the Yamal Peninsula system, another huge Siberian energy basin. Those hopes were to be dashed. After the Russian government signed a PSA agreement with Total of France, PSAs were banned. Resurgent nationalism was stymieing Western oil companies across Russia, and even Chevron's ambitions in Kazakhstan.

Chevron's fraught negotiations to develop Tengiz had been stabilized by John Deuss. The trader famous for Brent squeezes was representing the government of the oil-rich Gulf state of Oman. Seeking investment opportunities, Deuss, flying his Gulfstream between Almaty in Kazakhstan, Europe, Washington and Jackson Hole, Wyoming, brokered the "deal of the century" between Kazakhstan and Chevron. His intention was to profit through the financing and ownership of a new pipeline to transport Tengiz's oil to a port. Chevron's success depended on the pipeline, which, despite Kazakhstan's independence, was subject to the Kremlin's veto if the proposals were deemed to be unfavorable. Ken Derr appeared untroubled by that hurdle. Desperate to reverse Chevron's decline, and haunted by the company's loss of $1 billion in Sudan during the 1980s, he was prepared to gamble on securing the oil first, only afterward negotiating a pipeline's construction.

The preliminary agreement to develop Tengiz, an area twice the size of Alaska, had been signed by Derr and President Nazarbayev of Kazakhstan on May 18, 1992, in Washington. In the initial $1.5 billion investment the Kazakh government, advised by Morgan Guaranty Trust and Deuss, had persuaded Derr to reduce Chevron's share of the income from 50 percent to 20 percent. Over 40 years the Kazak government expected to make $200 billion.

In public, Chevron's success in Kazakhstan was credited to its technical superiority. The Kazakhs and the Russians, it was said, could not manufacture the special quality of steel pipes needed to resist Tengiz's corrosive crude oil, or provide the drills to reach 23,000 feet amid toxic hydrogen sulphide gas. In reality, Chevron's breakthrough to secure the oilfield had been due to a combination of risk and dubious practices during excruciating negotiations that were saved from stalemate by James Giffen and John Deuss. The two maverick traders were consulted by Chevron to fashion a deal with Kazakh and Russian politicians. Giffen, a 62-year-old New Yorker acting on behalf of other oil companies, was suspected of paying $78 million between March 1997 and September 1998 into Swiss bank accounts via the British Virgin Islands for the benefit of Kazakhstan's President Nursultan Nazarbayev and some ministers. Nazarbayev was alleged to have used some of the money to buy jewelry, speedboats, snowmobiles and fur coats. On March 31, 2003, Giffen was arrested at JFK International Airport under the Foreign Corrupt Practices Act and charged with bribing Kazakh officials. Two months later, J. Bryan Williams of Mobil pleaded guilty to evading taxes on a $2 million bribe connected to Mobil's purchase of a stake in Tengiz costing $1.05 billion. Mobil (before merging with Exxon) had paid Giffen's company, the Mercator Corporation, $51 million for work on the Tengiz deal, although Mobil insisted that Giffen was working for the Kazakh government and not them at the time. Giffen admitted depositing money in the Swiss bank accounts, but insisted that he had acted with the approval of the US government. The CIA, the State Department and the White House, he said, had encouraged his relationship with Nazarbayev. The prosecution remains in limbo, with Giffen on $10 million bail. The problem, as Chevron's executives acknowledged, was the immutable relationship between corruption and securing oil supplies in the Third World. Giffen was accused of paying an immediate $450 million deposit to sweeten Nazarbayev's interest. Ostensibly the payment was to finance Kazakhstan's share of the investment, but the FBI would subsequently allege that the money was a bribe.

Derr's success relied on pressure exerted by Vice President Al Gore and the White House on President Yeltsin and his ministers. Similarly, Andy Hall hoped that a visit by Ron Brown, the US secretary of commerce, would rescue some return from Phibro's $100 million investment in White Nights, which by 1993 had become a disaster. His

Russian partner had demanded extra money, which Hall called "out-right expropriation," and local government officials frequently "reinterpreted" the terms of the contracts and changed the law to demand extra taxes. Hall felt naïve and a fool for rushing in after Exxon had rejected the project. "They just raised the taxes whenever it looked like we were going to make money," he complained. "I didn't enjoy it." Brown's protests against arbitrary rules and taxes imposed on American investors did secure the Russian government's agreement to review taxation, but his announcement of success inflamed the nationalists, and Phibro would lose nearly all of its $100 million. In New York, Salomon Brothers wrote off $35 million, curbed Hall's trade in oil products and fired staff. Hall was not personally blamed. "He's made a sickening amount of money in Nigeria," rued a competitor, impressed that Hall had successfully speculated in Nigerian crude, buying at $12 a barrel and watching the price rise to $20. "He's an untouchable." Phibro had also, Hall acknowledged, earned "bucketfuls" of money trading Iranian, North African and Persian Gulf crude. But, he insisted, "We're always staying aboveboard. Nothing illegal or involvement with the rinky-dinky stuff." His trader's shrewdness did not prepare him for investment in Russia. "This is what happens when amateurs go into the oil business," chuckled an Exxon executive.

Exxon's aversion to risk benefited BP and Amoco in Azerbaijan. Azerbaijan, on the landlocked Caspian Sea, was regarded by Russians as the birthplace of the world's oil industry. Oil had for centuries seeped through the earth to the surface there and been used by locals for domestic fuel. Small refineries had been built before Robert Nobel, a Swedish industrialist, arrived in Baku from Saint Petersburg in 1873, searching for walnut trees from which to manufacture gun stocks for the tsar's army. Instead of wood, Nobel bought a refinery, and began to successfully compete against the kerosene sold locally by Standard Oil. Baku flourished as an oil town until 1945. After the Allied victory over Nazi Germany, Stalin abandoned the region and directed his engineers to explore in the virgin areas of western Siberia. Forty years later they returned to Baku, and in 1987 discovered oil beneath 980 feet of water in the Caspian Sea. In October 1990 BP signed an agreement with Caspmorneftgas, the Soviet ministry of oil and gas, to develop that reservoir. Soon after, the Soviet Union collapsed and Azerbaijan became independent. BP's choice was either to risk millions of dollars in Azerbaijan or to compete with Chevron in neighboring Kazakhstan.

Chevron was intent on betting everything on Kazakhstan, yet the Kazakh government was tempted to choose BP. In November 1991 President Nazarbayev arrived in London to meet BP experts to review the deal he had signed in June 1990 with Chevron. Tom Hamilton had been dispatched by BP to Tengiz. "The more I looked," he had reported to Browne, "the more I disliked. There's abundant crude but too much baggage including 10,000 local staff." Investing in Tengiz, he advised, was too risky. Critically, the building of a pipeline to transport the crude to a port across Iran or Russia had not yet been decided, and Kazakhstan's claim to oil from the Caspian Sea lacked clarity. With oil at $15 a barrel, Hamilton recommended that Azerbaijan was a better bet.

In October 1992, Ed Whitehead negotiated on BP's and Statoil of Norway's behalf to pay $40 million for the exclusive rights for a consortium of oil companies to establish whether Azerbaijan possessed commercially viable reserves. The license lasted for just 36 months. While three teams negotiated in Baku, Moscow and London, another was dispatched to establish the viability of the deposits. Across the Caspian's shallow waters it found leaking pipes, abandoned equipment and decrepit offshore rigs, visible relics of the bedlam of the Soviet era. Beneath the sea there was, according to Soviet estimates, 3.5 billion barrels of oil. To transport it, an existing pipeline called "the northern route" passed through neighboring Russia to the Black Sea. Hostile toward Azerbaijan since independence, Russian prime minister Viktor Chernomyrdin and the foreign ministry threatened to veto Azerbaijan's oil exports by limiting the pipeline's use.

Negotiating with Moscow was straightforward compared to the governments in Azerbaijan. Two presidents had come and gone since Ed Whitehead arrived in Baku before Heydar Aliyev, a former chief of Azerbaijan's KGB, grabbed power in a coup in June 1993 in which British agents were alleged to have offered weapons to Aliyev's supporters. To ease the third attempt to secure a concession, John Browne organized for ex–prime minister Margaret Thatcher to visit Azerbaijan, and BP offered the government's leaders $70 million as a "bonus" to finalize the $7 billion development. Having put itself in prime position to be awarded the license, in early 1994 the BP team awaited Aliyev's agreement to sign the contract, which since 1992 had been increasingly tilted in Azerbaijan's favor. Inevitably, there was a twist.

The successful negotiations, led by Al Gore and British prime minister John Major, to persuade Chernomyrdin to allow the consortium's use of Russia's pipeline to the Black Sea, prompted Aliyev to declare, as a negotiating ploy, that foreign help was no longer required. In a region infested by corruption, intrigue and wars, the demand by Marat Manafov, a pistol-waving associate of Aliyev's, for a final $360 million bribe to allow the Western consortium to continue negotiations was the last straw. Officially, the tendering process was halted until a new contract could be agreed. Browne calculated his response. Appealing to the dictator was pointless. The time had come, Browne decided, to call the government's bluff. "Circumstances change," Phil Maxwell of BP told journalists as he emptied his desk in BP's Azerbaijan headquarters, the mansion of a former oil baron, before flying back to London. Maxwell explained that BP was cutting its staff in Baku from 80 to 30 until President Aliyev resolved the uncertainty and was reconciled to competing on the world market. For some weeks the Azerbaijani government prevaricated. The president wanted a large number of investors in order to protect the new state from Russian aggression, and he wanted to play the oil companies off against each other. Unusually, BP, Statoil, Amoco and the five other minor partners in the consortium remained united. Aliyev's bluff was called. The country's financial fate and his survival, he knew, depended on producing oil within four years. Even if oil remained at $15 a barrel, Azerbaijan's income would be $100 billion over the field's lifetime, a phenomenal windfall. The Azeri government blinked. The agreement, dubbed "the Billion-Dollar Experiment" by

Exxon and the "Contract of the Century" by Aliyev, justified a celebration. One thousand guests were invited to the signing ceremony and dinner on September 20, 1994, in Baku's Gulistan Palace. The star guest would be John Browne. Others invited included William White, the US deputy secretary of energy, eagerly promoting Chevron's and other American involvement in the region.

Browne did not stay overnight in Baku, but left midway through the Azerbaijan government's celebratory banquet following the signature of the "Contract of the Century" to take his private jet back to London. "That was not good politics," President Aliyev later told BP's local representative, who concurred with his view. Browne was unconcerned. A done deal meant moving on. The political settlement, he believed, was best executed by others: Aliyev would be invited to London in 1997 to meet Queen Elizabeth at Buckingham Palace. The local settlement between Azerbaijan, Russia and Turkey was delegated to Terry Adams, BP's appointee to chair the consortium of 13 shareholders. "Without a pipeline there will be no development," said Adams, who had also advised against BP's investment in Kazakhstan after anticipating Chevron's problems with building a pipeline. While President Clinton unconvincingly posed as an "honest broker," Adams chose to negotiate in Moscow. With the help of British diplomats, he successfully arranged in October 1995 to use the northern route pipeline through Russia to the Black Sea, and began planning a new pipeline, avoiding Russia, through Turkey to the Mediterranean. Piping Caspian oil and gas to Turkey became BP's and Adams's recurring problem. Ignoring the cost and the political complexities, President Clinton was determined to wrest control of Caspian oil from Moscow, regardless of the anger this aroused in the Kremlin. Turkey had become critical to the West's strategic interests.

In Washington, Bill White, the deputy secretary of energy, and Rosemarie Forsythe, the Caspian expert on the National Security Council, urged Clinton to adopt policies to divert the region's oil to the West regardless of Russia's historic links. Rejecting those who urged the administration to act generously toward Russia, Forsythe displayed petulant anger at Russia's failure to provide a level playing field for Western oil companies. To outwit Moscow, she supported the construction of pipelines from Tengiz and Azerbaijan that bypassed Russia. Aggravating Moscow did not trouble Forsythe, who would be described as "Amoco's ambassador to the NSC." An alternative policy was advocated by Strobe Talbott, the president's special envoy to Russia. To encourage Russia's reformers to increase investment and to Westernize the country, he favored a conciliatory approach. Securing Russia's trust, he argued, would guarantee Russian oil supplies to the West over the long term. Forsythe rejected that measured approach. She was particularly irritated that ENI, the Italian energy company, seemed to enjoy favorable treatment compared to American oil companies. The Italian outsider had traditionally undercut the Seven Sisters' cartel during the 1950s, first in Iran, and then in North Africa and Russia. Now, the Italians once again seemed to be exposing the oil majors' vulnerability in the oil-producing nations. Clinton fought back. Unwilling to reconcile the contradictory policies among his staff, he pursued American interests regardless of the consequences during a meeting he and Al Gore held with Yeltsin soon after the signing ceremony in the Gulistan Palace. America's oil companies, he told the Russian president, were entitled to Caspian oil. Resolutely, Yeltsin replied that the pipeline and Azerbaijan's oil were Russia's and not America's interest.

As proof of his influence, there was an outbreak of violence, murders and bomb blasts across Azerbaijan. President Clinton's priority was to protect oil supplies, regardless of the background of those with whom he would have to deal to do so, and with American support Aliyev reasserted his authority. Clinton's success encouraged the administration to further humiliate Russia. Seeking allies around the Caspian to separate the oil-rich countries from Russia and pipe their crude to the Mediterranean, Clinton and Gore encouraged Exxon, Chevron and other Western oil companies to act under the "shield of government," blatantly antagonizing Moscow. HAPPINESS IS MULTIPLE PIPELINES read a bumper sticker handed out by American diplomats fizzing enthusiastically about "to the victor go the spoils."

To transport Azerbaijan's oil, Clinton had been urging BP to build the BTC pipeline from Baku to Ceyhan, a blue-water port on the Mediterranean, bypassing Russia. In Clinton's opinion, completing the pipeline would put the seal on Russia's defeat and American ascendancy in the region. BP refused the president's entreaties until its technicians had determined whether Azerbaijan's fields would yield five billion barrels, making it financially justifiable. That would not be established until 2001. BP's experts would discover that the reservoirs were better than anticipated: they expected not five but 9.5 billion barrels of oil to lie beneath the Azeri seas, a true elephant.

Clinton's demands to build a pipeline for Kazakhstan's oil would prove more difficult to fulfill. The ideal route to the Mediterranean, avoiding Russia, was through northern Iran. But American sanctions imposed in 1979 excluded that option. Classified as a rogue state, Iran, combined with Libya and Iraq, possessed 23 percent of the world's known oil reserves (923 billion barrels), but in 1996 contributed only about 6 percent of global production (3.6 million barrels a day). The sanctions had proven to be counterproductive. Iran relied on oil for 90 percent of its foreign earnings, yet was compelled to use 33 percent of its production for domestic energy and to import electricity from Turkmenistan. In an attempt to relieve the nation's poverty, the Iranian government was developing nuclear energy in order to release oil for exports, and was encouraging China to exchange nuclear and missile technology for oil. In 1997 Clinton was warned that China would increase its dependence on imported oil from 12 percent in 1995–96 to 40 percent by 2000, and would increasingly depend on Iran. That growth would inevitably impinge on America's needs. Over half of America's daily consumption of 18 million barrels of oil was imported, and about five million barrels came from the Gulf, which had 65 percent of the world's reserves. China's increasing consumption of oil could be accommodated if Western oil companies were allowed to develop Iran's natural gas fields in South Pars, an area bordering Qatar under 220 feet of water with an estimated 300 trillion cubic feet of gas and some oil. Initially, Clinton had agreed.

In 1995, Conoco signed a $1 billion agreement through a Dutch affiliate with the Iranian government to extract 120,000 barrels of oil a day from South Pars. Throughout the negotiations Conoco's executives had kept US officials informed. The company was offering no credits or special technology, and since no Iranian would enter America, it was assured that the agreement was legal. In fact, American companies were already annually buying 23 percent of Iran's oil exports, worth $3.5 billion, to supply US refineries calibrated for Iranian crude. Forty-four other countries were also using Iranian oil, especially Japan. In the integrated, and increasingly tight, international oil market, American sanctions against Iran, Libya and Vietnam were harming US oil companies. America's rivals had benefited from the sanctions against Vietnam and Libya, which forbade exploration in those countries by US companies. Nevertheless, ignoring logic and reality, on March 20, 1995, Clinton abruptly extended the sanctions by an executive order stopping Conoco's agreement with Iran. His ostensible purpose was to punish the country for supporting terrorism and developing nuclear weapons. The immediate casualty, besides Conoco, was oil exports from Kazakhstan and Azerbaijan. Both countries enjoyed "oil swap" agreements with Iran, exchanging oil shipped across the Caspian for Iranian exports into the Mediterranean. Spotting an opportunity, five months later Total of France combined with Gazprom of Russia (30 percent) and Petronas of Malaysia (30 percent) in signing a $2 billion agreement with Iran to develop the same reserves. To deter retaliation against Total, the French government announced that American sanctions would be "illegal and unacceptable." This anticipatory defiance was ignored in Washington. To curtail Total's ambitions, Clinton supported the Iran-Libya Sanctions Act in 1996, which empowered the US government to punish foreign companies helping proscribed countries to develop energy projects. Soon after, Total withdrew from Iran. Simultaneously, Gazprom's attempt to raise $1 billion on US markets with the support of the US Export-Import Bank faltered. Iran's ambitions were frustrated, and the operators extracting Qatar's natural gas were able to continue to siphon supplies from its neighbor's reserves.

President Clinton's decision to prevent the development of oil supplies from countries deemed to be enemy states roused little debate in America. Few perceived the contradictions miring his policies. Clinton's advisers spoke about oil remaining plentiful and cheap, yet his motives for aggressively seizing oil reserves along Russia's borders signaled America's growing dependence on imports.

In 1995, many Russians were convinced that the US government had drawn a map carving up Russia's oil wealth among American oil companies. To the suspicious, the drop in oil prices from $20 a barrel in 1991 to $15 in 1994 appeared to prove an American conspiracy to deny Russia value for its natural resources. No one in Moscow understood how much America's domestic oil industry had also suffered during that period. In July 1991, 251,000 out of 660,000 wells in America had been closed down. Russian prime minister Viktor Chernomyrdin's response to nationalist anger was to restore Russia's control over Kazakhstan's oil. A proposed pipeline to be built by the new Caspian Pipeline Consortium (CPC) to transport Tengiz's oil was forbidden until Russia's unspecified demands were met. Chernomyrdin also reneged on his 1993 promise that Chevron would be allowed to pump 120,000 barrels of oil a day through Russia's existing network. Ken Derr's trust in Russian assurances was dashed. The fate of the $1.4 billion pipeline appeared to be controlled by John Deuss, a man loathed by Derr, representing the ruler of Oman, a desert state at the entrance of the Persian Gulf.

In the battle to decide the CPC's negotiations with Russia, Derr, with the White House's support, was unwilling to compromise in any deal with Deuss about the ownership and the cost of the pipeline. The outcome of their personal battle depended upon Chernomyrdin's preference. Although the construction of the pipeline had been approved by the Duma on August 30, 1993, Chernomyrdin's cooperation wavered depending upon the personalities involved, the chance of bribes and disputes about funding. Chevron's fate looked bleak, and Derr panicked. Fearing that Chevron had lost billions of dollars, he slashed investment by 90 percent. Enviously, he noted that BG and the Italian oil company Agip had united with Gazprom on March 2, 1995, to develop the enormous Kashagan oil and gas reserves in Kazakhstan. The CPC pipeline project remained at a stalemate until, in December 1995, Chernomyrdin was persuaded to isolate Deuss and expel Oman from the project. Mobil and Lukoil, an independent Russian producer partly financed by Arco, joined Chevron as partners. Yeltsin and Kazakhstan's President Nazarbayev agreed in 1996 to build a 935-mile pipeline from the Caspian to the Black Sea. Derr was left on the sidelines while the fate of Russia's oil and natural gas became the central issue in a colossal struggle for power.

Chapter Seven

The Oligarchs

IN 1996, THE FATE of Russia's oil and natural gas, and of the state itself, had become entwined in the ambitions of seven bankers aspiring to become oligarchs.

Boris Yeltsin's prospects of reelection were endangered by the country's imminent financial collapse. Inflation was 200 percent a year, the ruble was weakening and the government was in danger of defaulting on repayment of its foreign loans. Scornful of Russia's nationalists, the pro-Westerners around Yeltsin feared that any impression of arbitrariness and paranoia would encourage Western companies to invest in Kazakhstan and other former Soviet republics, and ignore Russia. Based on advice by Boston Consulting and McKinsey, the Kremlin agreed to privatize Rosneft, Lukoil and Yukos, starting with the gas stations in Moscow. Gazprom, controlling about 25 percent of the world's natural gas, would also be partially privatized and become a quoted company. With 53 trillion cubic meters of gas reserves and 140,000 miles of pipelines crisscrossing Russia and crossing Ukraine and Belarus into Europe, Gazprom's value was incalculable—either worthless or invaluable depending on a market established by former outcasts, mostly Russian Jews and Western banks. Western analysts' estimates of its value ranged from $125 billion to $740 billion. The privatization was an invitation for the enrichment on a gargantuan scale of a small group of outsiders.

At the end of 1994 the seven bankers had hatched a plan to lend the government billions of dollars to stave off Russia's bankruptcy. As collateral they would receive shares in the companies owning Russia's natural resources. The plan's architect was Vladimir Potanin, a former employee at the ministry of foreign trade. Born in 1961, Potanin presented himself as a member of the Soviet elite who had been transformed into a cultivated banker akin to J.P. Morgan. His commercial career had been similar to those of the other aspiring oligarchs: he had started by consulting on import and export transactions after the collapse of the Soviet Union and had progressed to creating Onexim, or the United Export Import Bank, to broker export-import operations. His principal interest was Norilsk Nickel, a huge mineral conglomerate, but he also targeted a stake in Sidanco, Russia's third-biggest refining and oil-producing company, which owned oilfields across western Siberia, including the huge Chernogorneft field. He offered a third of the stake in Sidanco to Mikhail Fridman for $40 million.

Born in 1964 in Lvov in Ukraine, Fridman had been barred from Moscow's best scientific colleges because of his Jewishness, and was assigned to the dour Institute of Steel and Alloys to study metallurgy. In his free time he sold theater tickets, operated a courier service, imported photocopiers, sold Siberian wool shawls, and produced brushes to wash windows. Before ending college he was importing Western luxury goods and computers. Having mastered the Soviet bureaucracy, he graduated to profiting from the privatization of Russian industries. His Alfa Bank, opened with $10,000 in capital in 1990, earned millions of dollars trading the vouchers issued by acting prime minister Yegor Gaidar's government to the public to be exchanged for shares in Russia's major industries. In the rawness of the post-Soviet era, Fridman's $40 million commitment was recorded on one sheet of paper. "I trusted Potanin," he would explain. The shares, the two agreed, would be registered in Potanin's name.

Mikhail Khodorkovsky was the third putative oligarch plotting to keep Yeltsin in power. Born in 1963, like Fridman, as a Jew he had been denied a place in prestigious engineering colleges despite being top in his class at school. After a short career as a black-market trader he had created the Menatep bank to broker payments between government ministries, regional governments and industries. His ambition was to buy Yukos, the oil company. At the last moment two other aspirants, Boris Berezovsky and Roman Abramovich, combined with Yeltsin's approval to create Sibneft, to loan money to the government in return for shares in oil wells and a refinery. Two others made similar bids. Alone among the seven businessmen who sought a share of Russia's natural resources, Potanin was not Jewish.

Potanin, Fridman, Khodorkovsky and Berezovsky met regularly to pursue their conspiracy. In their agreements with the state, the former black-marketeers and traders undertook to auction the shares allocated to them as collateral for their loans. The highest bidder would have the right to manage the oil companies on the state's behalf until September 1996, when the state would be entitled to redeem its shares by repaying the loans. If it was unable to do so, the successful bidder for the shares became their legal owner. Since the country's finances were wrecked, Potanin and his collaborators anticipated that the state would be unable to repay the loans and recover the shares. Viktor Chernomyrdin, the prime minister and former minister of the gas industry and head of Gazprom, signed the agreement with the bankers in the Kremlin on March 30, 1995. Potanin's cool professionalism had persuaded him that the loans would finance the government and prevent the communists returning to power. Yeltsin gave his formal approval on August 31. Both the president and the prime minister were unaware that Potanin had inserted incomprehensible complexities into the documents in order to confuse potential rivals, beguiling even *The Economist* into commenting that the bankers "will not be allowed to sell shares to themselves on the sly." In reality, that was precisely their plan. Using rigged auctions, they intended to transfer the nation's wealth into their personal ownership.

Unseen by government officials and journalists, the auctions for

29 state companies on September 29, 1995, were bogus. Bidders found their access to the bankers' offices barred, or their bids rejected on technical grounds. Boris Berezovsky, a friend of Yeltsin, bought Sibneft for $100.3 million; Mikhail Khodorkovsky bought Yukos for $309 million—it would soon be worth $5 billion; Potanin become the owner of Norilsk Nickel for a pittance, and bought Sidanco for $130 million—two years later, Sidanco was valued at $5.3 billion. The IMF estimated that Russia's oil companies, actually worth $17 billion, had been sold for $1.4 billion. Gazprom was worth $119 billion, but in the sell-off the government received $35 million for 60 percent of its shares, one twentieth of 1 percent of their true value.

The new owners became masters of the system. Underhand means, including pliable local judges, were used to secure ownership of the oilfields. Ambitious to rank among the world's oil titans, Khodorkovsky began searching for Western experts to modernize his corporation. Vagit Alekperov, one of the seven and the new owner of Lukoil, with estimated reserves of 16 billion barrels, 15 percent of Russia's oil, had similar ambitions to end his company's stagnation and attract Western shareholders. Arco became the first Western corporation to buy a stake in a Russian company, taking a 6.3 percent stake in Lukoil (which was valued at $10 billion by Alekperov, and $850 million by Western banks) for $250 million, and pledging to invest $5 billion.

In those early days, the emergence of the oligarchs was less newsworthy than the success of their plan. On July 3, 1996, Yeltsin was reelected president. The Russian economy boomed while ex-communists, KGB officers and Soviet managers jockeyed to appropriate the country's major industries, criminalizing the country. Western bankers provided loans to the new owners of Russia's natural resources. Potanin, as a deputy prime minister in a government overtly helpful to the new businessmen, helped his co-conspirators in the loans-for-shares plot to divide the spoils. At every level, from the Kremlin down to the oilfields' managers in Siberia, insiders snatched a share, sometimes using violence, tipping the fledgling democracy beyond Yeltsin's control.

The chaos was broadly welcomed in Washington as providing an opportunity to replace Russia's historic control over the estimated 200 billion barrels of oil and gas around the Caspian Sea. Baku had become a boom town filled with Tex-Mex food, nightclubs and Texans smelling big money. "This is about America's energy security," announced Bill Richardson, the energy secretary. "It's also about preventing strategic inroads by those who don't share our values." Ignoring Russia's antagonism and the region's recurring wars over control of the Caspian, Richardson added, "We're trying to move these newly independent countries towards the West. We would like to see them reliant on Western political and commercial interests rather than going another way."

Foreign investors occasionally found difficulty identifying the right person to bribe, because there were so many. Officially, the Western oil majors repudiated kickbacks, but circumstances made refusal difficult. Many key officials in the food chain in Russia and around the Caspian asked contractors to pay for "services" supplied by intermediate companies especially established for skimming. The service fees were deposited in offshore bank accounts, especially in the Cayman Islands. While Statoil of Norway had been caught and expelled from Iran for bribing the son of President Rafsanjani, there were no supervisors in Russia or around the Caspian likely to cause such embarrassment. Nor was Washington complaining. Corruption was tolerated for the sake of democracy's future. Once Russians and the Caspian managers became property owners, Washington's reformers believed, the property rights of foreign investors would be protected. In that atmosphere, the CPC pipeline for Chevron's oil, running from Tengiz to Novorossiysk on the Black Sea, was finally approved, with completion set for March 2001—but, to President Clinton's irritation, on Moscow's terms. Obtaining Caspian oil without Russian interference preoccupied Clinton. He wanted the oil companies to build a 1,080-mile pipeline, avoiding Russia and Iran—from Baku to Azerbaijan and then through Georgia and Turkey to Ceyhan, a Mediterranean port. One obstacle was the estimated cost, $3.7 billion.

American oil companies, wary of low oil prices and endless political

shenanigans, wanted US government money to reignite their enthusiasm for the Caspian. Clinton refused, but dispatched Al Gore to
Moscow to lobby for more favorable treatment. As Gore traveled, a
stream of veterans including Henry Kissinger, James Baker, Dick
Cheney, and two former national security advisers, Brent Scowcroft
and Zbigniew Brzezinski, now a consultant for Amoco, elevated the
fate of Caspian oil into a new post–Cold War cause. Russia was condemned for seeking to control the region as part of the new Great
Game. The US administration, they urged, should support the Caspian nations against Russia. Several of those experienced Americans
were retained by the Caspian governments to promote their cause in
Washington. Caspar Weinberger, Ronald Reagan's hawkish former
defense secretary, spoke for the imperialists in May 1997 by decrying
Russia's attempts to "achieve strategic victory of its own: dominance of
the energy resources in the Caspian Sea region. If Moscow succeeds, its
victory could prove more significant than the West's success in enlarg-
ing NATO. The stakes in the Caspian are enormous." America's self-
interest to develop alternative oil supplies from the OPEC countries was
unconcealed. "Open access to the Caspian," continued Weinberger, "is
critical if the United States is to diversify its energy sources and reduce
its dangerous reliance on Middle Eastern supplies." Ignoring a thousand years of regional history, Weinberger, echoing Clinton's opinions,
portrayed Russia as meddling in a distant regime by supplying weapons
to Armenia, forging an alliance with neighboring Iran. He demonized
Iran for its hostility toward Azerbaijan. "If Russia and Iran succeed in
their designs on the Caspian," said Weinberger, "they will have potential leverage over Western economies, which will be left to rely on the
unstable Persian Gulf region for oil." The first step toward defeating
Russia, he advised, required Clinton to negotiate better terms for Azerbaijan from Congress. "Our long-term security interests are at stake,"
he concluded. Instead of the American oil companies engaging in a
dialogue with Russia's politicians, they were encouraged by the White
House to embarrass the Kremlin.

Clinton responded positively to Weinberger's exhortation. Eduard

Shevardnadze, the president of Georgia, was welcomed to the White House on August 4, 1997. His visit had been preceded three days earlier by that of President Heydar Aliyev of Azerbaijan. The former KGB boss was scheduled to spend 30 minutes with Clinton, but their discussion was extended to two hours. Azerbaijan's relations with America had been complicated by an official trade embargo forbidding loans to the country, imposed after an uprising in 1988 by Armenian Christians seeking self-rule in Nagorno-Karabakh, an enclave inside Azerbaijan created by Stalin. Aliyev also met William Cohen, the secretary of defense, to discuss "the development of strong US–Azerbaijani defense cooperation." The news footage beamed back to Moscow was calculated to incense Russia's nationalists and conservatives. The festering anger about the PSA agreements in Sakhalin re-erupted, with an immediate effect on all Western oil companies. Despite Russia's oil production having fallen from 12 million barrels a day in 1987 to six million in 1997, Western money and expertise were again regarded as suspect. Texaco, Exxon, Conoco and Amoco found Russia's unreliable and lethargic bureaucracy introducing new taxes and summarily reversing authorization for new projects. In September 1997 Al Gore, the oil industry's champion, returned to Moscow. In response to his complaints, Yeltsin restored tax and pipeline privileges to the early investors, but rejected requests by Exxon and Amoco for more PSA agreements.

Believing that Russia was stable, John Browne announced BP's intention to invest in its oil industry. His principal interest was Kovytka, a monster natural gas reserve near Irkutsk with sufficient gas to supply America for three years, owned by Sidanco. Browne's plan was to pipe the gas to China. In early 1997 Browne approached Vladimir Potanin, who had resigned from the government in March along with the rest of the Cabinet of Ministers. Enjoying good relations with the Kremlin and sympathetic to the West, Potanin owned two thirds of Sidanco; the remaining third had been bought by Mikhail Fridman, a director of the company, whose stake was registered in Potanin's name. Browne offered Potanin $500 million for a 10 percent stake in Sidanco. Since

Potanin and Fridman had paid $130 million for the entire company, the offer was warmly received.

Browne's proposed investment in Sidanco, and an alternative invitation to invest in Sibneft, were discussed at a quarterly meeting of BP's executives in New Orleans. Like Exxon and Shell, the corporation had been invited to invest in Russia during the 1980s, but the experiences had been frustrating. Secretive and suspicious, the Russian oilmen had refused to reveal any information. Now, Browne explained, it was different. He accepted the common wisdom that Russia would develop into a Western-style economy. The critics' argument during the meeting was familiar: the Russians considered the country's natural resources as their birthright, and would prevent capitalists profiting from their minerals. Browne's confidence in Potanin was shared by only a few, including David Jenkins. "It's a $500 million learning experience," said Jenkins, "and you'll love it." Browne understood the reasons for some of the reservations. The due diligence process to assess Sidanco's true worth would be imperfect, and BP, without Russian-speaking experts, would be unable to influence the company's management. Nevertheless, he trusted Potanin and agreed to pay $571 million (£389 million) for 10 percent of the company. Any doubts were swept aside by Potanin's sale of another 10 percent stake for about $550 million to the Sputnik Fund, financed by George Soros, the Hungarian-born financier, and Joe Lewis, a British currency trader based in the Bahamas. None of these foreign investors, including Browne, knew about Mikhail Fridman, or understood the background to Potanin's sale of 20 percent of Sidanco's shares.

Ever since the loans-for-shares deal had expired at the end of 1996 and Sidanco's shares had been auctioned by the government to the Onexim bank, Fridman claims that Potanin had tried to persuade him to sell his 33 percent stake, although Potanin denies this. Fridman had always refused, but Potanin eventually invoked his influence in the Kremlin, which Fridman says changed his mind. "He's the big boss," conceded Fridman, and accepted $130 million, a profit of $70 million, as compensation for the squeeze. Those shares were sold to BP and Kantupan Holdings. Unaware of the background, Browne asked Tony Blair to celebrate the deal at a reception in Downing Street. On November 18, 1997, the prime minister applauded while Browne and Potanin shook hands, and heard Browne announce that the venture was the precursor to investments totalling $3 billion in Russia over the next decade. Browne's optimism was based on the assurances of BP's political advisers that Russia was stable. No one realized that Russia's budget deficit was edging beyond the Kremlin's control.

Officially, Russia's crisis started in May 1998. Falling oil prices were wiping out profits and the collateral for Russia's foreign loans. Oil production was again falling, and crude was being traded inside Russia by barter and cash. On the international market, the cost of extracting Russian oil was too high—$8 to $10 for a barrel of Siberian oil compared to $2 in Saudi Arabia—and since 40 percent of Russia's exports was oil, the country's financial plight was dire. Viktor Chernomyrdin had been fired as prime minister two months earlier. "Where there's oil, there is blood," he quipped, complaining about the oligarchs' control over the industry and their destabilization of the government. To avoid the humiliation of defaulting on international debts, Sergei Kiriyenko, the new prime minister, wanted to raise $10 billion, partly by selling 75 percent of Rosneft, Russia's last wholly state-owned energy giant. The initial price was $2.1 billion. After Western companies shied away, blaming low oil prices, high taxes and turmoil in Russia, the price was reduced to $1.6 billion, and finally to $470 million. On auction day in May 1998, BP and Shell, the last prospective buyers, refused to bid at any price, blaming Russian instability and the plunge of oil prices. Mortified that Russia's financial structure was melting, Yeltsin appealed to Clinton for help. In July, the White House arranged for the IMF to grant Russia a $22.6 billion loan, but it was too late. In August, following an economic crisis in Asia, oil prices fell to $15, and then toward $10. On August 17 the Russian government defaulted on $40 billion of loans. Amid soaring unemployment and bank failures, Yeltsin agreed to devalue the ruble. Facing financial ruin, Russians lost their trust in the government and the banks. On August 23, five months after his appointment, Sergei Kiriyenko was fired and replaced as prime minister by Evgeny Primakov, a patently temporary appointment. Among the casualties as the oligarchs struggled to save their assets was Sidanco, which was on the verge of bankruptcy. John Browne was embarrassed, not least because Vladimir Potanin had disappeared. Deploying his presentational skills, Browne could have glided over the discomfiture of explaining the loss to BP's shareholders, had Mikhail Fridman not intruded into his grief.

By this time Fridman and his Alfa Bank had forged a partnership with two aluminum oligarchs: Renova chairman Viktor Vekselberg, who was worth an estimated $10.5 billion; and his partner Len Blavatnik, who had emigrated penniless from Russia at the age of 20 to the USA. After obtaining an MBA at Harvard, Blavatnik had worked for the accountancy group Arthur Andersen before founding Access Industries, based on Fifth Avenue. He had invested with Vekselberg in Russia's aluminum industry. Their joint company with Fridman, the Alfa-Access-Renova Group (AAR), had in June 1997 paid $800 million in an auction for 40 percent of the Tyumen oil company (TNK). Fridman, using the $130 million he had received from Potanin, had bought half of this, while Vekselberg and Blavatnik bought one quarter each. They would buy the remaining 60 percent of the company for much less in 1999. TNK's oilfields in western Siberia were adjacent to Sidanco's, and as the financial crisis developed during 1998, TNK's executives heard complaints from Sidanco's creditors about Potanin's attempts to avoid paying his debts by starting bankruptcy proceedings. In Moscow they also heard that Potanin, beleaguered by conflicting advice and complaints from Browne, Soros and Lewis, was struggling to survive. Still outraged by his double-crossing at the hands of Potanin, Fridman sought revenge.

Without Potanin's knowledge, he began buying Sidanco's debt at a huge discount. Besides all the Russian suppliers, he also persuaded the Westdeutsche Landesbank and other Western banks to sell Sidanco's debt to TNK for 10 cents on the dollar. Next, he applied to the bankruptcy court in Nizhnyvartovsk in Siberia to recover a notional $20,000 from Sidanco. The court agreed. Browne found himself caught in the cross fire. TNK, he discovered, was not seeking to recover just $20,000, but to strip Sidanco of all its assets, including the Chernogorneft oilfields. He also discovered that the judge in Nizhnyvartovsk was not inclined to protect BP's interests. On the contrary, Russia's bankruptcy laws allowed TNK to buy Sidanco's oilfields for a pittance, and to authorize the company to manage Sidanco's wells. Just one year after buying a stake in Sidanco, Browne embarrassingly reduced the value of BP's investment to $200 million, although in reality it was worthless. By summer 1999, Sidanco was an empty shell. TNK had become a substantial oil company. Outmaneuvered by Fridman, BP was regarded by rival oil companies as naïve and careless, while Browne was mocked by Fridman's manager for his unwillingness to cooperate.

Momentarily, Browne considered telephoning Fridman, but instead he prepared his retaliation in London and Washington. First he asked Tony Blair to dispatch a warning to the Kremlin. On September 7, 1999, the prime minister wrote to Yeltsin: "BP fears that what should and could be a healthy and profitable company will be manipulated into bankruptcy and collapse." He asked Yeltsin to "give this case your close personal attention," because "the case is being closely followed by other major investors in Russia." Browne's more important retaliation was in America. To finance its nascent oil business, TNK had received preliminary approval for a $489 million loan from the US Export-Import Bank. The money would buy equipment and technology from Halliburton to refurbish the Ryazan refinery near Moscow and to rehabilitate Samotlor, a giant oilfield in western Siberia. BP hired Patton Boggs, a firm of lobbyists, to persuade the bank to withhold the loan. Under pressure from Dick Cheney, the chief executive of Halliburton, James Harmon, the head of the Export-Import Bank, refused Browne's request "for the time being." The oligarchs were not surprised. "Jim's a good friend of ours," said one of TNK's owners, anxious to establish the company's credibility in America. In Moscow, Fridman was nevertheless concerned. TNK needed the money, and the company needed credibility in Washington. Repeatedly he tried to contact BP's representatives in Moscow, and Browne in London, to explain the story of his battle with Potanin, but his telephone calls were rejected. "There is nothing to talk about," Browne told an intermediary. Fridman sought the help of Andrew Wood, the British ambassador in Moscow, but was told, "BP doesn't want to talk to you. They say you're a hostile player." In London, Browne was ratcheting up his vengeance. The Republicans in the US Congress were protesting about soft and pointless loans to Russia. Browne joined the chorus, urging Al Gore and Madeleine Albright, the secretary of state, to order Harmon to abort the loan to a "criminal" company. In the battle of nerves, Browne feared he was losing—until Fridman blinked.

Fridman had been puzzled to receive a telephone call from Pinchas Goldschmidt, a Swiss national and Moscow's chief rabbi. "I hear you've got a problem," said the rabbi. "I've got a friend in London who knows Browne's mother, and I'll tell her that you're a good man and Browne should call you." Fridman was nonplussed. He never attended synagogue, and he could not believe that Browne would take advice about business from his mother. "Why not?" he replied with a smile. To his astonishment, two days later, Browne called from London. "People I respect tell me that you are a man of your word, that if you promise to do something you'll do it. Is that true?" "Of course it's true," replied Fridman. Meekly, he pleaded, "We don't want to harm BP. We're not against you guys. All we want is compensation from Potanin." "Come to London," said Browne, "and let's find a solution." At the meeting three days later in BP's private house in Hill Street, Mayfair, Fridman met Ralph Alexander, Browne's trusted Russian expert. "It's a big honor for us," said Fridman, "to have a relationship with BP. We are simple people and it's a big step in our lives." While the negotiations continued, Browne urged Madeleine Albright to order Harmon to withdraw the loan to TNK, on the grounds that it was against US interests. On December 21, 1999, Harmon obeyed. This exercise of power impressed Fridman. "It shows strength," he told a colleague, acknowledging BP's importance. Just hours before the order was formalized, realizing that BP could permanently damage his reputation,

Fridman finalized a settlement with the company. There was a mutuality of interests. Browne wanted a secure foothold in Russia, while Fridman needed technical help to rebuild TNK's and Sidanco's oilfields. Like so many Siberian oilfields, TNK's had been damaged by bad drilling and engineering. The agreement between the two would be the foundation of BP's activities in Russia. In return, the Export-Import Bank loan was approved. In London, the exploration section celebrated BP's decision to stay in Russia during their Christmas dinner in Trinity House. With Russian oil, everyone agreed, BP's growth was assured.

In the reconstruction of Sidanco, the Chernogorneft oilfields were returned by TNK to Sidanco. Fridman and the Alfa Bank took a 25 percent stake in the new company and a single blocking vote, while BP's interest rose to 25 percent. During the following year, the other foreign shareholders would be bought out by Fridman, and Potanin lost control of Sidanco. As a minority shareholder in TNK, BP was squeezed to collaborate with Fridman. The Russian, Browne concluded, was not dishonest, just very aggressive.

Browne and Fridman's peace agreement was unusual. President Clinton was encouraging the American oil companies to behave belligerently. In May 1999, six US ambassadors from the Caspian region toured the USA promising potentially huge profits for those who were "not fainthearted." *Oil & Gas*, the industry's trade journal, characterized the US administration's stance as "aggressive meddling" and "arrogance [that is] especially troubling" in the Caspian.

Clinton's forcefulness had been spotted by Vladimir Putin while he was Saint Petersburg's deputy mayor after 1992. Born in 1952 in Saint Petersburg, Putin had worked for 16 years in the KGB, including service for the First Directorate as a foreign intelligence officer in East Germany. In the gangster era in the early 1990s, he was mentioned in connection with Saint Petersburg's murky gambling syndicates, crooked businessmen and murdered tycoons. Less well known was his thesis, completed in the mid-1990s, explaining how Russia needed to recapture control over its natural resources to restore its status as a superpower. High oil prices were crucial to Putin's plan. His model for success was Saudi Arabia. Putin's nationalistic agenda appealed to an unusual coalition.

In August 1999, Boris Yeltsin was on the verge of dismissing Sergei Stepashin, his fourth prime minister in less than 18 months. Fearing Russia's disintegration, Boris Berezovsky and senior KGB officers advised Yeltsin that the 46-year-old Vladimir Putin was the ideal candidate to reform the country, and on August 3, 1999, he was appointed acting prime minister. Over the following weeks he aggressively prosecuted the war against the Chechens and established his authority after two bombs exploded in Moscow, destroying blocks of flats and killing 212 people. By the end of the year, a combination of former KGB officers and Putin's trusted officials from Saint Petersburg had steadied Russia. When Yeltsin unexpectedly resigned on December 31, 1999, Putin was appointed acting president. He immediately began executing his plan to take control of the media, to postpone elections and to appoint cronies from Saint Petersburg to key positions, especially in control of Russia's natural resources. Several of those who knew the details of his rise to power either went into exile or would die in mysterious circumstances. Beyond his tight circle of advisers, few anticipated Putin's intention to use the prices of oil and natural gas as levers of political influence.

Chapter Eight

The Suspect Traders

———◁◦▷———

FIXING THE WORLD'S oil prices had created raw hatred between Jorge Montepeque and Laney Littlejohn. Montepeque, the son of Guatemalan peasants, distrusted the oil industry. Every day, from his office in Singapore, Montepeque's small team of reporters employed by Platts, a specialist publisher bought by McGraw-Hill in 1953, recorded the prices for which producers and traders had bought and sold crude and its refined products for shipment across the world on tankers and railways and through pipelines, and published them as a guide for the world's oil trade. "I'm uncomfortable," Montepeque said, "because every day some traders are stretching the truth." Montepeque set out in 1991 to create a transparent, honest market. As the trade's self-appointed policeman, he was regarded by most traders as a necessary evil, but Laney Littlejohn saw only evil.

Four thousand miles to the west, within sight of the Gulf of Iran, Laney Littlejohn, an economist trained in Missouri, was employed by the Saudi government to fix the price of its oil. Littlejohn was housed in Box 8000, a three-story, reinforced building in the cramped Dhahran compound for 11,000 foreigners, and his daily routine was to collect market prices in Dubai, New York and London, estimate the amount of oil stored in ports and refineries around the world, and, after a simple calculation, inform the minister of oil the highest price customers could be charged for Saudi crude. Although Saudi Arabia supplied over 10 percent of the world's daily consumption, and through OPEC influenced the world's oil supplies, Littlejohn knew that his master's power to control prices was limited. Pricing Saudi oil, Littlejohn admitted, was two thirds science and one third "How much can we get away with?" The self-confident Saudi royal family, he knew, liked to imagine they were getting the best deal for themselves. After all, they owned Ghawar, the world's largest oilfield. Discovered in 1948, it was 174 miles long and 12 miles wide. By 1993, 34 billion barrels of oil had been extracted from it, and its unceasing production of 4.2 million barrels a day underpinned the nation's wealth. If the government needed more money, Littlejohn's recommended price was occasionally overruled by the minister. The extra nickels and dimes on five to eight million barrels a day were valuable. Accordingly, Littlejohn resented Montepeque's homily that a producer could control either volume or price, but not both. "Like quantum physics," said Montepeque, "you can't know the speed and direction at the same time." Oil prices, he believed, were a victim of the law of unintended consequences. Littlejohn resented anyone casting doubt about the credibility of his prices, and fumed at Montepeque's doubts about the market's honesty.

Saudi Arabia's finances had been crushed in 1986 by the collapse of oil prices, which had been unexpected by the Saudis, but with hindsight was inevitable. Until then, Sheikh Zaki Yamani, the internationally recognized oil minister, had given the impression that OPEC had replaced the Seven Sisters as undisputed price fixers. Littlejohn had been amused when Henry Kissinger mentioned a "fair price" for oil: "I thought Thomas Aquinas was the last person to talk about 'fair price,'" he said. The Saudis had assumed that the West was powerless to influence prices after Occidental's oilfields had been gradually nationalized by the Libyan government after 1970 and the oil majors had been restricted under US antitrust laws to collaborate and mitigate the damage. The Saudis and all the other OPEC countries concluded that the oil majors were vulnerable. But during the early 1980s oil supplies from the North Sea and Russia had overturned their assumptions. Fixed prices had been wrecked by the perception that the oil nations were producing between six and eight million more barrels of oil every day than required. Spare capacity and Saudi Arabia's subterfuge in under-cutting other OPEC members wrecked price-fixing. Littlejohn's former practice of confidently quoting the price of the sulphurous Saudi crude by checking the bids for West Texan Intermediate light on Nymex was wrecked as prices started falling from $26 a barrel during the last weeks of 1985. "The market's going to hell," Littlejohn told his confidants. "It's going down." His warnings were ignored by Saudi Arabia's rulers. Sheikh Yamani was unwilling to accept that the free market was destroying OPEC's cartel.

The disintegration had started in Nigeria. For nearly 20 years, the Nigerian government had sold light crude at fixed prices, until in the early 1980s its production was priced by traders against Brent oil. When North Sea oil unexpectedly became cheaper in 1985, Nigeria's crude was too expensive, and traders canceled their contracts. Simultaneously, the price of Dubai crude fell to $10. Tankers were sailing empty from Saudi terminals after American refineries refused to buy sour crude for high, fixed prices. King Fahd of Saudi Arabia was baffled. He had assumed that if Saudi Arabia produced six million barrels a day, the price would be $17.52. Neither he nor his entourage understood how international markets operated. John Kalberer, Aramco's chief executive, and Littlejohn explained to him that fixed prices and a fixed quota were inconsistent with a free market. Even Yamani spoke about the inconsistency of OPEC simultaneously pursuing higher output and higher prices. Troublemakers seeking to remove Yamani contradicted the Americans. Saudi Arabia, they insisted, could continue to produce six million barrels a day and sell it for around $18. "On Wednesday I was talking to Yamani," recalled Littlejohn, "and next day he was gone. I had no inkling. Those ambitious people who pushed him to the wall were fooling themselves about the formula. No one wanted to listen to me." Combining ignorance with the need for a high income, the OPEC countries refused to reduce production below 12 million barrels a day. "Sell my quota," Sheikh Hisham Nazer, Yamani's successor, told Kalberer. Obediently Aramco sold one cargo to Chevron for less than $5 a barrel. The company agreed not to reveal the price, but Aramco had no choice but to sack staff and produce less oil. Seven months after the crisis started, Littlejohn persuaded the new oil minister that the existing pricing formula was redundant. At the OPEC meeting in September 1987, the producers agreed to adjust to the new world. Rather than dictating prices, Littlejohn would discuss his new pricing formula with the "lifters" — his customers.

Like wine and cheese, the quality and value of crude oil varies significantly. The difference in price depends on factors including the amounts of sulphur and asphalts within the oil, and the distance from the oilfields to the refineries equipped to process that specific crude. Three principal locations determined the price. The first was Cushing, Oklahoma, the delivery point for WTI; the second was the North Sea, the pricing location for Brent, West African and Russian crude; and the third was Dubai, the pricing reference for oil produced in the Middle East for delivery to Asia. By 1989 a major distinction had arisen between buying oil in the future, speculating on paper, and trading actual cargoes of crude oil and products on "spot" — the sale of a single cargo, right here, right now, for cash, sometimes at a fixed price but often linked to an index published by Platts or Argus, a smaller British rival. In 1990, one third of oil traded was "spot" — for instant or future delivery for cash. Increasingly those "spot" trades were sold over the counter, or OTC, which meant they were unregulated in any market (only the futures market would be regulated). During the 1990s, the OTC market began to expand. By 1998, $80 trillion of oil was traded as OTC, compared to $13.5 trillion in exchanges. Each trade of the 120 types of other crude oils extracted in the Middle East, Russia, Africa, South America and elsewhere was determined by referring to the differential of the price of WTI or Brent, on Nymex and on the International Petroleum Exchange in London. Sixty to 70 percent of all crude oil trades were priced every day, reflecting the crude's quality against Brent, and the transport costs.

Considering all those variations, Littlejohn calculated the Saudi oil price by averaging the price of a basket of crude oils including Alaskan, WTI and Brent. Taking into account the amount of sulphur in Saudi oil and the freight costs from the Gulf, he would "netback" the price for delivery in New York to the cost for the customer at the terminal in Saudi Arabia. Littlejohn would hate to be accused of trying to influence prices. His unspoken quest was to discover the amounts of oil in storage in America, Asia and western Europe. If he could successfully manage the inventories of oil in the West, Saudi Arabia could influence prices over the long term.

Both Laney Littlejohn and Jorge Montepeque were aware of another complication. Crude oil prices were no longer solely dependent on supplies from the OPEC countries, the US and the North Sea, but were also influenced by the prices of products refined from oil, or the other way around. Trades of jet oil, kerosene, heating oil, diesel and the dozens of types of gasoline for cars sold "over the counter" by refineries to customers and between traders either in the future or for "spot" were all influenced by crude prices, and vice versa. The market was a vast, self-adjusting balancing act, reported by Platts and Argus. The principal users of those two agencies were the oil majors and traders, especially Glencore, Vitol and Trafigura. Registered in Switzerland and with offices across the globe, the secrecy surrounding the privately owned traders inevitably attracted suspicion.

Unlike the standardized Nymex quotation for the price of light sweet crude at Cushing, there was a vast variety of deals between the traders. Jet oil refined in Europe and sold to the US was priced differently from jet oil refined in Singapore and sold to Japan, or on a tanker heading from Rotterdam to Mexico, or from a refinery in Malaysia to Tokyo. The same jet fuels produced in America were priced differently depending on the refinery and which pipeline was used to transport it around the country. Even dishonest traders needed to know what prices their competitors had agreed to on a permutation of deals, and their only sources were Platts and Argus. The simplicity of Platts's and Argus's pricing exposed the market to unscrupulous traders.

In the early days, Platts employed 15 reporters in the USA, London and Singapore to telephone traders every day to discover the prices of their latest deals—the bids, offers and final contracts in every region. The work was not arduous. The Platts reporters asked simple questions, but, handicapped by the traders' superior knowledge and their occasional intention to manipulate the market, they had to rely on the information given to them by the traders. If only a few people were trading, creating an "illiquid market," the traders could easily hoodwink young reporters and secure profits on contracts based on Platts's report of a price. Good reporters tried to filter out the lies and balance out the assertions to establish whether the trader had simply invented a contract to be reported by Platts in order to increase his profits. "Some traders talk their books to manipulate the prices," grumbled a Platts executive. "Traders conspire to pull the wool over our eyes," agreed another Platts reporter. "We're fighting against distorted images in the market," declared Montepeque, who to his staff was the only barrier between honesty and dishonesty. "The best liar," said a Platts reporter, "is the trader who tells you the truth 95 percent of the time, slipping in one lie for his own purpose." An experienced reporter, Montepeque hoped, would "discount what he thinks is a phony price." But "interpreting" the market among the 20 traders of oil products and 15 traders of crude in America was chancy. Nevertheless, even the best manipulator could only fool the market for a couple of days. Beth Evans was one of the few Platts reporters feared by traders. If she suspected a trader of committing "hanky-panky" or making "five o'clock daisy chains" in the Brent market, she delighted in "wiping him out." "Death by Beth" was the fate of a trader nailed by her disapproval. Her anger was especially aggravated by British traders playing games on Thanksgiving Day, when the US market was closed, to manipulate unrealistic prices.

Montepeque's attempt to sanitize the system was resented by Littlejohn. "Jorge Montepeque is full of shit," he said. "There is no evidence of wrongdoing, but a lot of people bite on it." Montepeque was undaunted. Like the oil producers and oil majors, he knew that Littlejohn hated being told how to run his business, but some traders enjoyed flouting the rules. Prices offered to Platts on a P&C basis—private and confidential—were suspected of manipulation. "These traders," said Montepeque, "want a fantasy world of secrecy." "Game playing" had become endemic. "Resetting has to be done," said Montepeque, because "the oil trade is impenetrable for outsiders." He set out to create a transparent market with new rules for the messy trade in oil products—diesel, jet oil, gasoline and kerosene—being sold in different specifications, in constantly changing time zones, with freight prices influenced by pipelines and tankers heading toward dozens of destinations. Technically the traders were dealing in "swaps" on the OTC market or "futures" on Nymex.

The market's opacity suited Glencore, Vitol, Trafigura and others dealing on the margins in terms of market share. Marc Rich had reincorporated his company in New York as Clarendon, but once it had become notorious for lying, manipulation and secretly breaking sanctions to trade oil with Iraq, a group of his former employees decided in 1993 to break with Rich and the past by reestablishing the business in Switzerland as Glencore, which developed into the world's largest commodities trader, with annual revenues in 2008 of $152 billion. The successor company remained a colossus among mineral traders.

Playing "the basis game," certain traders and private corporations expanded their operations from beyond mere trading of petroleum products to renting pipelines, refineries and tankers for crude oil. Like scavengers in the African desert, ensuring oil was loaded and delivered for the best prices, traders crisscrossed the globe to dusty backwaters where they would wait endlessly for meetings that could lead to an offer to buy or supply cargoes of oil that the majors felt compelled to ignore. Refineries in eastern Europe, the Balkans, Colombia, Nicaragua, Ecuador and South Korea, blighted by shortages of money or shunned by the oil majors because their credit was deemed to be unreliable, were vulnerable to higher-priced cargoes.

In the aftermath of the Kuwait war, it was reported that some unnamed traders had noticed that vast amounts of crude oil, contaminated with sand and salt water, were lying on the desert, in the salt marshes and in damaged oil wells. At no cost, traders obtained the oil for treatment in Very Large Crude Carriers (VLCCs) anchored off-shore. After decontamination the crude was sold to refineries in Brazil and Tanzania, and a power station in Pakistan. The generators in the Pakistani power station, clogged up with sand and bitumen, had to be closed down. Several refineries in eastern Europe, especially in Romania, discovered after refining crude that it did not match the specifications in the bill of lading. Private detectives found that during the tankers' passage through the Suez Canal, the original loads of top-quality Iranian crude had either been transferred to another tanker or had been piped into a storage tank and replaced by inferior oil, which was undetected by the staff on the quay at the Romanian refinery. Unlike the major oil corporations, the Romanians had carelessly failed to employ an assayer to test the oil before acceptance. By the time the damage had been discovered, and the refineries' products—gasoline, diesel and heating oil—were discovered to be inferior or even worthless, so many shipments had been mixed together that it was impossible to determine for legal purposes whether the supplier was responsible for the culpable load.

In that sketchy world of squeezes and deception, the arrival of the bankers Morgan Stanley as oil traders shifted the balance. Under John Shapiro's direction, Morgan Stanley's oil business based on Sixth Avenue in Manhattan had grown from two traders in 1984 to 40 in 1990. Diversification meant greater risks and higher profits. Until 1990, the bank had predominantly dealt in crude oil, and did not fully understand the complexities of trading the refined products manufactured by the world's 620 refineries, each calibrated for different requirements in summer and winter. Among those recruited after March 1990 was Olav Refvik, a Norwegian employed by Statoil who revolutionized the bank's philosophy. In the volatile period after the Gulf War started in August, Refvik and Shapiro shifted the bank's emphasis from providing risk management for customers and primarily financing the trade of crude oil to risking money by trading products. Serious profits, Shapiro and Refvik realized, could be made by taking positions, or propriety trading, on the price of the so-called middle distillates—heating oil, diesel and jet fuel—where the "crack spread," or profits, were healthy.

Adopting financial instruments used by metal and foreign exchange dealers, Morgan Stanley would agree, for example, to guarantee the price of jet fuel for an airline in Chicago three months in the future. To protect itself from risk because of the inevitable change in oil prices, the bank would simultaneously buy another product, perhaps heating oil in Europe. Since the general price of jet fuel and heating oil, allowing for different specifications, would rise or fall simultaneously, the bank's profit was earned by "arbing on the difference in price," not by delivering the fuel. By protecting against price movements, or "hedging" relative prices, the bank offered customers the opportunity to protect themselves against every contingency.

Oil trading had fundamentally changed. First because the derivatives introduced liquidity or money into the market, and second because the market for oil products would eventually influence the price of crude oil, including OPEC's. Gradually, speculation among traders and the banks undermined OPEC's overwhelming power to control the world's oil prices. Morgan Stanley was becoming an intellectual powerhouse, "trading around the book." As the bank entrusted more money to Shapiro and the oil traders, Olav Refvik, regarded by one industry insider as "a piece of work with a very aggressive DNA," became known for "not being shy about an outsize bet, risking a big loss for a big profit."

Shapiro's success hurt Goldman Sachs. The founding team was blamed for failing to take risks and build the business. "They liked it when traders earned profits, but couldn't take the pressure of losses," noted Charlie Tuke. "They were so frightened of making the wrong decision that they were prevented from making the right decision." The new managers, Lee Vance and Frank Bosens, restructured the operation to take risks and follow Morgan Stanley's practice of making a market for anything and anyone. By then, those using Morgan Stanley's sophisticated methods of trading included BP and Shell. Their trading operations enjoyed a unique global footprint.

Establishing an advantage over rival traders depended upon possessing better information about events that would influence oil supplies and price movements. Noting how Marc Rich had earned his fortune by anticipating wars during the 1970s, every trader had sought unique sources of information to outwit his rivals. BP was becoming the industry's most aggressive trader. Since its expulsion from the Middle East and Nigeria, the company had focused more effort than others on trading from London, Cleveland and Singapore. BP's strength was its physical presence in every product and every region — producing, refining and marketing oil. This meant that BP's traders — employed on short-term contracts and paid bonuses to encourage aggression — received invaluable intelligence about shortages, surpluses and calamities. "A lot of people have lost money betting against BP, and made money following BP," said Peter Gignoux, a London trader. Even the best rival traders were by comparison working in a vacuum. "In trading you are either everywhere or you die," said Gignoux. "BP's culture was aggressive and arrogant, trading as a big name. Only when their conduct was so outrageous were they punished." During the 1990s, BP had often been suspected of manipulating prices of Alaskan and Brent oil. In June 1997, Trevor Butler and three other BP traders were finally accused of "frontrunning" Brent: Butler was selling small parcels of oil at a high price and then, having established a closing price on the futures market, selling a huge amount on BP's behalf at the high price. Inevitably, prices would subsequently fall. Both BP and Butler were found culpable. BP was fined £125,000 by the IPE, and Butler was ordered to pay costs. "It's called 'Grab a Grand,'" said one of the defending lawyers, who described manipulation as "systemic." The punishment of BP would be noted by rival traders as merely symbolic. Although BP issued a statement promising to "review our procedures," no one believed the company would behave less aggressively in future.

BP's use of "embedded optionality" — simultaneously trading paper and physical oil products, and monitoring activities at refineries to improve the flow of information about the business and mitigate the risk if it needed to store or supply oil — was also adopted by Morgan Stanley after the insolvency of Wyatt, an oil storage terminal in New Haven, Connecticut, in 1991. This would normally have been followed by the bank selling the assets at firestorm prices, but John Shapiro and Neal Shear spotted an alternative strategy of "going back to basics." After Wyatt was bought from the administrator, Morgan Stanley agreed to lease the terminal to store and supply four million gallons of heating oil and diesel every day to customers in the area. The physical ownership of oil gave the bank the flexibility to simultaneously trade real oil and speculate in "paper" oil, knowing that if a deal turned sour, it could avoid the risk of "having its feet held to the fire," and take delivery of physical oil. Shapiro had tested the idea in Oklahoma in 1986, and it worked. "Being in the physical business tells us when markets are oversupplied or undersupplied," he argued. The bank's traders were no longer the tail, but had become the dog. "I'm getting squeezed out by Wall Street's invasion into our territory," Phibro's Andy Hall complained.

Like all the traders, Hall had greeted the Iraqi invasion of Kuwait on August 2, 1990, as another opportunity for profit. Two oil producers at war and the prospect of a shortage provoked speculation. Prices rose during the first day from $21 a barrel to $27, and later hit $40. Unseen by the public, the traders were deploying their ruses and skills to make fortunes. None had taken into account the effect on the industry's image.

The oil majors' spontaneous hike of gasoline prices in the United States aroused public anger, and demands for a repeat of the windfall profits tax of the 1970s. None of the oil majors or the traders had anticipated that soon after the invasion Saudi Arabia would begin to recommission 146 old wells and drill 51 new wells, increasing production by 29 percent to 8.5 million barrels a day. Traders had not calculated that Saudi Arabia's spare capacity would compensate for the absence of Kuwaiti oil, or that stocks from America's strategic petroleum reserve would be released. Their expectation of a war premium evaporated. In early January 1991 oil prices fell back to $21 after Allied air raids on Iraq ended the threat to supplies. "Wall Street is ankle-deep in blood," smirked one trader in London. Andy Hall's speculative positions were hammered, and he withdrew from trading physical oil, concentrating on trading derivatives for high-percentage profits. He was relieved, like other traders, to be shielded from the growing conviction among the

American public that Big Oil was profiting from US soldiers dying in the Persian Gulf.

Protestors shouting about "blood for oil" and recalling Rockefeller's price-fixing cartels sought an explanation for the continuing high price of gas. Although the world was awash with low-priced oil, gasoline's price rise in the US following the invasion in August from $1.19 to $1.54 per gallon had not been reduced by early January. By contrast, the oil majors' profits, opaque at the best of times, had soared. Exxon's had risen by 300 percent, Mobil's by 45 percent, Shell's by 69 percent and Amoco's by 68 percent. All those profits, said critics, would have been higher if the companies had not used adroit bookkeeping to write off capital values, add to their reserves and settle tax disputes. One note in small print in Exxon's accounts embodied the seemingly disparate threads linking the public's dislike of Big Oil. At the end of 1990 it became apparent that Exxon had reduced its liability for taxation by setting off some costs caused by the *Exxon Valdez* disaster in Alaska against taxes. That calamity, and the reaction of Lawrence Rawl, the chairman, further damaged the industry's credibility.

Rawl had been telephoned at 8:30 a.m. on Good Friday, March 24, 1989, and told that the *Exxon Valdez*, a 900-foot tanker, had hit the Bligh Reef while maneuvering through Prince William Sound. By daybreak over 10 million gallons of Alaskan crude oil had leaked out of the tanker and was edging toward 1,100 miles of pristine coastline. Rawl was told that the tanker's captain had failed an alcohol test, and admitted, "You know it's going to be bad, bad, bad." By afternoon, he said, "I knew we were in a ditch pretty deep." Instead of flying to Alaska, he was persuaded to stay in New York. Even Exxon's chairman could be held on a chain by the corporation's lawyers. "You'll get in the way," he was told. "We've said everything that there is to say." His absence, the corporation's refusal to immediately admit liability or offer remorse, and the background to the incident fueled public loathing of Exxon and Big Oil.

The news soon emerged that Captain Joseph Hazelwood, in charge of the *Exxon Valdez*, had spent the previous afternoon drinking straight vodkas in a bar. Hazelwood had twice been convicted of drunk driving in New York, and Exxon knew he had been treated for alcoholism in 1985. Yet eyewitnesses among the crew would testify that he did not appear drunk, and that at the time of the accident the ship's third mate was navigating. Nevertheless, the effect of the collision was made worse by the captain's attempt to reverse the tanker off the reef, and by the failure to spray disinfectants over the oil. In a clash of testimony, Exxon would claim that the coast guards refused the company's request to spray the surface, and some transcripts of telephone conversations showed that local officials didn't know the location of the equipment and chemicals. In their defense, Alaska's officials would assert that Exxon was only trying to mitigate the horrendous publicity targeted at the corporation after the federal authorities began reporting the destruction of wildlife as a result of the oil spillage. The damage was colossal. Fish over a large area were destroyed, and the carcasses of 36,000 migratory birds and 100 eagles were taken to a freezer as evidence of Exxon's criminality. Television pictures of otters drowned in oil, dead seabirds and fish appeared to confirm the allegations by some of the 11,000 people employed by Exxon to clean the water and coastline that 85 percent of the dead birds had never been discovered.

The public's anger at Exxon was intensified by the news from New Jersey. On January 2, 1990, Exxon was accused of "stonewalling" an investigation about a spill of 567,000 gallons of heating oil from a pipe into the Arthur Kill channel, off Staten Island. Exxon admitted that its detection system had malfunctioned since 1978, but initially claimed that only 5,000 gallons of oil had leaked. Few were persuaded by a spokesman's denial that the corporation was "sloppy." The judicial process aggravated the anger felt toward Exxon. In Alaska, exactly one year after the *Exxon Valdez* spill, Captain Hazelwood was acquitted of navigating under the influence of alcohol and only convicted of a misdemeanor, namely the negligent discharge of oil. Two years later even that conviction would be overturned, and he emerged without any punishment whatsoever.

Exxon's progress through the courts was lengthier, but similarly controversial. Under the threat of a federal criminal trial starting in April 1991, Lawrence Rawl and Lee Raymond, his deputy, spent 57 days successfully negotiating a global settlement with Governor Walter Hickel of Alaska. Just before midnight on March 12, 1991, Exxon formally pleaded guilty to four federal misdemeanor charges and agreed to immediately pay a $100 million fine and a further $900 million over 10 years to settle civil claims. The following morning, politicians and environmentalists welcomed the agreement. By nightfall, the arrangement had been undone by Rawl's candid revelation about Exxon's concern for its shareholders. "The settlement," he announced, "will have no noticeable effect on our earnings." The worst legacy of the previous year, he confessed, had been the bad publicity. The settlement coincided with Exxon's highest-ever quarterly profit of $2.2 billion. The previous quarter, the corporation's profits had jumped 21 percent.

Judge Russell Holland, a federal judge in Alaska, was outraged. "I'm afraid that these fines," he said, "send the wrong message, suggesting that spills are a cost of doing business that can be absorbed...They do not adequately achieve deterrence." The fine, he felt, needed to be unabsorbable. The eccentric Judge Holland, a lanky, bearded Republican and a member of the Petroleum Club, was unwilling to allow Exxon to pay and move on. His intervention divided Alaska's politicians and ended Exxon's agreement to pay extra compensation. In a bold gambit, Exxon withdrew its guilty plea and challenged the government to prove its case in front of a jury. Alaskans, especially fishermen, who had suffered serious losses, feared they might receive no compensation at all after the stakes were raised. Under Judge Holland's direction, a jury ordered the corporation to pay $5 billion in punitive damages. A shocked Exxon appealed, launching a process that would only end in 2008.

To the public outraged by Big Oil in the midst of the Iraq War, it appeared that the oil majors had learned nothing since an oil spill into the sea off Santa Barbara in 1969 that had terminated offshore drilling in California and paralyzed Chevron's production in Point Arguello, off Santa Barbara, for three years. Negotiations to restart operations had finally been destroyed by the *Exxon Valdez* disaster and a contemporaneous oil spill off California. "These days," admitted George Babikian, the president of Arco's refining and marketing, "you mention oil company and people see some big, fat, greedy guy with a cigar and dirty fingernails counting his money. We don't have a hell of a lot of credibility." Big Oil's unpopularity was not only born out of Raymond's lack of obvious contrition in the aftermath of the Alaskan court's $5 billion award. Exxon was already vulnerable after a nasty battle between the oil companies and Chuck Hamel.

Hamel, a successful oil broker, had been ruined during the early 1980s after his customers found water contaminating the oil he supplied from Alaska. In 1985 he produced evidence that, he said, proved that the oil companies had deliberately added water to the crude, and sought compensation from Exxon and BP, two of the seven co-owners of the Alyeska pipeline. "I decided to expose the dishonesty of the oil industry," he later told a congressional committee. During his research, Hamel discovered that "the oil industry was turning Alaska into an environmental disaster." Employees of Alyeska, the owners of Alaska's pipeline, had provided him with internal documents proving the company's violation of environmental laws. "We are living in a conspiracy of silence," he pronounced, "waiting for an environmental disaster to happen." Over the years, he said, the pollution of the sea, earth and atmosphere around the terminal at Fort William Sound had been concealed by Alyeska's fabricated records, a charge Alyeska strenuously denied. "The oil industry isn't putting out anything but poison and lies," Hamel insisted.

Alarmed by Hamel's allegations, the oil companies discovered that their rebuttal, discrediting him as vengeful and insane, had encouraged Alyeska's employees to secretly supply him with more damaging information. Worse, Hamel's campaign coincided with the *Exxon Valdez* spill. "That goddamned, insane son of a bitch," Rawl was reported to have cursed. Fearful of Hamel's enhanced credibility, Exxon, BP and Arco hired Wackenhut, a firm of investigators based in Florida, to covertly identify the employees supplying the embarrassing information to him. Posing as an environmental lobbyist, Wackenhut's director of special investigations befriended Hamel, entered his home, unearthed his long-distance telephone records and secretly recorded his conversations. In September 1990, long after Hamel's sources had been identified, the undercover operation was exposed. Summoned to appear before a congressional committee on November 6, 1991, the directors of Exxon, BP and Arco were castigated by outraged politicians for corporate skulduggery.

Hamel was not assuaged. By then he had a third grievance against Exxon. Hamel had owned a lease at Point McIntyre, a stretch of land on Alaska's oil-rich North Slope. With apparent sincerity, Exxon's representatives had persuaded him that no oil reserves lay under this land, and, convinced by their tests and representations, he sold his lease to Exxon for a low price. Several weeks later the corporation announced the discovery of oil under the same land, and soon after, a rig began producing 158,000 barrels a day from it. Hamel sued Exxon. In 1996 he would agree to a settlement for compensation, though for far less than he expected.

Throughout these events, Lee Raymond betrayed no hint of concern about the Alaskans or Hamel. He appeared not to have thought about the potentially serious consequences for Exxon's future in Alaska. Yet the oil companies had pursued Hamel out of fear that his revelations of environmental damage would undermine their chance of drilling in the Arctic National Wildlife Refuge. Surveys by Arco's geologists suggested that 1,000 feet beneath the permafrost was an oil reservoir as big as any in the Middle East. Until then the passage of a bill through Congress to allow drilling had seemed assured. The *Exxon Valdez* spill challenged the entire project, and on November 1, 1991, the bill was rejected. "We have drawn a line in the tundra," declared Senator Joseph Lieberman, leading the opposition. Unexpectedly, the public outcry had also persuaded President Bush to terminate the proposed drilling off the Florida Keys and to buy back the offshore leases sold for $108 million to the oil companies in 1984. $200 million had already been spent on exploration to locate at least 10 billion barrels of oil, but that was ignored. Few outside the oil industry were impressed by the American Petroleum Institute's plea on behalf of "the free market" in the weeks after the Gulf War ended in February 1991. The hostility toward the oil majors was reinforced by their admission that profits in the fourth quarter of 1990 had increased by 77 percent, making them easy targets for vote-hungry congressmen. Politicians suspicious about the oil industry's manipulation of the market led the protests. Laws to control the corporations, limit fuel consumption and encourage fuel-efficient smaller cars poured out of Washington. Browbeaten, the editor of *Oil & Gas* commented, "Lawmakers are driven to tell otherwise free people how to behave."

Oil company executives and traders retreated from the argument. The Western economies were going into a recession, and the oil industry was in turmoil. Washington's hostility to Big Oil had hastened the industry's decline since the 1986 price collapse. Taxes and regulations were discriminating against small producers. Despite new discoveries in the Gulf of Mexico, Texas and Alaska, imports had surged in January 1990 to 54 percent, the highest in American history. Ninety thousand wells across America were still mothballed, dismantled and closed, leaving under 500,000 operating. Every year since 1986 domestic production had fallen by between 2.5 percent and 6.5 percent, and was now below seven million barrels a day. The domestic industry could have been expanded to produce nine million barrels a day, but it felt battered, especially by Senator Barbara Boxer of California's conservation campaign. The crisis gave the oil companies a chance to use the extra profits to invest. Instead, their profits were attacked. "We are now relying on military force rather than energy policy for our energy security," observed Senator Frank Murkowski of Alaska, unsuccessfully seeking more exploration in his state. "Americans will find it unacceptable to put American lives at risk because we have not made the hard choices to formulate a policy."

In the aftermath of Iraq's defeat, the industry's plight worsened. Most traders had expected prices to rise. Over 600 wells were blazing in Kuwait, whose oilfields were strewn with booby traps, unexploded bombs, land mines and huge lakes of crude oil. Yet demand was slipping just as record amounts of oil were being produced. Venezuela and Iran had increased production to earn more money at a time when oil stocks in the terminals were at their highest levels since 1982. Western fears of oil prices soaring above $50 a barrel had been replaced by Saudi Arabia's concern that overproduction of four million barrels a day would cause a slump in price. The price in December 1991 was $17, then rose to $30, but fell by the end of January 1992 to $16. In July it was below $16 and still falling. OPEC's overproduction and Iraqi supplies returning to the market aroused Saudi fears of $10 oil. Amid a mini-recession, the oil majors slashed their exploration budgets, decommissioned rigs and, in America, reduced crude production to 6.9 million barrels a day, the lowest in 35 years. In Prudhoe Bay, Alaska, Arco plugged a well that had been operating since December 1970. The site was marked by a Christmas tree. Following Russia, the nationalistic governments in Mexico, Brazil, Peru and Ecuador considered welcoming the return of the oil majors and the privatization of their oilfields to cure their budget deficits. By the end of 1993, oil was $15, a three-year low.

Injured, Andy Hall sold Phibro's two refineries. Pacing his office in Greenwich, Connecticut, he gradually shied away from squeezing his rivals, playing long instead. His withdrawal coincided with Laney Littlejohn desperately seeking to control oil prices and Morgan Stanley opening trading desks in Tokyo, Singapore, London, Houston, Denver and Calgary. Monitoring 24-hour trading from their New York headquarters, John Shapiro and Neal Shear understood that expansion brought greater influence over prices. New regulations to protect the environment and force the companies to manufacture cleaner fuels had created a shortage of tankers and pipelines and a scarcity of some oil products. Shapiro secured finance to construct new storage tanks—eventually the bank would have the capacity to store 40 million barrels of fuel—and in 1998 the charter of a tanker, the first of a fleet of 100 ships sailing among 40 locations. Transporting crude and its products from the cheapest to the most expensive markets became integral to the bank's operation. Speculation and profit from the world's lifeblood had become uncontrolled.

In the wake of the Gulf War, the chance of making huge profits from the volatility of the oil price had attracted Arthur Benson, the head of the "energy group" in Metallgesellschaft, one of Germany's leading industrial conglomerates. Benson was renowned for having made millions of dollars masterminding a "backwardation" strategy in the jet fuel market in 1991. "Backwardation" is when the "spot" or existing price is higher than the future or forward price. The opposite is "contango," when the "spot" price is lower than the future price. Predicting the shifts between "backwardation" and "contango" was the making and breaking of traders. Benson's "backwardation" success in 1991 emboldened his belief the following year that oil prices would continue to be in "backwardation."

The crucial event was OPEC's meeting in Geneva in September 1992. Benson noted that some producers, including Iran and Ecuador, wanted to restrict supplies to force prices up. Saudi Arabia rejected the suggestion, and the meeting ended in disarray. Believing that prices would rise, Benson sold derivatives involving 160 million barrels of gasoline and heating oil, roughly 85 times the daily output of Kuwait. Metallgesellschaft's risk was considerable, but Benson, believing that the company had sufficient finance to absorb the risk, calculated that it would profit whether prices increased or decreased. His belief that he was "hedging" the risk was mistaken, and Metallgesellschaft did not in fact have sufficient money to cover its trades, especially once oil prices fell further than he had anticipated. Instead of cutting his losses, Benson continued to buy oil derivatives. He was no longer hedging, but speculating that the "contango" would return to "backwardation." But the market remained in "deep contango." By September 1993, dubbed "a cowboy without cattle," Benson was ordered to unwind his positions, losing about $1.4 billion. Those who profited from Benson's folly were those traders who had bet against him. Among them was Castle Energy, a refining company that had sold Metallgesellschaft gasoline and heating oil for future delivery at the higher prices, on the assumption that prices would in fact fall. Their wisdom was Benson's disaster. He was fired, and Metallgesellschaft was ruined. Speculating about oil prices, insiders acknowledged, was not suited to amateurs.

The involvement of bankers and big traders complicated the pricing of Saudi oil for Laney Littlejohn. Regularly, Littlejohn was told by Jorge Montepeque that his pricing mechanism was being manipulated. "We're watching Japanese traders squeeze the market in Dubai," Montepeque declared. "Everyone's manipulating in Dubai. It's full of daisy chains." Aramco or Shell could have terminated the squeezes by releasing more oil in the region, but refused. Montepeque suspected their motives. Littlejohn was unimpressed. "Platts's prices are goofy," he complained, "because Platts's assessor is goofy."

As self-appointed policeman, Montepeque's ambition was to impose regulation and remove corruption from a trade crossing every national frontier 24 hours every day. "I'm a driver to bring up standards," he told traders. His cure in 1992 was to introduce "Platts Global Alert," an online system of bidding on the Platts screen in Singapore, starting at 4 p.m., for just 30 minutes. Within that restricted window, known as "market on close," traders could post the price they had bid or offered for a fuel. Montepeque chose that restricted dealing time in order to prevent dealers manipulating the price during the day, but compelling subscribers to use a screen also earned huge profits for McGraw-Hill. In that structured environment, Platts was not an exchange but a marketplace. "I'm bringing core transparency to the market and making the community more trustworthy," he announced. Littlejohn and others criticized Montepeque for creating more problems than he solved in order to generate profits for Platts. Thirty minutes at the end of the day, Montepeque was told, did not reflect a whole day's activity. "I like to trade market risk," said Charlie Tuke, "not what someone tries out at the end of the day."

As Littlejohn predicted, some traders invented new ruses. Artificially low or high prices were listed in the Platts window to establish the threshold of a dishonest trade. In a practice called "all-day capture" or "shower deals," a trader who was contracted to buy, say, six tankers of oil with four million barrels would sell 600,000 barrels at a low price into the Platts window to depress the published price early in the morning while showering in his bathroom. By fixing Platts's market until the end of the day, he hoped to manipulate a low price for the six tankers he was contracted to purchase the following day. Other dealers conspired together to "wash trade"—one publishing a price on the Platts screen for the sale of a cargo of oil to the other and agreeing to buy it back privately at a lower price. Their purpose was to set an artificially high price for a contract with a third trader; a variation is called "buying through the offer" or "selling through the bid" to set an artificially high or low price. Aggressive traders were also posting phony artificial deals in the screen window to increase their bonus payments from unsuspecting employers. The complications were compounded because Glencore and its major competitor Vitol, founded in 1966 by Dutch trader Hank Vietor and his son, were no longer mere traders but, like Morgan Stanley, had become owners of refineries, pipelines and storage tanks in Rotterdam and other ports. Occasionally Vitol controlled more oil afloat than any rival, albeit at most 1 percent of the total. The published price of oil products on Platts was critical to Vitol's profits because it could leverage—that is, earn profits—by speculating between crude and product prices. It could also trade the difference between paper and physical oil, using its ownership of crude in storage tanks to cover any position.

That was precisely the problem Montepeque noticed when Total, the French oil major, and Glencore were reporting sales of distillates. Platts reporters believed they were under pressure to publish reports of deals and prices to influence the pricing of a contract due to mature on that day based on the "actual" price in the North Sea. If all their business had been genuine, Montepeque knew, the whole region would grind to a halt with tankers. "It's suspicious trading," he concluded. "Each side wants to overwhelm the other to cover themselves." In December 2007, Platts would bar Total from trading until February 2008 for misconduct.

To prevent these ruses, Montepeque insisted that traders needed prior approval from a Platts reporter before posting their trade on the screen. The result, most complained, was as untransparent as before. There was a gap between the prices on the Platts window and the traders' actual deals. Small, last-minute trades caused prices to seesaw. At the last moment traders would offer huge amounts of a product like jet fuel in a remote location to distort the market. The system was as imperfect as previously. "The traders," Littlejohn realized, "are trying to pull the wool over our eyes." He blamed Montepeque's "market on close." Montepeque refused to explain how prices were set in the last minutes of trading if there had been no sale. Platts, the traders suspected, was simply inventing phony bids and offers so as to set a price. "Jorge Montepeque likes opaqueness to enhance his power and Platts's profits," Littlejohn complained. He telephoned Montepeque in Singapore and told him, "You're duping us by bizarre pricing." Montepeque rejected the criticism. "You're not the smartest man in the world," Littlejohn screamed at him. "I'm talking to a wooden post. I'm going to ignore you." The bankers in New York were similarly scathing. "Montepeque is unbearably self-promoting and self-important. He's running around with a fire hose but he's always getting screwed. The traders use him to make their profits. Each time he believes he has prevented manipulation, he is beaten by another ruse."

The proof was a trader's squeeze on Brent in 1994. The Brent market had become an impenetrable matrix of contracts developed among the traders through usage. Somehow the benchmark for oil prices worked so long as no one tried too hard to understand the complications, not least the anomaly that the decline of oil production in the North Sea had made Brent even more vulnerable to manipulation. Occasionally even the sharpest traders missed the signs. In 1994 the trader bought about 80 cargoes of Brent oil, although the field could only produce 30 cargoes a month. Leveraging between 15-day Brent and "contracts for difference" (CFDs), the squeeze was undetected as prices rose by $2 a barrel, earning the trader an estimated $30 million.

This was minuscule compared to the potential that Vitol's directors spotted once the United Nations approved the "food for oil" program in April 1995, allowing Saddam Hussein's regime to overcome an embargo on Iraqi oil by exchanging it for food. Embargoes were rich sources of profit for traders, especially as Iraq exported oil worth

$65 billion under the program. It was claimed that Vitol had made handsome profits striking oil deals with Serbia while UN sanctions were still in place, paying the paramilitary leader Arkan, a Serbian later found guilty of committing countless atrocities during the Balkan wars in the early 1990s, $1 million to act as an intermediary in 1995 to recover a debt. Vitol insisted the oil was delivered only after sanctions were suspended, and the deal was entirely legal. Just as Vitol hoped to escape censure for that trade, the directors did not anticipate any repercussions if their company bought Iraqi oil in breach of UN sanctions. Knowing that the sale of Iraqi crude was assured to those refineries that were calibrated to process it, Vitol's skill was to negotiate with Iraqi officials the amount of dollars to be deposited as "commission" in return for the oil. Like Oscar Wyatt in New York, Vitol knew that those payments, or "surcharges," broke UN sanctions, but none of the traders involved expected any legal consequences.

Faced with such ruthlessness, Jorge Montepeque's mechanisms to control the traders were, in the opinion of Carl Calabro, Littlejohn's successor, not "failsafe." Like Littlejohn, Calabro disliked any interference in his efforts to establish the best price for Saudi crude. The Kingdom considered Platts to be unhelpful. Limited markets encouraged "murky assessments," and prompted Calabro's question, "Is this price being manufactured?" Russian traders had entered the market, and were offering bribes to Platts reporters and threatening to steer the market. Real trades throughout the day were ignored, while those in the last 30 minutes were, Calabro thought, just bids and offers, not authentic trade. "The traders," Calabro realized, "were trying to pull the wool over our eyes." At the end of 1999 he decided to abandon the Platts window and trade on the IPE in London and through Argus, Platts's smaller British rival, which he described as "the real trade." Argus's prices were entirely based on 50 reporters around the US interviewing traders. "Argus was not vulnerable to traders crashing in on the close," Calabro explained. The British agency allowed traders and regulators to watch prices change in "real time" knowing that the reporters excluded those that had aroused suspicion because of their timing or size. Montepeque telephoned Calabro. "I didn't get a hosing from Jorge," Calabro reported, "I got worse, a one-sided lecture." Calabro replied in kind: "You're not the smartest man in the world, and the moment anyone thinks he is the smartest man in the world, it's the road to ruin. Never call me again."

Jorge Montepeque was bruised. Platts had been rebuffed by Aramco, and other traders were moving to Argus rather than Platts to price American crude oil. Platts was also rejected by BP, the industry's most aggressive trader. Montepeque's problems were drowned amid the industry's crisis as prices collapsed and Saudis' fears about their vulnerability were resurrected.

Chapter Nine

The Crisis

————◄O►————

LEE RAYMOND'S BELIEF in the free market had been vindicated. In August 1996 the accountancy firm Arthur Andersen had predicted that America's oil and gas industry was in its best shape in 15 years, and was poised for "double-digit" growth. Costs had fallen, spending on exploration had risen by 8 percent, oil prices had risen modestly and the oil majors' profits were assured. Even OPEC acknowledged defeat as the world's oil production increased by 2.9 percent but prices appeared fixed at around $20 a barrel. "These prices are here to stay," announced Rilwanu Lukman, OPEC's secretary general. During 1996, proven oil reserves increased by 11.4 billion barrels, and an additional 11.6 trillion cubic feet of gas had been found. The surplus of oil, the industry was convinced, could only swell.

Raymond's self-confidence reflected a disdain of Washington, the US government, Congress and the anti-oil lobby. The fear of oil shortages, rising imports and soaring prices symbolized by Jimmy Carter wearing a cardigan and declaring it was "the moral equivalent of war" to cut consumption had been forgotten. Size rather than the environment was again fashionable. Economy cars had been abandoned, and nearly a fifth of American motorists drove gas-guzzling SUVs, the biggest of which ran at four miles a gallon. Although oil imports were predicted by the US Government Accountability Office to rise to 60 percent by 2015, there was no fear of an oil shock. Smart technology that enabled drills to turn corners in rocks miles beneath the surface could be relied upon to produce more oil, and oil's contribution to industrial production was diminishing.

Raymond's sense of certainty enhanced Exxon's stature. He and Rex Tillerson were guided by their desire to satisfy their shareholders. They appeared to have no fear of failure. Exxon's reserves, compared to those of their competitors, were enormous. "Lots of molecules" was a familiar Exxon boast. Convinced that Exxon's balance sheet could withstand any cycle, Tillerson asserted, "We'll be the last one standing." The volatility of prices during the 1990s had convinced Raymond that Exxon should continue on its own terms, without departing from its preordained systematic process. Exxon invariably arrived late, not least in deep-sea drilling, and preferred to shun partnerships with other oil corporations in the hope of gaining unique access. Its reserves had fallen over the previous 30 years, but the corporation's haughtiness had not dimmed. It continued to cultivate a mystique of omnipotence, and governments were deterred from interfering in its interests.

Identified at the outset of his career by an internal code as destined for the board of directors, Raymond specialized in negotiating Exxon out of problems: selling a refinery in the Caribbean, or a nuclear plant in Washington State after a dangerously unfit chief executive had been removed; facing down threats from Venezuela to pay more for crude; and directing the cleanup after the *Exxon Valdez* oil spill. In the spotlight for the first time in Alaska, he had been asked by a lawyer during the preliminaries of a civil trial to assess the *Exxon Valdez* damages to describe his background. "I hope this doesn't get too boring," Raymond replied. "It kind of bores me." Asked later about his low profile, he explained, "I don't think much about it. I've never had a focus group to decide what my persona is out there." A good day, he added, was Exxon staying out of the news, and maintaining his own invisibility. His limited public appearances, displaying unusual self-control, aroused bewilderment about whether Lee Raymond was truly from the Exxon mold.

Raymond's insistence on limiting Exxon's civil and criminal responsibility, and his management of the *Valdez* spill and its aftermath, publicly redefined Exxon and Big Oil. Stoutly, he expressed no shame, and appeared to dismiss the emotional and financial distress the disaster caused to Alaskans. According to Exxon's Strategic Planning System, his priority was ensuring that shareholders' profits were substantially above the industry's average, and he combatively rebuffed challenges to that agenda. He embraced a culture that ordained that Exxon's rules and contractual agreements would be mercilessly enforced upon governments, employees and anyone else entrapped within the corporation's affairs. Unlike the employees of other companies, Raymond was trained not to "aim all over the sky," but to "beat opponents to death with lawyers and auditors." His opponents after March 24, 1989, were 14,000 Alaskans, especially fishermen, whose livelihoods had been destroyed by the oil spill.

In the weeks before the trial in Anchorage in May 1994, Raymond had single-handedly negotiated a settlement with the plaintiffs. At the final meeting, never raising his voice, he had confronted a room full of opposing lawyers and won the day. But the deal collapsed, and the plaintiffs, claiming $20 billion, had to prove Exxon's negligence before and after the spill. During the pretrial hearings Exxon's lawyers had successfully limited the corporation's liability to direct losses. The indirect losses, including the diminished value of fishing permits and the sharp drop in the numbers of salmon in the area—from 38 million to 7.4 million—were excluded from the claimable damages. Those legal distinctions aggravated the jurors, all local people. "This is corporate indifference and arrogance at its worst," Brian O'Neill, the plaintiffs' lawyer, told the court. "A company as large as Exxon thinks it is above the law. You need to take a substantial bite out of their butt before they change their behavior." O'Neill portrayed Exxon as an uncaring oil giant that entrusted its most modern tanker to the command of a known alcoholic, "causing misery to thousands of ordinary folk." In the courtroom sat Lee Raymond, said O'Neill, whose income had risen in 1990 to $909,000, while Exxon chairman Lawrence Rawl earned

$1.38 million, and both were refusing to help innocent victims. Highlighting Exxon's $111 billion revenue in 1993, O'Neill scoffed, "Exxon thinks its behavior after the incident was exemplary." Raymond retorted that Exxon had paid $3.5 billion toward the cleanup and compensation, a sufficient amount.

Exxon's defense was fortified by the rescue operation after the accident. Chuck O'Donnell, the senior executive employed by Alyeska, the owners of the pipeline running from the North Slope to the terminal on the seashore, was responsible for overseeing the response to any emergency. Awoken at 12:30 a.m., he was told that a tanker had run aground. Earlier that night his team had celebrated the completion of clearing a small spill from the *Thompson Pass*, an Arco tanker, and he was not in the mood to rush back down to the coast. After ordering a subordinate to head for the pipeline terminal, O'Donnell went back to sleep. The subordinate discovered that the equipment to contain oil spills was either missing or had malfunctioned. "I was shocked at the shabbiness of the operation," admitted a manager. Raymond argued that the damage following the spill was aggravated by Alyeska's negligence, and presented scientific studies financed by Exxon showing that Prince William Sound had suffered more damage over the past years from diesel spills than from the *Exxon Valdez*.

Despite his aggressive defense of his company, Raymond's arguments were rejected. The scientific evidence lacked credibility, and he was criticized for refusing to meet the fishermen, and rebutted by scientific witnesses who confirmed that the fisheries were destroyed, about 400,000 birds had died and the recovery could take 70 years. On June 13, 1994, the jury found Exxon negligent and reckless. The corporation was ordered to pay a further $5 billion in punitive damages. "The verdict," commented Raymond, announcing an appeal, "is totally unwarranted and unfair...The damages are excessive by any legal or practical measure." Raymond's insistence that Exxon should not suffer punitive damages was undermined by the emergence after the trial of his disingenuousness. In his testimony, he had described how in 1991 he had authorized the immediate payment of $70 million to destitute fish processors, ruined by the *Exxon Valdez* spill: "I said from New York, 'Forget the release, just pay the money. Get a receipt that you paid the money and some day we'll sort this out in court.'" He and his lawyers had not, however, revealed the conditions attached to that $70 million compensation: the fish processors had agreed to repay Exxon a share of any punitive damages subsequently awarded by a court. Raymond's secret arrangement, described by trial judge Russell Holland as "Jekyll and Hyde" tactics and "behaving laudably in public and deplorably in private," was criticized as "reprehensible," and "an apparent fraud on the jury."

Raymond dismissed the censure. He had vowed to contest the $5 billion damages award up to the Supreme Court, on the grounds that damages should be "reasonably predictable." Eventually, in June 2008, the Supreme Court reduced the damages to just $507.5 million, a year after Exxon made profits of $40.6 billion. The damages could be paid from two days' earnings. The fisheries remained destroyed. Experience had taught Raymond not to be cowed by any judge or to trust any government, especially after the recent hysteria in Washington about supposed price gouging and monopolies in oil retailing. Every appearance of an oil executive in Congress was exploited to win political points. Occasionally seeking government help was akin to supping with the devil. Raymond's contribution to the widespread distrust of Big Oil left him unconcerned. By advocating the shareholders' interests and alienating the public, he had reinforced the corporation's belief in his trustworthiness, and on April 28, 1993, aged 54, he had been appointed as Larry Rawl's successor. His conduct confirmed the sanctity of Exxon's values regardless of external circumstances. "Our culture eats strategy for breakfast," said one admirer.

"Play by the rules of the book" was the corporation's doctrine. "It started with JR," intoned Raymond's lieutenants, invoking Standard Oil's founder John D. Rockefeller to those selected to embrace the faith. "JR had integrity. He played by the rules. He never did anything illegal." New employees were told, "There'll be no cutting corners, corruption or lying. If you do, you lose your job." Deifying Rockefeller surprised outsiders, but within Exxon's secretive fraternity, agreement percolated from the top about the cardinal rules. By example and instruction, Raymond, like all Exxon's employees, had been nurtured to speak unemotionally, and only about facts. "When meeting outsiders," new arrivals were told, "either do not speak to them or be cautious and report back." Those lessons, impressed in youth, remained ingrained.

After *Valdez*, Exxon transferred its headquarters from New York to Irving, near Dallas. This move out of the spotlight had coincided with Rawl's and Raymond's courting of Republican politicians, especially George W. Bush. Ever since Theodore Roosevelt had declared in 1906 that the directors of Standard Oil were "the biggest criminals in the country," and unleashed the legal process that culminated in the corporation's enforced self-destruction in 1911, John D. Rockefeller's successors had downplayed their relationship with politicians and the government. Raymond gave the impression of hating Washington. "Exxon is squeezed too much by government," he complained, suggesting that politicians never did anything good. "I run the company on the basis of my long-term price assessments. I'm running a global business. Please stay out of my way." Exxon's rivals were unpersuaded. Oil titans only stayed out of Washington until they needed help, then they arrived with their hand out. Chris Fay of Shell regarded Exxon as the voice of the US government, while Chevron's senior directors suspected that Exxon had received government assistance to secure a natural gas concession in Indonesia, an oil lease at Upper Zackin in Abu Dhabi, the rights to Sakhalin 1 in Russia, and much more. Combatively, Raymond occasionally denied ever requesting or receiving government help.

Raymond had been trained to share Exxon's global outlook, but America was the center of its universe. Standard Oil of New Jersey, Exxon's original incarnation, had been the jewel of Rockefeller's empire. The conservatism of Standard Oil's senior staff, especially the international specialists, had transferred to Exxon. Created within an integrated market—the oil wells, refineries and customers were all adjacent—Exxon inherited more oil than the other Standard companies after the breakup, and Walter Teagle, Exxon's chairman in the 1930s, had bought more oilfields in the midst of the financial crisis in Venezuela and elsewhere, confirming Exxon as the world's biggest oil corporation. Accordingly Exxon had the least reason of any of the oil majors to take risks, and could balance losses in one region of the world with profits in another.

Watching Lawrence Rawl, a renowned strong-arm practitioner who used Exxon as a blunt instrument, Raymond had learned the technique of grabbing an inordinate share of the action regardless of sensibilities. Exxon, he knew, was sufficiently big to shun partnerships, but as Rawl and his predecessors had demonstrated, principles could be bent in order to accrue profits. An early partnership with Shell in Cuba had been unsuccessful, but with some political pressure from Washington, Shell had agreed in the late 1950s to admit Exxon as a partner in the spectacularly profitable Gröningen natural gas fields in the North Sea, and later in the Brent and other North Sea oilfields. Raymond had also learned that Big Oil did not need instant results. During that formative decade, he had witnessed both feast and famine. Uncertainty and gyrating prices complicated investment decisions, and oil companies could not turn money off and on. Exxon's custom was to sit tight, especially after the experience of the exploration and production executives, who had been ordered to spend money in the early 1980s, when high tax rates had encouraged drilling risky wells in the Rockies. "I need you to spend another $50 million," one exploration and production engineer was told in a telephone call. "Get rid of cash." Resources seemed to be unlimited. Prices rose and then crashed with the surplus of production, the development of efficient car engines and the shift to natural gas and coal to produce electricity. The ultimate blow was Saudi Arabia's decision in 1985 to maintain rather than reduce oil production. America's oil industry had been devastated. Half a million employees were laid off in a wave of hardship, aggravated by deregulated markets and denationalization.

Rawl and Raymond had understood the need to alter Exxon's

model. Costs were cut and the organizational structure was modified. But the fundamental Exxon way remained sacrosanct. Best practice was imposed from the center. Systematized checks and balances prevented individual initiatives. Raymond inherited a frictionless, problem-free machine, in which every transaction was religiously policed to extract economies of scale by standardization. Like God's, Exxon's gospel heralded the corporation's infallibility: "Accidents are impossible because no man makes mistakes, only a malfunctioning system." Unlike BP, which would be rejuvenated by haphazardly combining the remnants of a colonial era with three former Standard Oil companies—Sohio, Arco and Amoco—Exxon was purposefully designed by engineers over 100 years, gradually developing the perfect DNA to manage its assets. "The oil business," Exxon's staff absorbed, "is putting things down and living with them for a long time." Everything should work perfectly, without gold-plating. Improvements sprang from constant dissatisfaction and resistance to being first. There was no shame in arriving late and buying entry into the game. Exxon was designed to profit by watching others make mistakes. Smaller rivals, the wildcatters acting on gut feeling and shooting from the hip, would bear the risk of failure. Exxon took pride in being a "fast follower," carefully calculating if something worked before investing. "It takes me two months to train a specialist to operate in Angola," a Schlumberger executive proudly told an Exxon specialist. "Gee," replied the Exxon man. "It just takes me five minutes." "How so?" asked the puzzled executive. "'Cos that's how long it takes me to walk down the corridor to the Gulf of Mexico man and tell him that he's now assigned to Angola." The result was a "one size fits all" standardized Exxon expert, recognized across the globe. Consistent in his approach, he had a clear understanding of his role, the limitations of his responsibility and to whom he was accountable. Empowered by a "Triple A" rated corporation with low tolerance for differences, he was sharply aware of his restricted scope, but could boast on arrival in a new country, "We've got the best technology, the deepest pockets, the industry's best project management, and we will work with the government." The result produced neither spectacular successes nor crushing failures—just consistently the industry's highest profits.

Inflexibility was a mixed virtue. Raymond's obstinacy in Anchorage and later in Moscow was Exxon's strength. The intransigent culture—"My way or the highway"—was born from the tortuous decision-making process. Once Exxon's policy was fixed, its executives became suspicious of anyone who disagreed with it. "They're wrong," Raymond would say of those who warned about lead in gasoline or global warming, or, on a lower level, about a rival's interpretation of seismic data. Disagreement automatically aroused questions about motives. The technological revolution challenged the monopoly of Exxon and the other oil majors. Indoctrinated to believe that Exxon was better than the rest, and that any new idea must be mistaken unless it was invented by Exxon, Raymond begrudgingly acknowledged that Exxon's research, while the best, did occasionally fail to buy any noticeable advantage. The Standard research labs in New Jersey were closed, and the number of patents assigned to Exxon declined. Outside consultancies could occasionally interpret seismic data better than the oil majors, and operations were frequently subcontracted by Exxon to Schlumberger and Halliburton. Exxon was progressively becoming weaker. Unable to dictate the industry's agenda, the corporation nevertheless maintained its historic authoritarian pose in order to maintain the influence secured by its founders. The impression of ExxonMobil as a superb machine, engineered to clockwork perfection, was reflected by Lee Raymond's vainglory and the forbidding exterior of the corporation's new headquarters.

Visitors to ExxonMobil's hilltop mausoleum outside Dallas could have mistakenly interpreted the silent vacuum in the huge entrance hall as symbolic of the giant corporation's oppressiveness. Within that building, constructed over the Barnet natural gas reserve (America's largest, but not owned by Exxon), were tiers of seniority, a hierarchy based on status assessed by the "formal weighting" process, blessing the chosen ones with access to the executive dining room. The identically clad male inhabitants—starched white shirt, striped tie, gray suit and shiny black belt — appeared to be the products of a mold, akin to the Stepford wives. Fired by indistinguishable levels of conviction, they were subject to continuous evaluation to select the best and discard the rest. "We put them through a big distillation column," explained Raymond. "We have a very rigorous process of moving them around. Everyone in the world gets rated every year. It's a very, very rigorous process. Only the top of the column stays here." Those few, even though disciplined by Exxon's monoculture in how to conduct their discussions, incarnated the corporation's reputation as feared and revered. "Some wonder why they remain as Exxon people," observed a contractor working in the oilfields alongside those at the top of Raymond's column. "It's like they've been injected with forgetful drugs."

In a raw trade, the smooth self-presentation of Exxon's star executives was deceptive. Outsiders would carp that they "all sing from the same song sheet," but Raymond's closest aide countered, "We don't meet for Scripture meetings." In 2000, Enron's CEO Jeff Skilling, boasting about his company's supremacy, would compare the oil majors to dinosaurs. "They need to remember," replied Raymond, "that dinosaurs were around for a long time, and they were as mean as hell." Big meetings, encounters with strangers and social etiquette were avoided by the custodian of Exxon's unchangeable customs. The traditional lexicon of marketing — brand recognition, interface and image — barely troubled him. Proudly he presented himself as the personification of Exxon's parsimony and integrity, the chairman who like the lowliest employee was required to read "Standards of Business Conduct" every year, and to confirm by his signature his agreement to abide by that code. To others, Raymond symbolized Exxon's regimented "Arrogance Training."

Preceded as chairman by colorless, soft-spoken engineers wielding the power of statesmen, Raymond dented the mold. Those lacking his self-confidence were humiliated. "That's the stupidest idea I've ever heard!" he would bark at a subordinate or an opponent. Unglamorous, not least because of a harelip, he rarely concealed his conceit, or nurtured protégés, or made collaboration enjoyable. "Steamrolling by sheer intelligence" was the comment of a contemporary inducted with

Raymond into Exxon's "forced ranking system," an innovation introduced during the Second World War.

The combination of that culture, Exxon's obduracy toward the
environment and the growth of the oil market changed the relationship between the corporation and the government. Unlike the mid-
1970s, when the dialogue between the oil majors and the government
had been a civilized exchange based upon mutual interest, the mood in
the 1990s became sour. "We got bored with each other," said an Exxon
executive, "and less convinced that we needed each other."

America's politicians inflamed the distrust. Bush's successor Bill
Clinton's National Energy Policy Act, which aimed both to conserve
energy and to ban oil exploration in the Arctic National Wildlife Refuge, was criticized by the industry as representing "the destructive
power of exaggerated environmentalism." Although Clinton suppos-
edly checked his popularity every week against the oil prices, his energy
bills floated on an oil surplus were blamed by some for crippling the
industry. In 1998 he extended Bush's moratorium on drilling off the
US coastline by 10 years. Most Americans were content for oil to be
drilled in the Gulf of Mexico and parts of Alaska, but not off California, Massachusetts or Florida. Although Florida might have billions
of barrels of oil and trillions of cubic feet of gas offshore, and California had even greater reserves, those two energy-consuming states forbade offshore drilling. The US government's outlawing of Alaskan oil
exports to Japan was supported by the same people in California who
demanded that the government place pressure on Russia to increase its
exports of crude oil to America. As a relic of an era when America was
the world's biggest oil producer, the country had spawned 18 different
gasoline zones, meaning that licenses were required to move gas from
one zone to another. California's unique gasoline requirements rendered oil from Nevada's refineries inadequate for its purposes, so Nigerian oil was refined with special calibration in Pembroke, South Wales,
and shipped to California. The contradictions were unmentioned. "Oil
was surplus and there were other more important issues," admitted an
Exxon executive. The surplus neutralized some governments' interest

in developing energy policies. As an extreme example, the British prime minister John Major abolished the Department of Energy. "We don't need an energy policy," said Nigel Lawson, a former chancellor of the exchequer, convinced that Britain could rely on market forces, and could change taxation in the North Sea whenever additional revenue was needed.

The converse was the oil companies' perennial attempts to avoid regulations and taxation. Around 25 percent of America's oil and gas was produced from federal land, and since 1988, 18 American oil companies had paid lower royalties for oil drilled on federal lands, saving $5 billion by 1996. "The issue is simple," said Senator Barbara Boxer. "Big oil companies have knowingly and intentionally cheated the federal government and American taxpayers out of royalty payments for years. We cannot continue to subsidize Big Oil who list profits in the billions each year, at the price of our children's education." Most of the companies, including Chevron, BP and Conoco, repaid the government between $30 million and $153 million to end litigation. Only Exxon refused. It had successfully fought similar charges in California in 1992, and as usual expected to out-litigate the government. Exxon was among the oil companies Leo McCarthy, the lieutenant governor of California, had in mind when he commented, "The people of California must be treated respectfully as customers and partners, not as victims and targets for fleecing." In Alaska, the state government claimed that, to reduce the royalties owed to the state, producers were deliberately undervaluing the oil produced and overvaluing their transportation costs. Exxon did pay $154 million to settle this dispute, much less than BP's $1.4 billion and Arco's $280 million, but would refuse to pay a further claim, insisting that it had acted lawfully. Exxon's rivals were less assured in court battles, and similarly lacked its confidence in their ability to cope with unpredicted price changes.

In January 1997, fearing losses as oil prices unexpectedly fell, the industry dismissed some engineers, abandoned plans to recruit and train graduates, and canceled orders for new rigs. But in the spring prices stabilized, and fears of disaster evaporated. The oil companies began recruiting, paying bonuses and preparing for expansion. Some even forecast a boom. Over 1,000 oilmen crowded into the Hyatt Regency in New Orleans in early March 1997 to place bids for exploration leases in the Gulf of Mexico. In the electric atmosphere, everyone wanted to play, and the US government pocketed $824 million, 58 percent more than the previous year. Shell was the biggest bidder. Daily rentals for rigs soared to $145,000, compared to $49,000 in 1990. Convinced that oil at $20 signaled good times, OPEC ministers meeting in Jakarta in November 1997 raised production by two million barrels a day.

But after Christmas, news of crumbling Asian economies crushed the optimists. Brent fell from $20 a barrel to $12.80 in March 1998. The collapse confirmed Lee Raymond's certainty about oil surpluses and low prices for the foreseeable future. Although the demand for energy had risen by 3.7 percent and prices had fluctuated since 1996, Raymond and his competitors spoke of an "oil glut." Recently recruited engineers were dismissed, and drilling was terminated, prompting some to question whether the industry had become too smart for its own good. "Is technology too effective?" one expert asked. BP's horizontal well at Wytch Farm in Dorset had reached 6.28 miles, a world record. "Has seismic 3D run amok?" asked a trade magazine. Underwater exploration and enhanced production in established fields had discredited old truisms about oil shortages. There were new discoveries in the North Sea and off Newfoundland. Amoco had found a vast reservoir 35 miles off Trinidad, and Texaco was about to announce a new discovery 12,400 feet below the seabed off Nigeria. Venezuela's reserves were variously estimated at between 1.3 and 1.8 trillion barrels. The consensus was that demand would remain low in 2000, and the industry concluded that more oil was available than was required.

Saudi Arabia had good reason to fear economic mayhem. Ali al-Naimi, the petroleum minister since 1995, was in uncharted territory. The traditional weapons to control prices were failing. Two years earlier, an abrupt $6 price rise after the bombing of a building in Saudi Arabia used to house foreign military personnel had killed 19 American airmen, had rapidly evaporated. Instead of "supply anxiety" sparked by unrest in Saudi Arabia, traders assumed that the world could cope with any glitches, not least because an estimated 800 million barrels of crude oil lay unused in the Gulf's storage tanks. In that atmosphere, al-Naimi summoned an emergency OPEC meeting in Riyadh in July 1998. To cure its budget deficit, Saudi Arabia needed oil to be priced at $22. OPEC announced that production would be cut by 2.6 million barrels a day. Prices rose to $16, but after some OPEC countries cheated on their quotas and Iraq exported more than was allowed by the United Nations' quotas, they fell back. With full storage tanks, and with rival countries using technical innovations to increase production, OPEC was vulnerable. Little comfort could be drawn from the prediction of the Energy Information Administration (EIA), the statistics arm of the US Department of Energy, that OPEC would again dominate the world's supply once the sellers' market was restored. Despite OPEC's controlling 63 percent of the trillion barrels of oil already discovered, its old certainties seemed to be evaporating. In a masterly understatement, al-Naimi's experts reported, "The market is unbalanced."

To expel the demons threatening once again to revive the horrors of the late 1980s, the city of Houston hosted a party in September 1998 for 30 oil ministers and 8,000 officials and executives from 82 countries. Everything about the World Energy Congress was on a mammoth scale. Twelve tons of fireworks and the biggest dinner ever served in Texas, however, could not smother the smell of crisis. To survive low prices, some visitors suggested, would require reinventing the whole industry and its technology. By the end of November, the scenario was bleaker. Oil was trading at $11 a barrel, which barely covered the cost of exploration, drilling and recovery. In December, Brent slid to $8, the lowest level (accounting for inflation) since the 1930s Depression. The oil market's collapse was swift and devastating, and shares in oil and drilling companies fell by 70 percent. Thousands of workers were dismissed, and only 155 offshore rigs were working, a 60 percent decline in one year. Rig prices in the Gulf of Mexico fell by 70 percent. The remaining 500,000 marginal wells across America were jeopardized. America's independent producers and one-man strippers,

especially in Oklahoma, earned just $18 a day from their two-barrel-a-
day wells. The prediction by a banker from J.P. Morgan that oil would
"not see $20 in the foreseeable future" galvanized the terrified industry
to appeal for help to Congress. But the politicians were helpless. None
could understand how oil's value could halve in two years. By December 1998 there was talk of OPEC's imminent death.

The Saudis had survived the collapse of prices in 1996, but the new
falls resurrected their vulnerability. OPEC was powerless. Petroleum's
professionals had fallen victim to their own success. In an unusual
initiative, al-Naimi sought advice from the oil majors. Seven chief
executives were invited by the Saudi ambassador in Washington to his
home in McLean, Virginia. How, they were asked, could Saudi Arabia
develop its energy resources? Speaking first, Lee Raymond delivered
a dry speech, packed with statistics. Oil prices, he predicted with certainty, would remain at $10 for "another decade of hurt." The cost of
production, he said, had fallen from $16 per barrel in the early 1980s to
between $6 and $10. At the same time the cost of finding oil had also
fallen, from between $14 and $20 per barrel in the 1970s to between
$4 and $5. At a price of $10 a barrel, said Raymond, Saudi Arabia
would have ample profits, while Exxon would retrench. Raymond's
self-assuredness hung on his conviction that only those with a long-
term perspective prospered in the oil industry. Stability was Exxon's
mantra. "A useless, nonsensical meeting," said one of the invited chairmen. "I've just wasted a Saturday on bullshit. They don't know what
they wanted."

Raymond had behaved with unusual diplomacy at the meeting,
although his assumptions were privately questioned by others present.
The industry's economics, they believed, were fundamentally changing. As a tradable commodity, oil derivatives rose as fast as they fell
in an "aggressively self-correcting market." While Raymond and the
OPEC countries conservatively assumed that every new crisis would
duplicate past ones—Iran in 1979 and Iraq–Kuwait in 1990—Raymond's critics recognized that the new breed of traders responded more
quickly to changes, usurping Exxon's influence. John Browne, unlike

Raymond, had understood that survival in such turbulence meant recognizing that oil prices were beyond the producers' and the oil majors' control. Stability depended upon assessing the profitability of each operation as "return on capital employed" by managing and reducing costs, increasing profits and production volumes and avoiding self-inflicted wounds.

This mathematical modeling was a far cry from Ali al-Naimi's search for solutions to Saudi Arabia's predicament. The former shepherd, born on September 28, 1935, had abandoned his "many, many sheep" at the age of 12 to join Aramco as an office boy. His break came when he studied geology at Stanford in California, and his understanding of the shifts of oil's history was to prove invaluable as he rose to become Saudi Aramco's president. OPEC countries, he realized, could not uniquely determine the world's fate. In the cause of complying with market demands rather than dictating to the globe, al-Naimi decided to end Saudi Arabia's traditional power struggle with other OPEC members, especially Venezuela and Iran, and to apply some free enterprise to OPEC's management. OPEC's isolation, he decided, should end. Discreetly he summoned another conference of OPEC countries in Riyadh. He did not reveal that Luis Tellez, Mexico's energy minister, was invited to the meeting, although Mexico was not a member of OPEC. Tellez, a 41-year-old economist without previous experience in oil, had been stung by the news about one shipment of Mexican oil that sold for only $7 a barrel. The Mexican newspaper headline "Crude Prices Shipwrecked," echoed his fear of a national calamity. Studying his country's plight, he realized that, contrary to orthodox teaching about markets, the traditional formula of supply and demand should not apply to oil prices. Oil was different. With unlimited supplies, prices were determined entirely by demand. Creating "supply anxiety" among the traders would determine oil prices. "Oil trades on psychology," Tellez appreciated. OPEC's production, he believed, should immediately be cut by up to five million barrels a day, until prices stopped falling, and should thereafter be kept at a level below demand. The 800 million barrels of oil in storage should be reduced to remove that cushion. Before flying to Riyadh, Tellez met the energy minister of Venezuela, which had reached a similar crossroads to Mexico. Nationalism was hampering both countries' oil industries and damaging their economies. Both could either maintain their exclusive control of production, or invite investment by the oil majors.

Foreign investment in Mexico's oil — the country's lifeblood — had been complicated ever since nationalization in 1938. Expressing a mystical relationship between oil and sovereignty, Pemex was a bureaucratic social security agency rather than an oil corporation. In 1981 it had spent $6 billion on capital developments, but in 1988, with 213,000 employees, it had invested just $1 billion to develop new wells. Insular and inefficient, the Mexican oil industry stagnated, producing about 2.5 million barrels a day, denying the country the potential benefit of higher prices. In 1990, there seemed no prospect of change. Pemex's finances had been plundered to support socialism and corruption. "I want to confirm the fact," President Carlos Salinas told parliament, "that Mexico will maintain its ownership and complete dominion over hydrocarbons." The consequences in April 1992 were fatal. Gasoline fumes had seeped from a Pemex facility in Guadalajara into the sewers, causing an explosion that killed 205 people and injured 1,470. Mexicans linked the disaster with nationalization. In 1995 Joaquin Hernandez Galicia, the head of the Mexican oil workers' union for 25 years, admitted, "My oil workers were corrupt, drunken and courageous." His imprisonment for murder and stealing millions of dollars highlighted Pemex's need for foreign technology and independent management. The nationalistic barriers cracked. Financed by $20 billion in US loans, production increased in 1995 to 2.85 million barrels a day. Falling oil prices in 1998 encouraged the movement toward privatization.

Venezuela appeared to be heading in the same direction. Although PDVSA, the country's national oil company, was the same size as Mexico's and had been nationalized in 1976, it was eight and a half times more profitable, earning $2.7 billion in 1990 compared to Pemex's $320 million. PDVSA employed 47,000 people, one fifth of Pemex's staff. The United States bought two thirds of Venezuela's oil exports, and all the oil majors were entranced by the potential represented by the country's reserves of 66 billion barrels of conventional oil (5.8 percent of the world's total), 140 trillion cubic feet of natural gas and 268 billion barrels of bitumen oil in the Orinoco, a 54-square-mile area 120 miles south of Puero la Cruz. To meet the country's need for income from oil to repay the national deficit and reduce its austerity, President Carlos Andres Perez had spoken in 1990 about rapid privatization to increase production by 25 percent. Although a member of OPEC, Venezuela was defying demands by the more militant nations—Algeria, Libya and Iran—to cut production. The president sympathized with Washington's argument that rising oil prices would damage the world's economy, not least because his country needed American investment. To restore the impoverished economy, there seemed to Perez to be no alternative but denationalization. Extracting oil required knowledge and money rather than human labor, which only the oil majors could provide. With the promise that "Venezuela is the safe, sufficient and reliable bridge toward energy self-sufficiency for the Americas," Perez urged the oil majors to invest $48 billion over five years. In contrast to the "inevitability of confrontation" in the Middle East, he added, "The United States can trust, always depend on, Venezuela as a safe supplier."

In June 1992, Shell, Exxon, BP and Chevron were asked by PDVSA to bid for inactive fields that, with modern technology, could be revived. The chance of attracting investment was temporarily threatened by a minister's demand that PDVSA submit any plans to the government for approval, and after a failed coup attempt against President Perez in mid-1992 led by Hugo Chávez, a nationalist, socialist and professional soldier, the oil majors lost interest. Chávez was jailed for two years. In 1996, the same corporations whose assets had been nationalized in 1976 were again invited to look for oil on government sites. They were told that, despite OPEC quotas, Venezuela planned to double production by spending $60 billion. The inducement to sign PSA agreements was tinged by urgency. A bank had failed, and the country was mired in financial crisis. Perez's negotiations were even encouraged by al-Naimi, who feared that prices could fall from $11 to $5. Al-Naimi was also concerned that the Western oil majors would develop alternative sources of oil, and that non-OPEC countries, especially Russia, would undercut OPEC after consumption increased. He urged the governments of Iran and Algeria to copy Venezuela and invite the Western oil majors to search for oil in their countries.

The opportunity for the Western oil companies was unique. But they were confronted by a paradox. Although there was a surplus of oil on the market, and the promise that new technology would discover ample new reservoirs, over the previous nine years none of the major oil companies had found sufficient new oil to significantly increase its own reserves. Depletion in the North Sea and Alaska, and disappointment in Colombia, had frozen BP's reserves. Shell suffered similar problems in Nigeria. Stagnation haunted Exxon, Chevron, Texaco, Arco and Amoco. The producing countries' new vulnerability provided the majors with an opportunity to surmount their nationalism. Success depended upon nurturing sensitive politicians.

This was a decisive moment for the Western oil majors. In their search for diversifying supplies, breaking through the nationalistic barriers in Venezuela and Mexico was a foretaste of the security that would be guaranteed by developing oil reserves across the Third World. Successfully producing oil in those countries would protect the majors from contraction, and prices would remain stable. But oil executives were rarely subtle, or gifted with a sensitive understanding of the oil producers, especially at a time when President Clinton was politicizing oil in the states on Russia's southern border and maintaining embargoes against several major oil producers. Yet the arrival of Luis Giusti, the president of PDVSA, at the annual Latin American conference at La Jolla, near San Diego, California, where he reaffirmed that Venezuela's oilfields would be reopened to foreign investors, sparked the sort of excitement more usually generated by rock stars. Reflecting on the new mood to denationalize, the economist Daniel Yergin, the president of the respected consultancy Cambridge Energy Research Associates, confirmed that privatization was returning: "It's back to a high degree of interdependence," he wrote. "Everyone wants to be on the same team now." A wave of privatizations, including Bolivia's in 1996, convinced Yergin that the oil majors rather than the national oil producers would dominate the industry. The national producers, he suggested, would be akin to traffic cops and tax collectors rather than significant explorers and producers. The oil majors, he believed, would become transnational entities like the IMF and the World Bank, with the Third World producers accepting the reality of their superiority. Lee Raymond accepted that analysis, and Exxon, like Shell, Chevron and Conoco, began negotiations to return to Venezuela. The country's oil industry, they knew, was in a desperate state. Wells had fallen into disrepair, and the country was struggling to produce even 2.4 million barrels a day. Exxon, said Raymond, would fulfill its contract and would expect to be treated fairly. Venezuela's political turmoil terminated that hope. On February 2, 1998, Hugo Chávez was elected Venezuela's president. Ideology rather than economics propelled Chávez to reverse Giusti's agenda and deny American corporations the chance to develop Venezuela's oil. PDVSA was, he said, to be brought under his control and would cease to be a "state within a state." Three trusted supporters were appointed as directors, none with any experience in the oil industry, especially the army officer appointed as the director of finance. By the end of 1998 Chávez had halted the $65 billion renewal program, which he dismissed as an "irrational" expansion of production. The Saudis could silently smile. Less oil from Venezuela would increase oil prices, which was precisely what they wanted.

Chapter Ten

The Hunter

———◆◇◆———

THE NEWS FROM VENEZUELA confirmed John Browne's suspicions. Oil-producing nations were unreliable, and even if they possessed "elephants," the pace of discovery was too slow to fulfill his ambitions.

In a strategic review presented to BP's directors in Berlin in 1996, Browne had eloquently explained his desire for spectacular growth. As usual, his competitive benchmark was Exxon, which owned a huge backlog of unexploited assets. By comparison, he described what he called "BP's New Geography" as a mixed blessing. Disappointing production in Colombia was complicated by terrorism; drilling in Venezuela had proved to be too expensive; and corruption ruled out any return to Nigeria. Only Angola (Girassol), the Gulf of Mexico and Australia (the Perseus) offered unmitigated success. BP's production of oil and gas, he forecast, would increase by one million barrels a day to 2.5 million within a decade. But that was too slow and too little. To survive, explained Browne, BP needed alliances and a spectacular acquisition. His targets were in Russia, China and America. One year after his presentation, his ambition had crystallized. "We want to overtake Shell," he said in a newspaper interview. "Who's BP?" was Shell's waspish reaction. Those two words galvanized Browne's competitiveness. His target was Mobil, the former Standard Oil Company of New York.

Browne had been telephoned by Lou Noto, Mobil's gregarious chairman and chief executive. Over the previous months Noto, the son of Italian immigrants, had pondered Mobil's fate. As a middleweight, the corporation's chance for growth was limited. Ever since he had started in the industry in 1962, and managed Mobil's refinery in Saudi Arabia, Noto had understood the danger of "hitting our heads against the wall." Unless there were certain profits, Mobil would stay away from any investment. By the mid-1990s the financial scenario was not encouraging. The losses suffered by all the oil majors in the so-called "downstream" — the refineries, gas stations and marketing — had accelerated. Supermarkets in Britain and France were selling cut-price gasoline, while the refineries, producing excessive amounts of unprofitable gas rather than profitable distillates, were losing 60 to 70 cents a barrel. To reduce the losses, the majors swapped products and shared pipelines. In America, Shell, Texaco and Aramco had forged a $10 billion union of six refineries, 80 crude oil and products terminals, 17,000 miles of pipeline and 8,800 gas stations, using some oil supplied by Saudi Arabia.

Mobil, Noto knew, also needed a partner. John Browne's rescue of BP from near death had impressed him. A combination of the two middleweights, he calculated, could rival Shell and Exxon. He proposed a trial marriage in Europe to Browne. The merger of BP's eight refineries, worth $3.4 billion, with Mobil's six refineries, worth $1.6 billion, plus their marketing operations would capture 12 percent of the European market, the same as Shell and Exxon. Browne readily agreed. BP would sell two refineries, and staff would be cut by combining the companies' headquarters. Noto had anticipated saving $400 million every year, but the savings rose to $600 million. He was only distressed by market research showing that BP was more popular among motorists than Mobil. "It still wasn't easy yanking the signs off 23,000 of our gasoline stations," he admitted, "and wasting $200 million just to replace the signs." Encouraged by the collaboration, in late 1996 Noto and BP's

David Simon discussed the next stage, a merger of equals, over dinner in New York. "I'll talk to John about it," agreed Simon.

For Browne, the news revived an idea he had discussed 10 years earlier while working in Cleveland, Ohio, except that he wanted BP to take over Mobil. The hunted was transformed into the hunter. Browne began attempting to persuade Noto to accept his vision. "We need to find a way into Noto," he ordered. Research was commissioned by BP's bankers to understand the Mobil chairman's life and his passions. One discovery was his interest in the work of the novelist Joyce Carol Oates. Before a crucial meeting at the Hay-Adams hotel in Washington, a first edition of one of her books was bought for Browne to present to Noto as a goodwill gesture. The gift was unmemorable for Noto.

Huge mergers were unknown in the oil industry, but Browne's arguments were, Noto knew, irrefutable. The existing business model, Browne explained, was obsolete. Threatened by decline, the majors needed lower costs and higher profits. His arguments were supported by unpublished research produced by Doug Terreson, a Morgan Stanley analyst. Mergers among the Seven Sisters, Terreson predicted, would be necessary "to avoid being competitively disadvantaged for years to come...If without a partner you need to find one."

Noto's skepticism appeared to evaporate. Not even their companies' different cultures, he agreed, were a barrier. By February 1997 the framework of a deal had been secretly agreed upon. Noto ordered his team to scrutinize the terms. They reached a gloomy assessment. "This isn't a merger of equals but a takeover," Noto was told. "It's transatlantic with tax problems. There's got to be value." The chairman agreed. "It's beginning to ungel." Noto called Browne and told him, "This deal's got to be priced differently. Mobil's going to disappear. Our shareholders need a premium." Browne agreed to reconsider the price. During March the paperwork was completed, and the two sides agreed to sign the contracts on March 28, Good Friday, at the Four Seasons hotel in New York. Browne arrived on Wednesday from London to inspect the stacks of documents awaiting his signature. The following morning his aides heard whispers that Noto had changed his mind. The obstacle, some speculated, was not the price but the danger that the Federal Trade Commission (FTC), the regulator, would resist a foreign takeover of American assets; others suggested that Lee Raymond had offered Noto a merger with Exxon. In reality, on the eve of the signing Noto simply wanted more money. Browne did not respond, assuming that Noto feared for Mobil's future without BP. He was mistaken. "It's coming apart," Noto told his staff. "There's not enough beef in the patty to do it." Entering Browne's suite the following morning, Noto told him to his face, "No, there won't be a deal." Shortly after, Browne flew back to London in his private jet. His failure did not sadden everyone on that flight. Some felt Browne was too radical and his pace too hectic. He dismissed their reservations. "You'd better think what we now do," he ordered his trusted associate Nick Butler as he stepped into his car. Anticipating a rush of corporate activity, he did not want to be left behind.

"Our next best bet is Amoco," Browne told his fellow directors. His verdict was confirmed by John Thornton, the acquisitions king at Goldman Sachs. Browne had astutely pinpointed a more vulnerable target than Mobil. "Buying Amoco at the bottom of the cycle," said another banker, "is a good move because the company's valuable assets are cheap and we can squeeze more out of them." Sharing the oil industry's conviction that prices would remain around $10 or $11 for the next five to 10 years, everyone agreed that the risk was minimal.

Based in Chicago, Amoco was North America's largest private natural gas producer and, as the refining arm of the former Standard Oil, the owner of five refineries across the country. Under Larry Fuller, an undynamic chief executive, the company was floundering. Since 1992 it had drilled 190 wells in four continents at a cost of $9 billion; its reward was a record collection of orphans. Burdened by that failure, the company was selling more oil than it was finding, and stubbornly remained America's fifth-largest oil company, increasingly seen as a country-club company. Foreign governments were rejecting Amoco's bids for exploration licenses in favor of Exxon and Shell, who were using the same technology.

Among Amoco's attractions for BP were its superior operations in the Gulf of Mexico, especially its engineers and 3D computer programs. Other assets complemented BP's. Amoco's chemical plants were not identical, and its 9,300 gas stations in the US were located in different areas from BP's 18,000, diminishing the regulator's possible objections. The combined company would be the world's third-largest oil major and North America's biggest producer, even bigger than Exxon. A deal was also important, stressed Browne, to camouflage the fall in BP's pre-tax profits from $8.8 billion in 1997 to $4.8 billion in 1998. In 1999, if the deal was completed, he anticipated dismissing 10,000 staff.

Browne's directors were impressed by his masterly presentation. Some carped that he appeared like a perpetual star school pupil, regurgitating information in the form required to earn praise, but everyone agreed that the hunt for an acquisition was over. That was particularly true of Peter Sutherland, the company's chairman since May 1997. The 51-year-old Irishman's career as a commissioner in the European Union, chairman of the World Trade Organization and latterly chairman of Goldman Sachs International gave him impressive credentials to scrutinize Browne's performance and character. Circulating around BP were accounts of Browne's close relationship with his mother Paula, with whom he shared his home in Chelsea. She regularly accompanied him to formal dinners, and had lived with him during his period as a mature student at Stanford University. Wry smiles were exchanged about the anxious plea after the Brownes' departure from Stanford that a BP employee fly to California and rescue John's teddy bear, which had been forgotten in his apartment. It was returned to London in a Federal Express carton. On a personal level, Sutherland and Browne had little in common. "There was no natural glue between a jowly Irish Catholic with no background in oil and a partly Jewish atheist gay seeped in the industry," observed a Browne aide. The chief executive's manner and dominance swept aside any critical discussions about the bid for Amoco.

Browne's approach to Larry Fuller in February 1998 was pitched to avoid any repetition of Lou Noto's painful rejection. In full flow,

Browne's faultless salesmanship as he enthused about the advantages of marriage was irresistible to the marooned oilman, enthralled by a suitor of impeccable merits. "Amoco had for long been the best partner to dance with," Doug Ford, the company's vice president responsible for downstream, volunteered. Rival oil companies had considered the prospect. Lou Noto had probed and walked away, while Ken Derr of Chevron, probably the ideal partner, had plotted but not moved by the time Browne approached Fuller. Browne's offer of shares, not cash, for Amoco was sugared with an enticing comparison. While BP shares had risen by 90 percent over the past five years, reflecting the company's transformation, Amoco's share price had fallen. The "merger," with a 15 percent premium, would be an offer of BP's shares valuing Amoco at $49.8 billion. BP, Browne explained, would not consider a takeover of Amoco, but a merger of equals into a corporation quoted in New York as "BP-Amoco." Browne was deliberately vague on the subject of whether Fuller or himself would be the merged corporation's senior executive. He did not mention that BP's code name for the offer was "Project Belgium," a clear reference to a weak and divided country susceptible throughout history to effortless invasions and occupations.

Browne claimed the credit for the possibility of a cashless offer. His endless cost-cutting and focus on efficiency had been rewarded. Assiduously, he had also cultivated his reputation in America as an inspired visionary, single-handedly softening the public's prejudice against dirty Big Oil. In a carefully researched speech about the environment he gave at Stanford, his old college, in May 1997, he had confessed to a Damascene conversion. He admitted his mistake, shared by the chiefs of Exxon and Shell, in disparaging the link between fossil fuels and climate change. In front of his carefully selected audience, he rejected his past opinion. Acknowledging the existence of global warming, he urged the world to "begin to take precautionary action now" and engage with Greenpeace to discuss the reduction of greenhouse gases. The acknowledgment had won Browne and BP considerable plaudits around America and Europe; but not from Lee Raymond, who was outraged by Browne's exhortations to his competitors to join BP and reverse global warming. In the same year, at the World Petroleum Congress Raymond caused uproar during a discussion about climate change by proudly describing himself as prepared to "stand up in public and point out some of the scientific uncertainties and damaging economic consequences of the proposals under discussion." Speaking as chairman of the influential American Petroleum Institute, he sought to halt policies limiting carbon dioxide emissions, because one consequence would be a reduction of oil consumption. Anyone who annoyed Raymond scored points with Larry Fuller, and Browne would do so again.

In the wake of Shell's imbroglios at Brent Spar and Nigeria, Browne publicly broadened his commitment to the environment to include health and safety, ethics and human rights. His timing astutely anticipated the announcement days later by Chris Fay, Shell's chairman in Britain, that Shell was converting itself into an energy company focused on promoting sustainable development. Fay's admission that the increase in Shell's oil spills from seven tons in 1995 to 111 in 1997 was "completely unacceptable" amounted to a confession that Shell could no longer unconditionally say to society, "Trust us." Browne elucidated the same ideas in a more sophisticated manner. "BP," he said, "is becoming a company for all seasons. Behaving ethically and with concern for people and the environment isn't something new for us... We have to be clear what our standards are, and we have to have a process in place to ensure that those standards are being met, consistently and universally." Those sentiments beguiled Fuller, and Browne hoped they would have the same effect on any American politicians prejudiced against a foreign takeover.

The merger was announced on March 29, 1998. Over the following months, Rodney Chase, BP's deputy chief executive, drove through Browne's true agenda. Tough, chippy and giving the impression of being burdened by complexes, Chase was Browne's "enforcer," the type of henchman always useful to authoritarians. The combined new company's aim, Chase revealed to Fuller, was to save $2 billion annually by more focused exploration, improving procurement and cutting staff.

Among the unsuspecting casualties would be Fuller himself. At first he failed to understand Browne's true intentions, or to grasp the significance of Chase's disarming smokescreen. "The participants had been unable to pinpoint who first suggested the merger. It just happened." It was not until July that he perceived his misunderstanding. Browne was not a generous team player, but orchestrated BP's deeds around himself. He was not interested in a merger of equals, only a takeover. Finally, on August 2, Fuller succumbed to Browne's pressure and agreed to become BP-Amoco's deputy chairman, pending his retirement. "A brilliant twist to BP's advantage," observed a rival.

On August 11, 1998, Browne watched the stock exchange screen in London register the formal announcement of the "largest ever industrial merger." BP-Amoco's shares valued the company at $52.41 billion. The new company was capitalized at $120 billion compared to Exxon's $163 billion. With combined reserves of 14.8 billion barrels of oil, its oil reserves were second only to Shell's. BP's share price immediately soared by 11 percent, and continued rising, apparently endorsing trust in Browne and the acknowledgment of a new American idol. Thrilled as the new company's value inched toward $140 billion, Browne, despite his lack of physical stature, was hailed for casting a giant shadow across the global industry.

But after the accolades, the gloom returned. The takeover, Browne knew, would produce hardly an extra drop of oil, or provide any new guarantees for the supply of energy. He hoped to win applause from investors attracted by higher profits, but knew that critics were writing the oil industry's obituary, cynically quipping, "It's cheaper to drill on Wall Street than to drill in the mountains and backwoods." Browne could fend off that prejudice by speaking about "secure reserves for the future" and crucially, to balance the complaint that the oil majors had "stripped themselves down to just making money," pledging that "supply can always meet demand." The guarantee sounded authentic. The oil industry was confident about the healthy profits it would reap from the growth of the world's population by one billion over the next decade—or an extra 250,000 energy consumers every week. Phil

Watts of Shell echoed Browne's mantra, "We can buy our way to the future."

The Amoco takeover gave Browne a taste for more. One advantage had already manifested itself in Azerbaijan. In 1998, BP produced the first crude oil from the Caspian Sea. To enhance its income, the Azeri government had planned to threaten BP that the contract would be switched to Amoco. The takeover sabotaged that ploy. Size was power, and Browne's restlessness to please Wall Street became infused into BP's culture. Some of his old colleagues and the new directors retained from Amoco were disturbed by his preoccupation with finance. Engineers were either retrained or displaced by accountants speaking about "portfolio management" rather than concrete, drills and metal in an attempt to win investors' confidence. But pleasing Wall Street's analysts proved counterproductive. Convinced that there was a surplus of oil, they penalized companies that reinvested their profits to find more oil rather than using the cash to reward their shareholders. While Exxon's balance sheet, supported by high reserves, could withstand any cycle, BP's fate was still precarious if additional reserves were not found, or if oil prices collapsed.

Anxious to maintain the analysts' support by spreading optimism, Browne spoke the jargon they understood. "I think the logic at these prices is for a focus on margins rather than volumes," he told a meeting at the Institute of Petroleum. "It is clear that investors will examine the way in which this industry performs against other sectors and, in particular, the way in which we use capital." Browne's macho forecasts of increased performance, profits and production growth displayed his pride in BP's dramatic transition compared to its rivals. While Shell's profits fell from $8 billion in 1997 to $5.1 billion in 1998 — the industry's average fall was 18 percent — BP's rose 13.6 percent, exceeding the analysts' expectation of 9 percent. Browne credited his savage cost-cutting, and anticipated that profits would soar from $4.6 billion in 1997 to $6 billion by 2002, even if oil stayed at $14. "We're beginning to create a new company," he announced modestly, "but I think we've only just begun to understand quite how much it could deliver." His self-assuredness destabilized BP's competitors. Mark Moody-Stuart succumbed to Wall Street's pressure to make an acquisition to avoid Shell becoming a target itself. Shell's most attractive suitor was Texaco, its partner for refining and marketing in America, but the technical obstacles presented by Shell's dual ownership remained insuperable. That did not prevent analysts, encouraged by Browne, from seriously speculating about the possibility of BP merging with Shell to become the world's number one. Browne's reward was Lee Raymond's reaction. Sitting imperviously in the catbird seat, Raymond had assumed indifference to BP's fate, but any challenge to Exxon's primacy was intolerable. Fortunately for Raymond, Lou Noto finally conceded that Mobil could not survive in a $10-a-barrel environment.

In the months after Noto rejected the merger with BP, Mobil's fortunes had declined. Although it was a clever marketeer, the company had run out of growth. At a crossroads, with an aging portfolio and aging staff, Mobil was unable to compete with Shell in Turkmenistan or the Gulf of Mexico. Its hopes in Russia had been derailed. Even in Qatar, Mobil's successful natural gas project was handicapped by the corporation's limited finances. "He's so full of himself," griped one of Noto's bankers, "that he's overextended in Qatar and Australia and too aggressive toward Kuwait. He's got to sell."

Between 1990 and 1997, Mobil had replaced only 83.6 percent of its reserves. In the fourth quarter of 1996 the company's net income had fallen by 13 percent. "The world," Noto admitted, "has changed. The easy things are behind us. The easy oil, the easy cost savings, they're done." Over the previous seven years, Mobil had fired 33 percent of its workforce and sold billions of dollars' worth of assets. By mid-1998 the downturn had become longer and more entrenched than Noto had anticipated. "Now we're getting down to the muscle and not fat," he said. Despite following John Browne's example of cutting costs and generating shareholder value, the reasons for a merger had not changed. "Ours is a great business," Noto conceded, "so long as you're number one or two...The status quo just didn't work." Mobil, he decided, was too small, and while the oil price was low, the company, according to an assistant, was "sliding south." Not long before, Noto had damned those who left the company as "traitors." In the new circumstances, that was forgotten. Without contemplating that Mobil might be more valuable in five years, he became fixated on a single agenda. "Lou's gone hunting for a marriage," observed Nick Starritt, a Mobil executive who had recently moved to BP. The prospect of merging with Shell was discussed, but Mark Moody-Stuart's lack of enthusiasm, reflecting the divided shareholding and the millstone of Shell paying in shares and not cash, excluded that option.

The only suitor was Exxon. Ever since the breakup of Standard Oil, the suspicion had lingered that Mobil and Exxon had forged "a conspiracy or a common understanding" to fix prices. The discovery that after the 1950s the two companies had secretly funneled profits to Saudi Arabia through the American tax system reinforced the conspiracy theorists' beliefs. Regardless of the past, Exxon, blessed by a high market multiple, was a financial engineer whose shares could be recommended by Mobil's directors. In June 1998 Noto was speaking on the telephone with Lee Raymond. The FTC had recently allowed a merger between Shell's and Texaco's downstream operations in America. The combined company controlled 15 percent of the market, demolishing the previous limit of 7 percent. "I'm really pissed off about this," Noto told Raymond. "Other industry members are not going to sit around. We can all benefit from consolidation." Noto allowed the conversation to touch on the idea of a merger. He emphasised his aversion to ceding complete control: any union was to be "a merger of equals." After all, the two corporations had started from the same tree, and their cultures were similar. Raymond appeared to agree.

During the first weeks of negotiations, Noto consulted his board, while Raymond delayed telling his directors what was happening until the next regular meeting. Naturally, he did not mention his impatience with John Browne's ambitions. Even in the privacy of the boardroom, he maintained the pretence that Exxon was nonchalant about size and trends, and was only concerned with opportunities and satisfying shareholders. Nevertheless, the advantages of a merger were overwhelming.

Ever since Standard Oil had been broken up into 34 parts by a Supreme
Court decision in 1911, the two biggest parts, Exxon, about half of the
original corporation, and Mobil, the second-largest part with 9 percent of net value, had clashed with the regulators. In 1953 President
Eisenhower contended that two former parts of Standard Oil and three
other companies were participating in an unlawful conspiracy; and in
1960 Exxon was compelled to divest itself of an oil company in the Far
East. If Exxon and Mobil could remerge, the major pillars of John D.
Rockefeller's empire would be restored, albeit without Standard Oil's
power to oppress its competitors. But most of all, having previously
ridiculed the notion of an oil company involving itself in extracting
natural gas, Exxon would inherit Mobil's invaluable expertise in Qatar,
one of the world's biggest and most profitable natural gas sources.
Combined with Mobil, Exxon would be better placed in the Gulf of
Mexico, West Africa and Kazakhstan. The new company would also
become a major petrochemical supplier, with 12 percent of America's
refining capacity. But Exxon's fundamental weaknesses would remain.
Like Mobil, its huge reserves were declining, and with oil at $10, profits
would suffer until OPEC regained some control over prices.

That was a short-term problem. In the long term, the arguments in
favor of a merger were irrefutable. Raymond agreed that Exxon would
pay $81.2 billion for Mobil, compared to the company's market value
of $74.9 billion. Valuing Mobil's oil reserves at $8 a barrel, Exxon was
paying a 24 percent premium. Indisputably, the new company would be
the world's biggest, worth $275 billion compared to Shell's $205 billion.
Despite their size, Ron Chernow, the biographer of Rockefeller, observed
that the oil majors were "pitiful, helpless giants." "Giants" was questionable. Even combined, ExxonMobil, with 120,000 employees, possessed
only 3.8 percent of the world's oil production. The merger merely created a bigger minnow. In the final negotiations, Noto acknowledged
that talk of a "merger" masked surrender. Mobil's headquarters in New
York would be shut down and incorporated into Exxon's office in Irving,
Texas. Mobil's creative employees were alarmed. "I've served in the army,
and I'm not going back," was a common sentiment among many Mobil

employees who feared suffocation in Exxon's sterile atmosphere. To sugar his pill, Noto was promised $6 million if the deal succeeded. His share options were worth $58 million.

The announcement of the merger on December 1, 1998, was marked by a joint public appearance by Raymond and Noto. Raymond, despite his efforts to remain invisible, wantonly antagonized analysts and journalists. His comment about the "awesome" size of the new organization, and his claim that it represented a "seismic shift" in the industry that would enable the new ExxonMobil to stare down Saudi Arabia, lacked credibility. His description of the savings envisaged — he mentioned cutting capital spending by $400 million in 2000 — confirmed that he favored retrenchment rather than expansion. No one took seriously Noto's remark that "This is not a combination based on desperation." The next hurdle was Congress. Both chairmen expected that the $10.5 million they had spent during 1997 on lobbying in Washington would be rewarded. The FTC presented a greater problem. Raymond and Noto knew that the obstacles to completing the partial reunion of Standard Oil would be substantial, but felt that by hard bargaining they could be reduced to merely irritating. "I'm not interested in your views of how we ought to deal with the FTC," Raymond abruptly told a questioner during his presentation.

Robert Pitofsky, the FTC's thoughtful chairman, was finalizing his requirements for BP's merger with Amoco. He had decided to be obliging. Although there was an overlap in the wholesale and retail selling of gasoline in 30 states and nine metropolitan areas, Pitofsky only ordered that 134 gas stations and nine terminals be sold, and that 1,600 gas stations be given the opportunity to switch suppliers. Nothing, Browne could smile, had gone wrong. But merging Exxon and Mobil was on a different scale. Raymond had already fired an intimidatory shot. Creating the world's biggest private corporation, he knew, would provoke alarm. "Big is not necessarily bad," was a favorite defense he deployed, and on this occasion, considering Shell's size and BP's expansion, his appeal to the "national interest" appeared unassailable: "The merger will mean that ExxonMobil will be able to compete in the same scale as the largest foreign firms." Creating the world's biggest private oil corporation, however, offended Pitofsky's belief that competition was critical to the industry. He began the largest investigation in the FTC's history, knowing that Raymond offered no concessions. Nothing, Raymond insisted, should be sold. Inevitably, Exxon hired the best lawyer. Rick Rule, the former head of the antitrust division of the Justice Department, knew that Pitofsky had the power to seriously meddle in the merger, but left him in no doubt about the consequences. Exxon, said Rule, would engage in a battle. Shrewdly, Raymond endorsed the threat, so that Pitofsky "knew the problems if we went to court." By the end of the negotiations the combined company agreed to sell many assets, including 2,431 gas stations, especially in California, but clung to pipelines and refineries that appeared to be in jeopardy. "The level of concentration in the US," Pitofsky concluded, "will set off anti-trust alarms." Frustratingly, he separated America from the world market.

The two mergers, BP-Amoco and ExxonMobil, had set the herd heading in the same direction. "To survive in this new situation," said Thierry Desmarest, the chairman of France's Total, "all companies are struggling." Total, ranking ninth, merged with Petrofina of Belgium. Peter Bijur, Texaco's chairman, and Ken Derr of Chevron admitted that they were considering mergers. Shell was particularly interested in Chevron, whose reserves, after success in West Africa and the Caspian, had increased by 60 percent since 1995. A major obstacle for the herd was Pitofsky, who had become suspicious of oil chiefs speaking about efficiencies while "their motive may be to eliminate competition." John Browne was not listening.

BP's purchase of Amoco was completed at 9 p.m. on December 31, 1998, and celebrated at a party in Britannic House, BP's London headquarters. Browne and his team were euphoric. "We're on a roll," exclaimed one. "An upwards trajectory." "The magic years," agreed another. The Amoco executives invited to join the celebration were less exuberant. Surprised by the modesty of Britannic House, they had become suspicious of Browne, and suspected that BP had underestimated the strength of Amoco's natural gas reserves. Mindful of

Browne's scorn for engineers, some recalled Shell's complaint that BP had appeared uninterested in the quality of engineering when the two companies had collaborated on the Mars rig in the Gulf of Mexico. Financial engineering seemed to be Browne's priority.

Privately, Browne was grateful that the hiatus around the merger concealed BP's vulnerability to the unexpected fall of oil prices. The company's profits fell by 23 percent in the first quarter of 1998, from £755 million to £582 million, and they would also drop in the second quarter. To those criticizing his preoccupation with BP's share price, Browne replied: "People say we are too focused on immediate results, I say to you there are 40 quarters between me and retirement and every one of them matters." Like all the oil majors, he spent another $2 billion buying back shares instead of investing in additional exploration. BP's high share price was critical to his next coup. The prelude had been performed in the midst of the Amoco negotiations.

In January 1999, Mike Bowlin, the chairman of Atlantic Richfield, or Arco, had telephoned Browne, whom he knew from American Petroleum Institute (API) meetings. "I think we should get together and see if we can combine." "Why are you interested?" Browne asked. "We're a takeover target," replied Bowlin, "but the shareholders will only agree to a sale if they're paid a premium and the employees receive generous provisions." Bowlin's explanation was precisely what Browne had expected. America's oil majors had reached a crossroads. With oil at $10, the weak could not cope. Just like Texaco's chairman Peter Bijur, Bowlin, a human resources director promoted to run an oil company, was vulnerable. For years Arco, ranking as the seventh-largest oil company, had drifted, not least when it bought *The Observer* newspaper in London. Shareholders had received high dividends only because the directors approved unsustainable loans. With declining funds, Arco's investment in exploration had fallen from $1.6 billion in 1998 to $950 million in 1999. The losses had increased over the previous three years after the nationalization of its oilfields in Venezuela and a succession of dry holes across the world, especially in Algeria. Millions of dollars had been lost by Chuck Davidson, Arco's legendary explorer, drilling dry wells in the Mars field in the Gulf of Mexico; Shell's subsequent success in the field humiliated Arco. The company's problems had been aggravated by its loss of an 8 percent stake in Lukoil after expectations did not materialize. Battered by that litany of woes, Bowlin feared that if the merger frenzy evaporated before Arco's fate was resolved, the company's shares would be squeezed and the full extent of its disarray would be exposed. Politely, Browne did not question Bowlin's unconvincing pitch that if BP didn't bite, Arco had an alternative suitor. "I've got to finish Amoco and think about it," he said. "I'll come back to you."

Before returning Bowlin's call, Browne summoned a meeting of BP's executives and professional advisers at a hotel in Hampshire. The industry's predicament appeared to have deteriorated. Philip Verleger, oil's guru, would call it the "Third Oil Shock." The first had been the aftermath of the Arab–Israeli war in 1973; the second followed the fall of the Shah in 1979 and Iraq's invasion of Iran, which paralyzed 6 percent of the world's oil production. The third, according to Verleger, started at the end of February 1999, when WTI, the light sweet crude sold from Cushing and used for pricing on the New York Mercantile Exchange, was selling at $12 a barrel. The common theme of all three "shocks," he explained, was a period of economic growth stretching oil supplies to the limit. Then suddenly the boom stopped and demand for oil shrank. The mood was encapsulated by *The Economist*'s immortal front cover on March 6, 1999: "Drowning in Oil," it read, and the magazine forecast that the price of oil would fall to $5 a barrel. During that month, OPEC ministers were due to discuss an increase in prices. In a choreographed performance, they announced that daily production would be reduced by 2.1 million barrels a day. The cut would be welcomed by the oil majors, which were equally anxious to earn a higher income. Economists calculated that the inevitable price increase would cost consumers $480 billion in one year, but that the "New Economy" of advanced technology, flexibility of labor, increased efficiency and low inflation could absorb higher oil prices. Days later, Brent oil rose to $15. Experts speculated that the oil majors might have overreacted to the low prices by dismissing thousands of their workforce. Adding to the confusion, Mark Moody-Stuart explained that Shell would only consider projects "which fly at $14," and would let them "sleep easy in their beds at $10." If prices rose to $40, he said, there was a danger of recession. Oil at $50, he concluded, was "highly unlikely."

John Browne suffered from no uncertainties. After the ExxonMobil merger his lust for BP to rank as number one had grown. By taking over Arco, he had previously explained to his intimate circle, BP would once again be America's biggest oil producer. Capitalized at $187 billion, it would outrank Shell's $175 billion to become the world's second-largest oil major. Shell, he added, would be vulnerable to a takeover by BP. Combined, BP-Shell would overtake ExxonMobil, his ultimate ambition. Chevron and Shell were identified as rival bidders for Arco, but as both owned gas stations and a refinery on America's West Coast, they would incur costly divestitures to obtain the FTC's approval, while BP had already sold its only refinery in the region. After reading a schedule of Arco's assets compiled by Byron Grote, BP's vice president of exploration and production, Browne declared the prize to be breathtaking. Arco's two refineries and 1,700 gas stations on the West Coast would complement BP's 16,000 service stations in the Midwest and the East. In the upstream, Arco's wells in the Gulf of Mexico and the North Sea were profitable. The company was a critical supplier of natural gas to Asia from reserves in Malaysia, Thailand and especially at Tangghu in Indonesia, with 20 trillion cubic feet of natural gas. The jewel was Arco's share of Alaskan oil. The huge reserves of the North Slope were split between BP in the west and Arco in the east. Uniting the two areas would save substantial sums and offer an additional financial advantage. After BP had failed to honestly declare the value of natural gas from Arco's Alaskan fields it had used to pump crude oil out of the ground, Arco had been awarded $40 billion in damages. Buying Arco would remove that liability.

The discussion of BP's executives in the Hampshire hotel was not harmonious. Alaska's economy depended entirely on oil, and the ownership of their resources was a particularly sensitive matter for the local population. Browne was warned that Alaskan politicians were opposed to a single operator on the North Slope. The merged company would control 860,000 acres there, but Alaska's law forbade a single company leasing more than 500,000. Although Browne conceded the possibility of having to relinquish some acreage, he expressed confidence that Tony Knowles, Alaska's governor, would change the law out of self-interest. Several voices cautioned that after Exxon's merger with Mobil the FTC might object to further consolidation: Robert Pitofsky had warned that further mergers would be minutely scrutinized. "You're tempting fate," one voice pronounced. Browne ignored the warning. Having won the FTC's approval on Amoco in a record four months with little divestiture, he was committed to an encore. He wanted control over nearly all Alaska. American law, he believed, would bend to BP's desires. In any case, hinted Browne, it was a win-win situation. Either the merger would be approved, or in the unlikely scenario that it was not, a rival would be blocked from doing the same. Doug Ford, the former Amoco director appointed to the BP board, did not voice his doubts, merely observing, "Everyone thinks it will be a piece of cake because the FTC let Amoco through." With the dissidents silenced, the executives agreed that BP's opening bid for Arco would be 26 percent above the share price, valuing the company at $26.8 billion. "We're buying at the bottom of the market," smiled Browne as he contemplated exchanging BP's highly valued shares for an underperforming company.

All of Arco's directors would be expected to resign. That condition suited Bowlin. Like his fellow directors, he was also focused on the prospect of a large payoff. Before leaving Los Angeles, he had admitted to his anxious board, "BP is the only game in town. I've got no cards to play." Steeled for difficult negotiations, he was taken aback by Browne's attitude. Within minutes of sitting down alone with him at BP's headquarters in London, he realized that Browne was "hot to do the deal. It was like Sherman riding into Georgia." Bowlin asked for a 45 percent premium. Browne counteroffered 26 percent. To Bowlin's surprise, Browne did not even question the huge severance payments for Arco's

executives. Bowlin enjoyed another surprise after Browne arrived at his
Los Angeles office near the Beverly Hills Hotel on March 24, 1999,
to finalize the details of price and timetable, and to sign the heads of
agreement. "I don't want to let them get out of the deal," Bowlin had
told his lawyers. Bowlin's lawyers had anticipated that Browne would
object to the contractual terms committing BP to pay an onerous penalty if the deal fell through. Instead, he signed with a smile. BP was
locked in.

Over the weekend, news about the deal leaked into the media,
and negotiations were accelerated. Browne was untroubled. The public announcement of the bid on April 4 raised Browne's profile into
the stratosphere. He accepted the invitation to stand in the spotlight.
By any measure, he was leading a global juggernaut ranking alongside Silicon Valley's giants. "Its emphasis on growth and shareholder
value," commented Philip Verleger, "has paid enormous benefits to
the BP shareholders." BP's flexibility and lack of bureaucracy, added
Verleger, outshone its competitors. Unmentioned in Browne's public
pronouncements was ExxonMobil, which was still substantially ahead
of BP at $243 billion. Hailing the creation of "the largest oil output
of any non-state company," Browne spoke about "a compelling strategic and geographic fit of quality assets" and "the immense potential it
offers for future growth. In Alaska in particular, the synergies we can
achieve from combining our operations will greatly increase the competitiveness of the state in the face of uncertain oil prices and provide a
strong incentive for significant investment in existing and future fields."
Browne staked his reputation on executing the merger. Not everyone in
Britannic House was sure he had taken Alaska's history into account.

Richfield Oil had established Alaska's first commercial oilfield at
Swanson River in 1957. Renamed after a merger as Atlantic Richfield
(Arco), the company discovered oil in the east of Prudhoe Bay. The
next hole it drilled was dry, and the directors offered Exxon a 50 percent
share of all their remaining leases across the North Slope in return for
financing drilling of the next well. Exxon's profits in return for the risk
were probably the biggest in oil's history. BP, which owned a substantial

portion to the west of the bay, struck oil soon after. To limit costs, Arco became the operator for all the owners on the east of the Slope, and BP on the west. They shared the ownership and maintenance costs of the Trans-Alaska Pipeline System (TAPS), which opened in 1977, linking Prudhoe Bay with the tanker terminal at Valdez. For the first 10 years, the oil companies prospered. Production in 1979 reached 1.5 million barrels a day, the authorized limit, from 218 wells. To sustain that production, the companies drilled more wells, until in 1988, the peak year, 690 wells were needed to produce the same amount. Ten years later, producing the same volume required 870 wells. By 1999, production had fallen to 35 percent of 1988's output. The state's income, which had risen from $793 million in 1982 to $5.7 billion in 1991, fell to $2.1 billion in 1998, with the certainty of continuing decline. Under the banner "No Decline after 99," Arco used improved technology to extend the field's life beyond its expected death in 2000, but, contrary to the original forecast that the North Slope contained 13 billion barrels of oil, the companies anticipated extracting only 11.4 billion by 2018.

Alaskans hoped that the oil majors would find new reserves. Anticipating failure, BP refused to search, but Arco spent $300 million drilling a succession of dry wells, aggravating its financial distress. Alaskan politicians feared that the state's economy, dependent on oil, was jeopardized. Before the *Exxon Valdez* spill, Congress had been on the verge of approving a lease to explore for oil in the protected Arctic National Wildlife Refuge (ANWR), but thereafter the attempts were repeatedly blocked, and finally vetoed in 1995 by President Clinton. Without access to the ANWR, the oil companies refused to build a pipeline to transport Alaska's 25 trillion cubic feet of natural gas to the United States. Browne's solution was a merged company controlling 78 percent of the resources of the North Slope, which would save $200 million annually. "We must do more for less," Julian Darley, BP's director in Alaska, explained. Browne hoped that his pledge to finally build the natural gas pipeline costing at least $5 billion would sway the doubters.

But local politicians were unconvinced. Arco had a good reputation on environmental issues and for its care for workers' welfare. BP's reputation, by contrast, was not enhanced by squabbles about maintaining the pipelines and contributions toward local charities. In October 1999 the company had admitted the illegal release of hazardous material and the inadequate supervision of a contractor on Endicott Island off Prudhoe Bay, and had had to pay $22 million to settle the criminal and civil charges. Infractions were common among all oil companies' operations, but Browne had not considered local sensitivities. To move on to the next target he needed rapid completion of the takeover, and he refused to contemplate any delays. On September 1, 1999, he told shareholders that his discussions with the "various regulatory authorities are moving constructively ahead." He added, "All this gives me great confidence that we can meet our target to complete the deal before the end of the year." Shareholders of both companies overwhelmingly approved the plans, although Browne ought to have realized that his optimism was misplaced, as he knew from his conversations with Pitofsky that the FTC was hostile.

Bruce Botelho, Alaska's attorney general, had raised the critics' flag on June 30 during a hearing in the US Senate. BP, he said, was an aggressive cost-cutter. Workers complained about poor pipeline maintenance and feared job losses. "The potential impacts [of the merger]," he said, "on the oil and gas industry in Alaska and thus on the Alaska economy and its citizens are major." Smaller companies would be discouraged from investing by the merged company's control of the infrastructure, tanker space, and access to all the seismic data and the pipeline. Even the price of oil supplied to the US West Coast, Botelho continued, would be influenced. At the end of October, Governor Knowles agreed that the enlarged BP would be too dominant, and demanded some divestiture. Browne brushed these negatives aside. Alaska, he assumed, wanted a deal, albeit with slight modifications; as he saw it, the real hurdle was obtaining the FTC's approval.

For years, Browne had prided himself on being a master of American politics. After living in Cleveland, studying at Stanford and mixing with Washington's power brokers, he would take no lessons about America's financial culture. Robert Pitofsky, the FTC's erudite chairman, he believed, would give his blessing to the merger with minimal conditions, in the same manner as the Amoco deal. The FTC's criteria were to prevent the elimination of competition in marketing and refining. The Commission had never shown any concern about upstream — production in the oilfields — because prices were set by the world markets. Convinced by his lawyers that Pitofsky could not change these criteria, Browne hoped that the FTC would approve the Arco deal at the beginning of December, although the ExxonMobil merger could delay his timetable. The first warning of problems ahead was Pitofsky's declaration soon after Exxon's merger was announced, "There's now a pronounced trend towards concentration and we have to be concerned about that." Browne ignored that comment, but could not disregard another observation from the FTC's director of competition that BP's merger with Arco would "cement market power and harm competition" for gasoline on the West Coast.

The negotiations with the FTC were entrusted by Browne to Larry Fuller as "BP's man in Washington," assisted by John Gore, a BP lobbyist on the Hill, and Byron Grote, another of Browne's favorites. For weeks the three did little, and reported nothing. Fuller in particular was hapless, but Browne was unconcerned — until Pitofsky delivered a bombshell. The merger, he ruled, would not be approved unless Arco's Alaskan interests were sold to another oil company. His reason was that the US West Coast, especially California, relied on Alaskan oil, and that competition would be jeopardized by BP's dominant position in Alaska. Browne's whole rationale for buying Arco was threatened. He was seen to shudder.

Browne and Bowlin were assured by their lawyers that Pitofsky was legally and factually wrong. He was conjuring an image of the oil market on the West Coast as sealed off from the rest of the world. That was nonsense, the lawyers explained. The oilmen believed that Pitofsky could be persuaded that the West Coast would never be a victim of a squeeze from Alaska, which only supplied 28 percent of its oil, and oil could be imported from anywhere in the world.

Pitofsky's decision, Browne knew, could be challenged both politically and legally. While an appeal was prepared, he ordered his staff to start lobbying against Pitofsky in Congress and the White House. John Gore returned from London to offer Fuller, Bowlin and Bob Healy, Arco's experienced lobbyist, "a list of talking points to push on Capitol Hill." "One idea is that we should urge swift approval," said Gore, "because 2000 is coming and we must be ready for the Millennium Bug." Healy roared with laughter. "The FTC could say then wait until after 2000." Gore was unamused: "We have to do this because that is what John Browne wants." No one, Byron Grote reminded them all, said "no" to John Browne. "We're going to spin it in this way," he ordered. Healy became disenchanted that while Bowlin and Arco's directors waited impatiently for big payoffs regardless of the company's fate, the British refused to listen to any American advice. He experienced this obstinacy again during Browne's next rushed visit. After cruising through the offices of senators and congressmen involved in the oil business, Browne sat in BP's Washington office expressing his frustration. Although he was greeted as a star, the politicians were unwilling to help. "Don't they know who I am?" he shouted at Healy. "Why don't they do what I want?" Healy was nonplussed by his truculence. "I can't understand why this can't be done," Browne continued. "Why is this a problem?" Browne did not grasp, Healy realized, that despite his fame he remained an outsider in America.

Browne decided that there was no alternative to confronting Pitofsky himself. The genial lawyer welcomed the chance of meeting the oil aristocrat and Peter Bevan, his legal adviser. In an unfailingly civilized atmosphere, Browne adopted what one observer called "The olde William the Conqueror view that 'We know something you guys don't know because we're taking over.'" "You're acting against the law," he politely admonished Pitofsky, claiming he had misinterpreted the West Coast oil market. Smiling, the regulator replied, "I wrote the laws." Browne had failed to glance at the bookshelves. Had he done so, he would have noticed that the fourth edition of *Cases and Materials on Trade Regulation* was edited by Robert Pitofsky. Browne was arguing with the author of America's standard textbook about monopolies, competition and trusts. In a landmark case, Pitofsky had persuaded the Supreme Court that takeovers should be judged by the trend toward a monopoly in the market. No one had briefed Browne that Pitofsky had been appointed as head of the FTC by Bill Clinton, an admirer, who had used his textbooks to teach the antitrust course at Arkansas's law college. "Be moderate but aggressive and undo the FTC's dormancy during the Bush years," the president had urged Pitofsky during their meeting to confirm the appointment.

John Browne's crisp and forceful manner did not offend Pitofsky as much as Peter Bevan's limited understanding of the law. His visitors' ignorance about the opposition to the takeover on the West Coast was as damaging as their tactless referral to his approval of the Amoco-BP merger. Much, Pitofsky explained, had changed since then. He had welcomed the first merger, which would introduce BP into the market as a strong competitor. "Regulators realize that bigness isn't necessarily bad," Philip Verleger had recorded. But in 1998 there had been 14 oil companies. Now there were 12, and with the certainty of more mergers there was a problem, explained Pitofsky, of preserving competition. Browne barely listened. Convinced that nationalistic prejudice was deciding the fate of his cherished deal, he expressed his anger about formal advice given by the FTC's senior lawyer to Pitofsky: "We should not allow BP to control Alaskan oil," he had written. "I don't think the proposed combination of BP and Arco is good for consumers." Browne challenged his objectivity. "We offer concessions and they change the rules in the middle of the game," he complained. Unless the FTC gave formal approval, he threatened, the deal would be consummated and the FTC would have to fight BP in court to secure its termination. Irritated that Browne had not noted the views he had previously expressed publicly, Pitofsky regarded Browne as a bit of an opportunist. "That's why we build courts, Mr. Browne," he smiled. "If you think we're wrong, go and try to persuade a judge." The comparison with Lee Raymond was not flattering. The American had always listened to Pitofsky's questions and, unlike Browne, gave unambiguous answers.

Browne returned to BP's office enraged. Looking for culprits, he blamed John Gore for ignoring the FTC and "the political dimension." Next, he cursed American nationalism and suspicion of foreign oil companies. "They've gone too far," he griped. "They don't see the world as a single market." He scorned Bowlin for focusing on his payoff and ignoring Washington. Finally, he took comfort from Bowlin's quip, "Pitofsky's looking at shadows," and flew home.

Searching for a way to turn the tide, Rodney Chase telephoned Healy from London. "John Browne has spoken to Tony Blair," he reported, "and Blair is fully engaged to talk to Clinton, who will speak to Pitofsky, and this will be sorted out." Healy laughed. "Clinton will never call Pitofsky. John Browne needs to get real and stop tearing his hair out." Turning to John Gore, whom Bowlin and Healy rated as a substandard lobbyist, Healy complained, "The problem is you guys. It's always 'We know best.' You have a bizarre British fantasy about the way this city works." The news from the White House was predictable. Sandy Berger, the national security adviser, warned that the president would refuse Blair's request to intervene. "John's a terrific articulator and a lousy closer," Rodney Chase eventually admitted to Healy. "I'm always fearful when he does things alone."

During November, Browne became anxious. Since July oil prices had risen from $20 to $27. Stocks were dwindling and production had fallen to 72.3 million barrels a day as OPEC's cuts materialized and Iraqi supplies diminished. The price volatility, wrote Joseph Stanislaw, a cofounder with Dan Yergin of the energy consultants CERA, was caused by sharp changes in the world's economy, new technology and more consumers. New sources of oil and deregulation, Stanislaw foresaw, would increase competition, expand privatization among the oil producers and eventually reduce prices. Tony Hayward, BP's director of exploration, echoed Stanislaw's optimism that Russia and the Caspian, like the North Sea and Alaska in the 1980s, would be "the new salvation against OPEC"; and these new supplies would be augmented by previously hostile countries including Algeria, Libya, Kuwait and Iran "reengaging" and welcoming the return of Western majors. Those mistaken assumptions mirrored John Browne's similar misunderstandings about BP's position in America.

After contemplating threats about deadlines, over Christmas Browne decided on a showdown. On January 13, 2000, BP announced that the merger would be finalized within 20 days. Pitofsky retorted that the FTC would block the deal in court. On January 30, Browne backed down and offered concessions. His tactics backfired. BP, the FTC's staff told Pitofsky, had never proved that the merger would boost efficiency, or disproved their conviction that BP's monopoly would "eliminate the greatest threat to the perpetuation of that power," that is, competition. Two days later the FTC voted by three to two to block the deal by applying for an injunction in San Francisco. The first hearing was due on March 10. Browne vowed to fight in court. His tactics drew no praise, and his plight deteriorated.

To Browne's dismay, although Alaska's politicians had finally approved the takeover subject to conditions, BP's rivals were encouraging West Coast politicians to question the company's motives. Lee Raymond was heard commenting about Browne's ambition to decide the fate of a Californian company and voicing concern about BP being poised to become America's biggest oil producer. The state attorneys general in California, Washington and Oregon, anticipating elections, were encouraged by local refiners and, some suspected, Exxon's lobbyists, to file lawsuits in early February 2000 opposing BP's activities. Half of the West Coast's refineries, the three argued, were geared to higher-priced Alaskan oil, and BP had been accused for years of squeezing the market by selling Alaskan oil to Japan. BP's assurances, they claimed, could not be trusted, if only because Browne's promise to Larry Fuller about a partnership of equals in the Amoco merger had proved to be worthless. These antics infuriated Browne. Emotion, he believed, was overwhelming the facts. The majority of the West Coast's crude was not imported from Alaska, and BP had no refineries on the West Coast, which would give it the power to control prices. Pitofsky was unpersuaded. If BP's complaint ever came to court, he told Browne, he would show that BP had exported Alaskan crude at a loss to Asia and to other regions in the US to squeeze refiners on the West Coast.

The prospect of a protracted court battle alarmed Alaska's Senator Ted Stevens and Governor Knowles. Pitofsky's "single-handed effort" to scuttle the deal, said Stevens, threatened Alaska with "bankruptcy if this drags on for three or four years." Alaska's fate, said Knowles, would be put into "a deep freeze" by "a distant court." But Knowles was outmatched by the three West Coast state governments. "You have to do what Pitofsky wants," Bowlin told Browne the following day. "Pitofsky sees it my way and the merger will go through," Browne stubbornly replied. "Are you sure?" asked Bowlin. Puzzled, Bowlin called Pitofsky. "You do it my way or it won't go through," Pitofsky warned him. Bowlin was flummoxed. Despite having spent so much time in America, Browne did not understand the system. The reason, Bowlin assumed, was Browne's global sense of himself, his belief in his ability to define the problem and find the solution. Bowlin nevertheless remained loyal to Browne, until he read newspaper reports that the delay in the merger was driving Arco toward bankruptcy. "That's a blatant lie," stormed Bowlin furiously. "Our credibility is damaged." Summoning Byron Grote and John Gore to Arco's office, he told them, "That's damned untrue and a dumb thing to do. You're stupid to start something fallacious and I'm going to correct it." Gore says he cannot recall the episode, while Grote would not comment. Around Washington, in Bowlin's opinion, the two men had become tagged as arrogant and of "low intelligence." "They're no good at mixing and mingling," decided Bowlin, "and that's cost good relations with the government. Now it's gotten worse. Their integrity is in question." Finding a solution to the stalemate was easy. Without telling Browne, Bowlin called Pitofsky to arrange a meeting. Abandoning any loyalty to Browne, he directed Pitofsky to a clause in the merger agreement that allowed Arco's Alaskan interests to be sold to another company. "Will this break the deadlock?" he asked. Pitofsky nodded.

Ten months after the agreement was signed, Bowlin told Grote, "Time to offer up Alaska." In London, Browne recognized that Alaska, which had formerly been seen as the jewel, had become the albatross. Reluctantly he acknowledged that under the agreement, BP was liable to pay billions of dollars in penalty fees if the merger was abandoned. "We're fucked," Browne told aides in his London office. "We've lost the prize." Pitofsky had won. There was no choice but to sell half of Arco's Alaskan production and a share of the pipeline. Soundings had already identified that James Mulva, the president of Phillips, would buy Arco's turf for about $7 billion. At $3 a barrel, Mulva had snatched a bargain and the resurrection of his sliding career. "It's one of the most attractive transactions that we've seen in a long time," he commented. Blessed with that windfall, in 2001 he could merge Phillips as an equal with Conoco and become chairman of the world's sixth-biggest oil company. Mulva's triumph intensified Browne's pain.

Shortly after the Alaskan deal with Phillips was announced, ExxonMobil notified Browne that proceedings would be launched to block the sale. Unknown to BP, in 1964 Arco had granted Exxon first refusal if its Alaskan fields were ever sold. ExxonMobil wanted to exploit that lever to assert control over the whole Alaskan operation. Harry Longwell, ExxonMobil's respected executive vice president, met Bowlin to discuss how Exxon rather than BP could operate the North Slope. "No one has spoken to us about operatorship," said Longwell. "This is of enormous importance and no one is speaking to us." Bowlin was unsurprised by the familiar Exxon "mistake-free holistic approach," but was jolted by Longwell's hardball demands. "When a Brit farts, the sun shines out of his face," was a favorite Exxon expression. Longwell's measured antagonism appeared to confirm the misgivings that the American giant had encouraged opposition to the merger. Old suspicions about Exxon were reawakened. Over 30 years earlier, after oil had first been discovered in Alaska, BP had suspected that Exxon was deliberately delaying production to avoid competition with its other American oilfields, and was encouraging conservationists to protest about the environmental damage that would be caused by a pipeline, mentioning in particular interference with the migration of the caribou. Nothing had been conclusively proved, except that fighting Exxon was a thankless business. Now Exxon was seeking an injunction in a Los Angeles court to block the Arco-Phillips deal. Capitulation was Browne's only choice. The ownership and operation of Prudhoe Bay was realigned among BP, Exxon and Phillips. Exxon would become the biggest oil producer, while BP received the largest share of natural gas.*

"We didn't really need Alaska," Browne told his board of directors at the next meeting in London. He did his best to gloss over any impression that he had behaved rashly. One member privately recorded, "John is not sulking in a corner. He concealed any impression that he had behaved irrationally." Others concluded that his explanation resembled the emperor's new clothes. "Everyone knows it's a mistake," one director admitted, "but he was not confronted with dismissal. There was no postmortem." The giant was certainly diminished. BP was relegated to rank as number three among the oil majors. The value of the deal would be rescued as oil and gas prices rose. Unnoticed by the board was the FTC's warning that BP's inheritance of Arco's pipeline and storage interests in Cushing could excite suspicion about undue influence over worldwide crude prices.

In Los Angeles on April 14, 2000, Mike Bowlin read a news agency report: "FTC Approves BP-Arco Merger." While the negotiations had been dragging on, oil prices had risen from $10 to nearly $34 in March, but had then gone down to $22. By October they would rocket back to $36. This volatility, Bowlin concluded, proved Arco's inability to survive. He did not make a final call to Browne before bidding farewell to a few people and handing over his office to Byron Grote. The time for champagne had long passed. He drove home to Pacific Palisades. "Will you put Miracle-Gro on the roses?" his wife asked. With a $27.6 million payoff, Bowlin had no reason to work for the remainder of his life. Most of Arco's executives had also departed with handsome payoffs.

---

* Previously ExxonMobil had owned 42.9 percent of the natural gas, Arco 42.6 percent and BP 13.8 percent. In oil, BP owned 51.2 percent, ExxonMobil 23.8 percent and Arco 21.9 percent. The field would produce 564,000 barrels a day in 2000.

The integration of Arco into BP, they knew, would be ruthless. Eventually, Browne visited Los Angeles to host a reception for the city's civic leaders. "You should move your headquarters from London to LA," he was urged by the mayor. "Maybe we could," he replied. Browne did not reveal that Chicago rather than Los Angeles would be BP's center of operations in America, although curiously he decided that neither the Windy City nor any of the trophies inherited from Arco and Amoco merited a stopover during his visit.

John Browne (left) enjoys his moment of triumph. BP's successful bid for Amoco in 1998 propelled BP to rank as a world-class player. Amoco's chairman Larry Fuller (center) and Peter Sutherland (right), BP's chairman, would both become disenchanted with Browne.

Browne's glory was tarnished in 2005 after the giant Thunder Horse platform in the Gulf of Mexico tilted because of an engineering fault (above), followed in 2006 by oil leaking from a corroded pipeline in Alaska (below).

Shell's fortunes soared after the 1960s thanks to huge oil discoveries in Nigeria, but attacks by armed gangs in the Delta region (top) and pollution (left) fractured Shell's reputation. The Nigerian government's show trial and execution of Ken Saro-Wiwa (above) for campaigning to curtail the exploitation in 1995 transformed Shell into an international pariah.

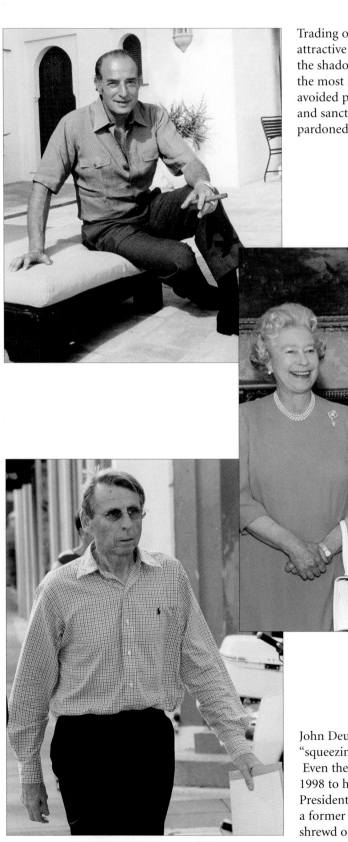

Trading oil on the margins is attractive to those occupying the shadows: Marc Rich, among the most infamous oil traders, avoided prosecution for tax evasion and sanctions-busting but was pardoned by President Clinton.

John Deuss (left) is infamous for "squeezing" North Sea oil supplies. Even the Queen (above) was asked in 1998 to help BP secure sympathy from President Heydar Aliyev of Azerbaijan a former KGB chief who became a shrewd oil supplier.

n 1997, Russia's fledgling bankers offered to save their country from bankruptcy if President
Boris Yeltsin pledged the nation's oil and mineral reserves to the bankers in exchange for
oans. (Left to right: Rosprom's Mikhail Khodorkovsky, Media-MOST's Vladimir Gusinsky,
BS-Agro's Alexander Smolensky, Onexim's Vladimir Potanin, Inkombank's Vladimir
'inogradov, Alpha Bank's Mikhail Fridman, Yeltsin)

The following year, Mikhail Khodorkovsky (above left), Mikhail Fridman (above right and below left) and Vladimir Potanin (below right) became oligarchs and billionaires by securing shares in Russia's vast oil fields for minimal amounts.

Chastened by John Browne's success in buying American oil companies, Lee Raymond (above right), Exxon's chairman, persuaded Lou Noto (left), Mobil's chairman, to agree to a merger and stymie Browne's ambition for BP to become number one in the world.

Raymond (left) hoped in 2003 to persuade President Putin to allow ExxonMobil to buy Yukos, owned by Mikhail Khodorkovsky. Rex Tillerson (above), Raymond's successor, witnessed the destruction of that strategy.

Chasing Browne's success in Russia (left to right: ExxonMobil's Rex Tillerson and Lee Raymond, Chevron's David O'Reilly, ConocoPhillips's James Mulva, Vladimir Putin, interpreter, Putin's adviser Sergei Prikhodko, Lukoil's Vagit Alekperov, Ambassador Yuri Ushakov).

Raymond (left) and Khodorkovsky at the World Economic Forum in Moscow in October 2003

Oil speculators, many trading through the New York Mercantile Exchange (Nymex), pushed oil prices up from about $30 a barrel in 2003 to $147 in July 2008.

The most admired trader was Andy Hall (above), the head of Phibro, who personally earned about $125 million during 2008. Recruited to Phibro in 1982 by Tom O'Malley (right), Hall emulated his mentor's discretion and savvy.

After meetings with Putin, Phil Watts (above right), Shell's aggressive chairman, reinforced Shell's foothold in Russia but his success unraveled as Walter van de Vijver (left), his deputy, accused him of deception over Shell's oil reserves.

Walter van de Vijver exposed the company to accusations by Oleg Mitvol (above) of environmental destruction in Siberia, which required Jeroen van der Veer (below, second from left), Watts's successor, to humiliatingly apologise to Putin.

Ironically, the Russian oil industry ranks among the world's worst offenders as enemies of the environment, as indicated by this photograph, taken in Siberia.

Oil creates strange bedfellows: (above) US Vice President Al Gore chased Nursultan Nazarbayev, the suspicious president of Kazakhstan, to secure more supplies for America; while (below) President Vladimir Medvedev of Russia courted Hugo Chávez (standing, left), Venezuela's left-wing president, to challenge a historic source of America's oil supplies. German Khan (seated in front of Chávez) of TNK signed the deal on Russia's behalf.

John Browne was thrilled in 2003 that his deal with Mikhail Fridman was blessed by Putin (above). Browne imagined that Rodney Chase (right), his enforcer, had tied up all the loose ends but he failed to account for Igor Sechin (below), the Kremlin's "Mr. Oil," and his sympathy for the oligarchs. The efforts of Tony Hayward (bottom, right), as Browne's successor, to resolve BP's crisis with Alexei Miller (bottom, left), Gazprom's chief executive, at a football match, failed.

Ali al-Naimi, Saudi Arabia's 74-year-old oil minister, shrewdly talked up oil prices during an early-morning run in Vienna in May 2009 before an OPEC meeting.

Chapter Eleven

The Aggressors

——◄◦►——

PAUL ADAMS MOVED BP's oil trading headquarters from Cleveland to Chicago soon after the takeover of Amoco. BP's senior trader understood the company's remarkable inheritance from Amoco better than others. BP had become the dominant owner of oil storage tanks in Cushing, the epicenter of pricing WTI oil on the New York Mercantile Exchange. In theory, with 20 percent of the storage, BP's power to influence Nymex's prices by choking or flooding the market had become overwhelming. By controlling the amount of oil entering and leaving Cushing's storage tanks, and managing access to the pipeline network across America, BP's traders could profit at their rivals' expense. To BP's surprise, Robert Pitofsky had failed to comprehensively examine the Cushing connection during his review of the bid. Probably he had not understood oil trading, and assumed that BP would give all the oil companies free access to Cushing on the same terms as its own giant refineries at Texas City and Whiting. After trading oil for 20 years, Adams and his superiors in London were unlikely to ignore their new trump card.

Soon after the move to Chicago, Chris Moorhouse, BP's chief of global trading, summoned a three-day conference in a London hotel to discuss "the Prize," as he called the acquisition of Amoco. "The key," Moorhouse told the group, which included Adams, "is to look for chances to arbitrage BP's assets and give BP an advantage." The priority, he continued, was to "optimize the system; to use trading to get 'the Froth'"—his expression for profits from simultaneously producing, refining, trading and controlling Cushing. Byron Grote agreed. "Trading must make a bigger contribution to BP," he said. Grote, like Moorhouse, understood the power of the "Cushing Cushion" and the profitable insanity of Nymex maintaining the landlocked storage complex in the prairies as the key to the world's oil prices. In passing, he mentioned "What We Stand For," BP's recently published code of ethics. The section headed "Vision and Values" made no mention of squeezing the market. Neither Grote nor Moorhouse mentioned John Browne. Pertinently, the chief executive remained perpetually invisible to the traders. Nor did Moorhouse mention Doug Ford, his superior on the main board. Ford, a former Amoco executive who specialized in refineries, knew practically nothing about oil trading.

BP's trading advantages had been maximized since the 1992 Gulf War. Over the following years, personal fiefdoms in the regions, products, upstream and downstream had been demolished. Using constantly improved communications, traders had been amalgamated into a global 24-hour organization. Automatically, any event anywhere in the world prompted a reaction from some of the 575 traders and 5,000 support staff. "BP had real country insight at the top level which Glencore, Vitol and Trafigura didn't have," admitted Mark Crandall, a founder of Morgan Stanley's oil trading who later moved to Trafigura, one of the large private traders registered in Switzerland. The most experienced rival traders in New York and Houston were nervous about BP's institutionalized advantage.

Jimmy Dyer, a squash-playing BP trader in Chicago, became renowned for exploiting the advantages. "I wouldn't be long [or short] in the spot this month," Dyer would say, and thousands of contracts would be influenced. "Dyer's fast," Moorhouse told his colleagues. "He sees opportunities others don't." Renowned for his energetic trawling of the Internet for information, and for always demanding answers, Dyer had a reputation for pushing to the edge but not beyond. Motivated by the bonus culture, his rivals were not enthralled. "He's loaded the bullets in the gun in Cushing and he's socking the rest of the world," complained a London trader. "He's coining it going up and coining it coming down." Charlie Tuke was a wary fan: "If he emptied his tanks, his players on the other side didn't like him." The walking wounded called Axel Busch, now at Energy Intelligence, a publishing and information company, and whispered their venom. By owning the physical infrastructure at Cushing, and producing crude oil and oil products, while trading crude and derivatives on Nymex and as OTCs, BP's rivals suspected the corporation of manipulation of WTI's prices. No other oil major was similarly empowered or involved: ExxonMobil only traded its own oil; Conoco's trading was minor; Chevron was still building a trading business after buying BP's refinery in Singapore, which bequeathed BP's trading secrets; while Shell's trading was modest compared to BP's—only two Shell traders were known to earn over $1 million a year. BP's operation was also the most aggressive. "BP's impossible," its rivals cursed in an agonized chorus. In New York, one medium-size trader spotted BP's traders bidding against each other to squeeze rivals. "They're playing WTI to Brent," he gasped, before noting how they effortlessly reversed the cycle. "Just learn how to navigate around them," he ordered, dismissing the idea of a formal complaint: "We're big boys and letting the regulators in is too complicated."

Vitol's senior trader in Houston spotted the same pattern, and told his office: "Never worry about WTI's fundamental price. Just think, what is BP doing? They can turn Cushing into shortage or surplus. Forget prices, focus on spreads." BP's traders, he realized, were even trading against BP itself by speculating against the company's refineries: "It's bonuses and easy money for them to take money off their own company." The Platts manager in Houston watched as BP filled up Cushing and waited. In that city, he quipped, "The rivals earned income trading off the crumbs dropped on the floor by Dyer." Traders at Morgan Stanley often retreated rather than compete with BP, and Goldman Sachs completely withdrew from some trading to avoid losses. Even the legendary trader Andy Hall regarded his old firm warily during the "loosey-goosey market" between 1997 and 1999. He had earned a fortune buying long and squeezing the shorts, disproving Morgan Stanley's conviction of the benefits of asset possession to glean market intelligence, but in the months after the Amoco and Arco mergers even the master of the squeeze felt BP's heat. The halcyon days of squeezing small traders had disappeared. Like his rivals, Hall identified Jimmy Dyer as having "a big impact on prices." As one trader put it, BP was challenging its rivals: "If you don't like what I'm doing, I'll hang you out to dry, but if you like, you can come into the water and play with me." Despite appearances, the Chicago traders were operating beyond London's control.

Chris Moorhouse in London faced a dilemma. Appointed at the last moment in 1998 as a "caretaker" head of trading for six months because the incumbent had fallen ill, Moorhouse was ill suited to challenge his subordinates in Chicago. "Do you intend to take an active role?" Paul Adams asked him. "Because if so, I'm off." Old-fashioned and dazzled by the "sweep of new trading conditions," Moorhouse recalled the huge losses he had suffered in 1983 after taking a bad position on crude oil in Rotterdam. "No, I don't intend to be involved," he replied. He was hardly equipped to master BP's squeeze. "You should look on the screens for immediate profits which can be locked in," he told Adams, "but BP should not be taking long or short positions." Privately he hoped that his traders knew how to behave, and not to take undue advantage. "The Chicago team know that they have to be careful," he told his superior after his "caretaker" role was converted into permanency. Too timid to instruct BP's 70 traders in Chicago and Long Beach not to exploit their advantage, he was reassured by Adams that they could trade without supervision, although Moorhouse would be personally accountable. "We're on the same wavelength," he told a colleague in London with evident relief. On past experience, Moorhouse did not expect to be troubled by America's regulators. Normally, any laws against market manipulation would require compulsory disclosure and supervision, and the relevant laws of that time imposed no such requirements. In any event, there was no evidence that BP's traders were other than successfully aggressive, but within the law. "I had no alternative but to have complete confidence because I didn't want to lose Adams," he would tell a superior. "I knew that there were insufficient controls but I had no power and no responsibility." Moorhouse could explain that, despite the allegations about BP manipulating prices in the US, nothing was ever proved. On the contrary, an investigation in 1999 of BP's alleged manipulation of prices in Alaska discovered no evidence of wrongdoing.

Paradoxically, at the same moment that BP was suspected of influencing the market by sophisticated methods, its own traders fell victim to a familiar ruse. In 2000, traders employed by Arcadia Petroleum,* a London company owned by the Japanese conglomerate Mitsui, spotted that the widespread use of derivatives since the 1990s to hedge oil offered another chance to squeeze Brent. Although supplies from the North Sea field had fallen to a trickle, traders still used "paper barrels" of Brent as the benchmark to set prices for their contracts across the world. As previously, if Brent's price was artificially squeezed higher or lower, the manipulator could earn a fortune from related trades across the globe. In 2000, Tom O'Malley, more sensitive than others to the risks of untransparency, noticed some familiar signs. Tosco, his new company, was America's largest independent refiner and oil distributor, owning 4,500 gas stations. O'Malley had swum against the tide. Battered by the volatile oil prices during the 1990s, the oil majors had abandoned refining as unprofitable. John Browne, uninterested in engineering and aware of refining's perilous finances, had ordered the sale of some of BP's refineries, including Grangemouth in Scotland. Exxon and Shell had followed, improving some refineries and selling others. O'Malley had capitalized on these fire sales, paying 10 cents on the dollar, especially for BP's East Coast refineries. He reasoned that the banning of the construction of new refineries in America since 1976 for

---

* Arcadia Petroleum has no connection with Arcadia Group Ltd.

environmental reasons would cause a shortage of capacity. Anyone who could overcome the bottlenecks in refining would be guaranteed profits. But just as a potentially lucrative shortage materialized, O'Malley became vulnerable to a squeeze,

In an earlier era Arcadia Petroleum had stored huge quantities of North Sea oil in South Africa and, as prices moved up, sold consignments to India for huge profits. "A really good play," commented one rival. In September 2000, O'Malley noticed that Arcadia Petroleum had bought more Brent crude than could be produced or delivered during that month. By that time Brent, the production from multiple fields in the North Sea operated by Shell and Exxon, had declined from 60 cargoes per month to 18. Prices were rising by $3 a barrel along the entire Atlantic basin — in Europe, Africa and East Coast America. Arcadia Petroleum's plan, O'Malley's staff believed, was to create a "shortage" to influence the price of WTI in New York. "The most extreme example of an artificial price being created," Philip Verleger called the conspiracy. "The idea that one could corner, could encompass an entire benchmark market and consequently manipulate potentially other prices is absolutely fascinating and, in my view, astonishing." BP, O'Malley believed, was also affected. As a result of the squeeze it went short, and Grahame Cook, the head of the company's European trade, had to compensate Arcadia Petroleum for BP's failure to deliver a consignment of crude. After this financial punishment, BP became more cautious in trading Brent. O'Malley did not surrender. In March 2001 he filed a $10 million claim in a New York court, accusing Arcadia Petroleum of using "illegal and monopolistic conduct" to force Tosco and other companies "to pay substantially more for crude oil than they would have if Arcadia had not manipulated the market." The case was settled out of court, and Arcadia Petroleum was not permanently damaged. In 2005 the company was sold after registering an annual post-tax profit of $57 million. By then, Jorge Montepeque had single-handedly taken steps to prevent a similar squeeze. In 2002, "Brent" was expanded to include three North Sea fields — Brent, Oseberg and Forties — which fed into the pipeline terminating at Grangemouth.

BP had been damaged by the Arcadia Petroleum squeeze. Before it had taken place, Browne had asked whether the company's trading was bringing it into disrepute, but he was deterred from interfering because of the high profits. A telephone call in 2003 from Adams to Moorhouse changed that. "The CFTC is having a go at us," said Adams. The Commodity Futures Trading Commission (CFTC) suspected that BP's trading on Nymex was illicitly seeking to influence the market. Moorhouse did not ask for an explanation or a report. There was nothing, he believed, that London could contribute. "I'll let you get on with the lawyers and the CFTC," he said, unaware that the regulator was scrutinizing the recorded conversations of BP's traders. Soon after, Vivienne Cox was appointed to supervise Moorhouse. "The investigation is taking up a lot of time and effort," was her only comment. Traders, she knew, often boasted excitedly about their influence and their positions. Frequently they exaggerated, occasionally they said things that were untrue. Illegal trading, she assumed, would not be discovered on the sound tapes. The unresolved question was whether BP's traders, much of whose pay was made up of bonuses, had become greedy. Rivals suspected there were many casualties in Browne's race for growth.

Chapter Twelve

The Antagonists

———◆⟨◦⟩◆———

THE NEWS IN AUGUST 1998 that BP was merging with Amoco had ruined Mark Moody-Stuart's sailing holiday in Greece. "That's put the cat among the pigeons," sighed Shell's stoic chairman, concerned that BP was poised to overtake Shell. Exxon's purchase of Mobil 16 weeks later compounded his grief. His company, Moody-Stuart realized, appeared directionless, introverted and risk-averse. Even Shell's technical superiority was slipping. For a year Moody-Stuart had searched for potential purchases, but Shell's convoluted share and corporate structure obstructed mergers. Chasing a string of pearls rather than the crown jewels, in his words, had exposed the company as slow and unwieldy. Even hostile takeovers for small companies had failed because the financial inducements were unenticing. "Shell," smiled a rival, "is the plain girl no one has got engaged to."

Accepting defeat, Moody-Stuart calculated that Shell's salvation required a merger of itself. Royal Dutch and Shell Trading should abandon the 60–40 structure created by their founders 91 years earlier. The notion, he knew, was controversial. The problem would be the older, conservative Dutch directors, who feared the eradication of Dutch control. Secretly, Moody-Stuart approached younger directors to seek their support. Two of them, Phil Watts and Steve Miller, the president of

American Shell, immediately agreed. The support of Jeroen van der Veer, a young Dutch director, and Maarten van den Bergh, the president of the committee of managing directors, was crucial. To Moody-Stuart's relief, both appeared to be favorable, although van der Veer would later deny that he gave any pledge of support. Moody-Stuart now appeared to have five votes, which meant the conservatives would be defeated. The principal obstacle, he believed, was Lo van Wachem, the former chairman who had remained on the supervisory board as the guardian of Dutch interests. Despite the split board, van Wachem had continued to dominate the company. Moody-Stuart's appointment as chairman had been approved by him on condition that he held the position for only three years, and he had vetoed a proposal to extend Moody-Stuart's tenure. The distrust between the Dutch and the British board members had spurred Moody-Stuart's initiative. The contrariness among the directors needed to be expunged.

At a meeting of Shell's directors in London in October 1998, Moody-Stuart launched a conversation about the company's future. As anticipated, Lo van Wachem and his supporters were cool, but agreed to continue the discussion at the next meeting in December. As usual, after the meeting the Dutch reverted to their native language, incomprehensible to non-Dutchmen, in the corridor outside the boardroom to protect the privacy of their conversation. Moody-Stuart remained unsuspicious, trusting his four pledges of support. He failed to grasp the intensity of van Wachem's dislike of British culture, and did not anticipate that Jeroen van der Veer and Maarten van den Bergh would succumb to van Wachem's entreaties. Both felt that Moody-Stuart was in too much of a hurry, and while they were prepared to discuss changes, they would not be rushed into a merger. They were vulnerable to van Wachem's eulogy about "respect for the past" and warnings about "betrayal of the greatest talents." "This is too sensitive," said van Wachem, appealing to the younger directors to "protect national pride" by "putting the idea on the shelf."

An unsuspecting Moody-Stuart arrived at the December meeting to continue the debate. To his surprise, the response was icy. Royal

Dutch's directors, van Wachem was satisfied to see, subsequently voted unanimously against the merger. In a private meeting in his office later that day, van Wachem warned Moody-Stuart never to raise the subject again. The Dutch, he emphasized, would always use their 60 percent share to assert control—the tail would not wag the dog. Biting his lip in the funereal atmosphere, the humiliated chairman remained silent. "I came close to being fired," he would later acknowledge.

Days later, on December 14, 1998, Moody-Stuart stood humbled before a group of analysts at Plaisterers' Hall in the City of London. Van Wachem's deadline for his retirement by the end of 2000 undermined Moody-Stuart's credibility. Shell, he knew, had become the underdog to Exxon and BP. The company's market value had fallen by $43 billion since April. Criticized for allowing the company to have become distracted by woolly sidetracks like publishing a worthy manifesto, "People, Planet and Profits," Moody-Stuart acknowledged that without a mega-merger, Shell would struggle. "I'm absolutely sure that our group's reputation with investors is on the line," he admitted. To create an upbeat mood he pledged that he would be "clearing the cupboard" with sales and dismissals, and that the corporation would reinvent itself. Shell, he said, would be restructured over the next five years, and profits would increase by 50 percent. "We are large enough to be the leading company on our own without any merger," he said with limited conviction. "Either I deliver or not. I'm sure there will be consequences if I don't."

Still scorched by Brent Spar and Nigeria, Moody-Stuart sought to rebrand Shell as honest and transparent. The company, he acknowledged, needed to establish "public trust," values of "sustainable development" and "social responsibility." Acknowledging that such matters concerned Europeans more than Americans, he added that "Analysts do not value those things highly, but some major shareholders definitely do." Exxon and Chevron, he knew, did not suffer the same burden. Doggedly, he persevered. To hasten the "internal merger," Shell's country chairmen had been disempowered, and responsibility for taking major decisions on exploration had been centralized in The Hague and London.

The turbulence was not welcomed by the 8,000 dismissed employees, particularly those who had been enjoying perks in the abolished hierarchies, or by the axed country barons in Britain, Holland, Germany and France, but the reforms did improve the company's effectiveness. The profits in 1999–2000 would increase by 38 percent to $7 billion, compared to 1998, when profits in the third quarter had fallen by 56 percent. But the financial improvement glossed over Shell's failure to find new oil and gas. Negotiations had started with the Nigerian government for a five-year development program worth $8.5 billion in more stable areas away from the delta to produce an additional 600,000 barrels a day, and there was also the Brutus field in the Gulf of Mexico, which would produce over 200 million barrels of oil and gas; but those projects were insufficient to replace Shell's annual depletion.

Brutus, a cutting-edge development that involved drilling 20,000 feet into the rock from a rig a mile above on the sea surface, was being supervised by 43-year-old Walter van de Vijver, Shell's chief executive in the United States. Although talented, "the Tall Angry Dutchman" had been summoned to The Hague to receive informal warnings about his manner. Moody-Stuart had ordered him to negotiate a collaborative venture between Shell and Enron, a major international electricity and gas supplier. After lengthy meetings in Houston with Ken Lay, Enron's chairman, van de Vijver reported to Moody-Stuart, "I can see no sense in their model." Renowned for spontaneity and impetuosity, van de Vijver concluded that Moody-Stuart's enthusiasm for the Enron deal exposed him as another Briton prone to pursue unreliable market trends and following the last man's flawed advice. Van de Vijver was equally prejudiced against Phil Watts, Shell's head of exploration and production. Brazenly, he told John Hofmeister, the director of human resources, that he regarded Watts as an obstacle rather than a boss; and he similarly dismissed Jeroen van der Veer, responsible for Shell's chemical industries, as "useless." His contempt for his colleagues mirrored the fractured relationships among Shell's directors during 2000 as they decided on Moody-Stuart's successor.

Jeroen van der Veer was favored by van Wachem, but was regarded as too young. The second candidate was Paul Skinner, the head of downstream, criticized by some as "a legend only in his own mind." Resistant to the new requirements of diversity and inclusivity, Skinner was rejected on Moody-Stuart's suggestion in favor of the 55-year-old seismologist Phil Watts. Watts's candidacy was opposed by Chris Fay, the recently retired chairman of Shell UK, who mentioned his suspicions of a "geological mafia cover-up" to Moody-Stuart. Fay suspected that Watts's reduction of the exploration budget, his unsuccessful search for new oil and gas, and Shell's recent reinterpretation of its existing reserves to increase the amounts of oil and gas under its control was concealing the actual depletion of the company's assets. His protest was ignored, and subsequently Watts took issue with Fay's comments. Moody-Stuart had no doubts about Watts's integrity. "He is a good Christian and very thorough," he said. Although Watts could occasionally be a bully, Moody-Stuart felt his energy and ability were what was needed to destroy "the bureaucratic monolith." Others believed he was too aggressive. Contrary to Shell's gentlemanly traditions, Watts enjoyed playing hardball, and could be abrasive with those who failed to match his standards.

Those who doubted the wisdom of Watts's appointment as chairman in December 2000 were equally nervous about the promotion of van de Vijver to head of exploration and production. The intemperate Dutchman's dislike of Watts was unlikely to foster a collegial atmosphere, especially as he spoke of himself as having been chosen to inherit the chairmanship. Moody-Stuart decided to ignore how the mutual dislike between van de Vijver and Watts would be aggravated by the inherent tensions among Shell's management. According to Shell's unique constitution, Watts, although the chairman, lacked any authority over the Dutch directors, including van de Vijver. Unlike Lo van Wachem, Watts's personality would undermine his attempts to impose his authority, which was certain to give rise to difficulties in the future, considering van de Vijver's disdain for Watts. The mutual distrust was interpreted by van de Vijver as confirmation of his eventual succession. Initially, Watts was uninterested in van de Vijver's dreams. His own ambition was to revive Shell, and he could cite his success in Sakhalin, the huge development on Russia's Pacific coast, as evidence of the repairs that had been made to the company's international reputation since Ken Saro-Wiwa's execution.

In the early 1990s Watts had identified the production and sale of natural gas as profitable. His "Walls of Gas" strategy was built on Shell's mastery of the complicated technology shared only by Exxon-Mobil. Years earlier, Watts had been enthusiastic about Shell's LNG ambitions in Brunei, Malaysia and Australia, but nationalization in Malaysia and government-inspired delays in Australia had frustrated the company's schemes. But Shell's development at Sakhalin 2 offered potential profits from LNG.

The bid for a license to develop the oil and gas deposits at Sakhalin 2 had been won by Marathon Oil in 1990. Knowing that the costs of exploiting the reserves were too daunting for his company, Vic Beghini, Marathon's president, identified Shell as the only major oil company that was suitable to become its partner. Teo Oerlemans, a senior Shell engineer, stressed Sakhalin 2's principal advantage to Shell's directors: LNG could be shipped directly to Japan without using Russia's pipelines, and Mitsui had contracted to buy the natural gas. Accepting Oerlemans's advice, Shell became Marathon's junior partner in 1992. By 1994, Shell's engineers had condemned Marathon's technical competence as "bad" and their relationship with Marathon's employees as "terrible," but Beghini resisted pressure from Sakhalin's governor to replace Shell with Exxon. On June 22, 1994, witnessed by US vice president Al Gore and Russian prime minister Viktor Chernomyrdin, the $10 billion contract to develop Sakhalin 2 was signed in Moscow. The agreement, said Gore and Chernomyrdin, proved that Russia would abide by international law and would regularize its tax policies. Under the first PSA agreement with the Russian government, Marathon had 30 percent of the venture, Shell 20 percent, and the remainder was divided among other Western companies. The development of the project was slow, expensive and beyond Marathon's finances. On December 13, 2000, after Beghini's retirement the previous year, Watts flew to New York and bought out Marathon's interest. Shell now owned 51 percent of the company, with Mitsui and Mitsubishi owning the remainder.

Watts's success was significant. Since finding oil in 1958 on the West African coast, Shell had barely added to its portfolio. It had gained a share of the oil in the North Sea and the Gulf of Mexico, but had failed miserably in South America. $300 million had been wasted drilling dry wells in Alaska, forcing Shell to abandon some exploration in the Gulf of Mexico and reluctantly to accept BP's investment to continue exploring in the Mars field. Sakhalin 2 ended those barren years. Within the company there was jubilation. Potentially, Sakhalin was a cash cow, but only after engineers had overcome unprecedented difficulties to build the first $3 billion "train" to produce the LNG. Veterans of the North Sea boom were recruited to build two offshore platforms and pipelines to produce 90,000 barrels of oil a day, all year round. Limited production of oil had started, albeit restricted to the six warmest months of the year, and the first tanker had left Sakhalin for South Korea on September 20, 1999. Watts had good reasons to celebrate.

Watts's self-congratulation concealed his artlessness about Russia. He had bought Marathon's share of Sakhalin 2 without adequately considering the reaction of the Russian government, assuming that his negotiations and Marathon's payments, as requested by Valentin Fyodorov, Sakhalin's governor, to improve the region's infrastructure, had secured Shell's position. Disdainful of Moscow's control, Fyodorov encouraged Watts's conviction that any deals made locally were beyond the Kremlin's control. Russia's internal turmoil seemed to encourage that belief. The chiefs of Gazprom expressed no interest in Shell's plans, not least because, without any money or expertise, the Russians could contribute nothing to the project. Observing the unlimited powers exercised by Fyodorov until 1993, and by Igor Farkhutdinov, the governor after 1995, Watts assumed that everything would be agreed upon locally. Shell was providing Farkhutdinov with the finance to split from Moscow, but neither Watts nor the company's directors in Europe considered the implications of encouraging his autonomy. Unlike in Nigeria, where Shell could rely on the British Foreign Office for advice, it was short of the diplomatic support in Moscow enjoyed by Exxon, Chevron and BP. Shell's executives thought a close relationship with Farkhutdinov would suffice. "We don't need to consult anyone in Moscow," said Watts, slightly troubled that the contracts agreed to with Farkhutdinov were "a little too splendid." Critics of Shell mentioned their impression that Russia appeared to be being equated with Nigeria, and that the Russians were regarded as vodka-drinking idiots. The more malicious suggested that the appearance of a black Nigerian engineer in Sakhalin, and of a female project manager, had irritated those Moscow bureaucrats who did not appreciate Shell's diversity programs. More seriously, the bureaucrats in Moscow did resent Shell flying local Sakhalin officials to The Hague for "indoctrination." To Watts, distant Moscow was easily forgotten. He had found penetrating the small clique governing Russia impossible, and the PSA agreement was watertight.

To protect Shell's $10 billion investment in Sakhalin, the LNG had been presold in Asia as a guarantee for loans from banks, to order the special tankers and hire thousands of skilled engineers to build the production facilities. Watts knew that Shell's directors would never have committed the company to a $10 billion project before "placing all those ducks in a row" and without owning 51 percent of Sakhalin. He also knew that Shell's investment was impossible without a flawless PSA agreement, guaranteeing profits and limiting taxes. The Russians had no wiggle room in the future. But most important, with oil prices between $10 and $15 a barrel, Shell's skeptical directors had only approved the project after being reassured that the costs would be low. Forcefully, Watts had persuaded his budgetary staff to shoehorn the project within the $10 billion budget. Few involved in the calculations visited Sakhalin, and even those who did derided the higher estimates by the "elephant riders" in the Far East before producing a precise estimate of $9.624 billion. Watts's assertion that Sakhalin 2's profits would be considerable, and the costs contained within the budget, had assisted his selection as chairman. With unconcealed self-praise, he rated Shell's PSA agreement as superior to Exxon's for Sakhalin 1. After five years of negotiations, finalized between 1995 and June 1996, Exxon had been compelled to accept a 30 percent stake, and to include two Russian companies with a total stake of 40 percent and other partners in the $15 billion project. Further down the food chain, after five years Mobil, Texaco and Exxon were still waiting at the end of 1999 for an agreement to develop Sakhalin 3.

The mega-mergers and low-priced oil had placed additional pressure on Watts, when he was head of exploration and production, to meet Moody-Stuart's targets to increase Shell's profits and size, especially after the horrors of 1995 and a poor exploration record, which included Shell having successfully bid for the worst acreage in Angola. Like BP, Shell was focused on increasing dividends to improve its share price. The stock market value of oil companies was partly dependent on the amounts of oil they owned and had available for production. To achieve growth without risking money meant producing more oil, and the cheapest way to do that was to improve engineering at existing wells. An alternative was to increase the amount of oil and natural gas categorized as "reserves." The valuation of reserves was determined by a combination of industry guidelines and experts' estimates, and a formula devised by the American Securities and Exchange Commission. The guidelines for booking reserves were first published in 1936 by the American Petroleum Institute, and changes to reassure shareholders about their company's value, requiring "reasonable certainty" of production, were implemented in 1978. A company could not book reserves until money had been allocated for drilling and production was anticipated. As technology improved through horizontal drilling, the oil companies agreed that the SEC's interpretation was unreliable, and that its interference should be rejected. Presuming that the SEC would accept the companies' own estimates of their reserves, each oil major adopted its own individual criteria.

Shell's standard changed in 1997. Moody-Stuart was under pressure to prove the company could grow without a merger, and his desire to "improve efficiency" extended to enhancing its reserves of oil and gas. Shell's "Leadership and Performance Group" was ordered to issue a directive to "create value through entrepreneurial management of hydrocarbon resource volumes." In future, Shell would book reserves without allocating any money for production: the reserves would be increased simply by relaxing the guidelines. In the past, reserves had been decided locally by each country chairman, and it seemed each country had used different rules. To achieve uniformity, Watts endorsed the company's decision to reinterpret the rules, assessing reserves upon a "reasonable expectation" of an available market rather than using the SEC's gauge of "reasonable certainty" of producing and selling them. The change of policy had started in 1995, when Nigeria's rulers, to improve their ability to raise loans and channel more money into their Swiss bank accounts, had encouraged Shell to increase its oil reserves by four billion barrels. To justify the increase, Shell and other oil companies were offered more acreage for exploration. This had coincided with Shell's own initiative, the Reserves Addition Bonus. Embarrassed by its falling share price, Shell's directors had asked Tim Warren, the head of technology in exploration and production, to investigate whether its reporting of reserves was too conservative, and could be increased by relaxing the accounting guidelines.

By that time Watts had become director of strategic planning and sustainable development. Although he was not directly responsible for Nigeria, he and the local country chairman were induced to increase Shell's reserves, and to agree to the Nigerian dictators' request even though chaotic conditions in part of the delta prevented exploration. No one in The Hague questioned the abandonment of the company's natural conservatism and its policy of "If in doubt leave it out." In 1997 Shell booked the Gorgon field in Western Australia as reserves in its official accounts, although the plan for production had not been approved by the government and no sales contracts had been signed. For those reasons Chevron, Shell's partner at Gorgon, had not booked the reserves. Similarly, in Oman Shell booked some oil reserves where extraction was not yet possible. At Ormen Lange in the North Sea, off northern Norway, Shell booked reserves in 1999 after drilling two exploration wells but before complicated technical problems had been resolved. BP, Exxon and Statoil delayed booking the same reserves. All those bookings assumed that the reserve numbers would eventually prove to be accurate and would be approved by the SEC. Shell was renowned as the gold standard for honest reporting, so no one doubted that its oil and gas reserves were growing. Nor was a paradox noticed: while the reserves increased, the exploration budget was cut. The conundrum was how, at a critical moment, Watts could deliver his promise of 5 percent growth. Combined with incentives, it appeared that "self-interested optimism" alone would increase the reserves. "Oil men can play to any rules they are asked to play," observed John Jennings, a former chairman of Shell. "Oil breeds arrogance because it's so powerful."

The increase in reserves, and Watts's arrival as chairman in January 2001, coincided with an unexpected surge of oil prices from $11 to $30. Across Europe and America, cheap oil had become an unchallenged assumption. OPEC had been dismissed by many as impotent, and oil as irrelevant as cotton. In 1980 the US had spent nearly 9 percent of its GDP on oil; by 1999, it had fallen to 3 percent. Oil, it was assumed, was not a growth business. Investors were unconvinced that the oil companies needed reserves, and felt that rather than invest in finding new oil, they should return their profits to shareholders. Fearing repeats of oil gluts, OPEC countries, burdened by debt, had also not invested. The cushion of spare production had dwindled from 15 million barrels a day in the mid-1980s to 2.5 million (18 percent to 5 percent). World production of crude had fallen from 67.2 million barrels a day to 63.8 million, mostly in OPEC countries, which, despite abundant reserves, could not instantly expand production.

Reports of a new shortage rattled those who had supported America's embargoes against Iran, Libya and Iraq. Fearing a threat to the economy, President Clinton sent Bill Richardson, the energy secretary, to tour OPEC countries and appeal for a 10 percent increase of production. Richardson was partially rebuffed. To reduce prices to the $22 to $28 range, Saudi Arabia's petroleum minister Ali al-Naimi agreed to increase production by 5 percent. Next, Clinton protested that Mexico's prices were too high. He was told that the increase of prices from $8 in December 1998 to $24 was good for the country. Washington warned President Tellez not to be "too pushy. The big players of the oil market have already said there needs to be more oil. The message has gotten through. Heavy lobbying could backfire." By September 2000, some were speaking of an "oil crisis." Middle East oil was being sold to Asia rather than the West, and OPEC was resolutely united to keep prices high. In Congress, which had been asleep about oil for many months, there were sudden demands for punitive action against OPEC and Mexico.

On October 12, oil hit $36 a barrel, sufficient to persuade Shell that extracting oil from the Athabasca tar sands in Canada would be profitable. This prompted OPEC to speak of producing more oil to reduce prices and make the tar sands uneconomic, but prices remained high in early 2001. America's energy policies had become incoherent. Using fuzzy phrases about comprehensiveness, environmental friendliness and integration of energy, George W. Bush, the new president, talked about his intention to "modernize conservation," meaning that drilling would be allowed in Alaska. But the energy industry's high expectations of Bush sank as California plunged into darkness. No new power station had been built in the state for over 30 years, and deregulation, which Bush supported, had lurched the state's energy companies toward bankruptcy. Rather than simplifying the importation of energy from neighboring states, California's electricity-generating corporations discovered that it was either loss-making to provide local energy, or that they had become the victims of Enron's manipulation to demand uneconomic prices to supply power from outside the state. Although the 38 days of blackouts during 2000 and 2001 were caused by financial shenanigans, they profoundly awakened America's sense of its need for secure energy supplies.

Then, in mid-2001, oil prices unexpectedly fell. Russia's oil production rose, undermining OPEC's squeeze, and concern about shortages disappeared. Proven reserves had increased by 12.3 billion barrels in 2000, and would rise by a further 3.7 billion in 2001. Despite low investment and increased consumption, the world's reserves of oil and natural gas had risen consistently since 1985. "I am positive," Sheikh Yamani predicted from the sidelines in July, "there will be a crash in the price of oil." New discoveries and cuts in consumption, he forecast, would neutralize OPEC's 13 percent cut in production in 2001. The oil glut had returned. After the terrorist attacks on September 11 in New York and Washington, the Western economies dived, and oil fell in November to $17 a barrel. "Has there ever been greater uncertainty for the direction of oil prices?" asked Bob Williams, a researcher for PennEnergy. No one could foretell whether prices would soar or collapse. The "crisis" days of peaking at $32 were mere memories as al-Naimi begged Russia to cooperate with OPEC and cut daily production by two million barrels to force prices up. President Putin refused, causing al-Naimi to fear that prices could fall to $14 in 2002, draining Saudi Arabia's cash reserves. Fear of recession gripped the market. In *Global Oil Trends 2002*, CERA predicted that falling demand would cause prices to fall to $6. In Congress, the same politicians who the previous year had demanded punitive action against OPEC sought to renew sanctions against Iran and Libya for another five years. Libya's ambition to increase production of its high-quality sweet crude by three million barrels a day was blackballed. Even ExxonMobil ran a protesting advertisement: "Unilateral economic sanctions are rarely effective but do discourage development of non-US energy supplies that would add to global supply diversity. This sanction is ripe for revision."

During 2001, Watts's self-confidence grew. The temporary surge in oil prices had increased the company's profits to $1.5 million per hour, enabling it to deposit $11 billion in the bank, part of the estimated $40 billion in cash accumulated by the oil industry. Shell shared a problem with its rivals—how to spend that money. Compared to the late 1980s and the 1990s, when the oil-owning countries controlled

90 percent of the world's reserves, encouraging the oil majors to explore for new reserves, those countries had become restless and resistant to the majors. In Venezuela, the Caspian and Russia, resistance was growing. The paradox of the oil majors declining during an oil glut frustrated Harry Longwell of Exxon. "The challenge," he admitted while leading Exxon's search for an extra 1.6 billion barrels a year to restore the corporation's depleting reserves, "is production replacement. That's becoming more difficult." While Exxon, despite its difficulties, found more oil than it sold in 2000, Shell's oil and gas reserves, despite the new relaxed booking criteria, actually fell, relegating the company to joint second in the league. Unable to invest all its allocated funds to find new oil deposits, Shell spent $4 billion in the first quarter of 2001 buying back shares, emulating Exxon's $2.35 billion buyback in 2000. Big Oil, it had been suggested two years earlier, could survive with fewer assets. Opinion now swung in the opposite direction.

One option arose in Saudi Arabia, where the king offered Shell and ExxonMobil the chance to invest $16 billion in a natural gas processing plant, to spend between $5 billion and $10 billion on exploration in the Red Sea, and to build pipelines and petrochemical, electric and desalination plants. Exxon's share would be $8.5 billion. After examining the offer, both companies regarded the price to participate in Saudi Arabia's exploration ambitions as "the entrance fee to the poker game." The chance of profits in the near term was fanciful, and 9/11 put an end to the idea. Claiming to be outraged by Western media criticism of their country's connection to the terrorists, the Saudis withdrew their offer. By then, Watts had decided that Shell needed instant growth. A spending spree, he told his directors, was essential to restore the company's reserves. Endlessly, Shell's planners number-crunched in their efforts to strengthen Watts's presentations. Despite the Dutch directors' 1995 veto of the proposed purchase of British Gas, he reopened the case for a bid. Eventually, in 2001, the Dutch directors agreed to offer $17 billion, valuing BG's oil reserves at $16 a barrel. Watts disagreed. "We'll only get it with a slam-dunk bid at $20 billion," he countered, valuing the reserves at $22 a barrel. Doubting that oil prices would rise, the Dutch directors refused even to consider offering more than 40 percent above the share price. In 2009, after the stock market crash, BG would still be worth about $50 billion.

Reducing his ambitions, Watts next sought the board's agreement to bid for Enterprise, a British oil company floated in 1984 with five oilfields in the North Sea. Managed by Graham Hearne, Enterprise had additional oilfields in Norway and a share of the Tahiti field in the Gulf of Mexico, and was operating the Bijupira and Salema offshore projects in Brazil, which produced about 250,000 barrels a day. Considering the company's potential 10 years earlier, shareholders had lost confidence in Hearne. Like Lasmo, another British oil company, Enterprise had failed to use its North Sea fields to expand and to gain a place among the major players. Together, Lasmo and Enterprise symbolized Britain's squandered opportunity.

Enterprise had been created in 1983 by a Conservative government ideologically opposed to the state producing oil and gas, and keen, regardless of Britain's long-term energy security, to immediately receive a large amount of cash. Convinced that there would always be a surplus of oil supplied by BP and Shell, Margaret Thatcher privatized the North Sea's reserves in 1986, just as prices crashed. This was followed in July 1988 by an explosion on the Piper Alpha platform, killing 167 workers. By then, civil servants had fled from the shrinking Department of Energy. Floundering without any considered policy, over the following years the government repeatedly altered the tax rates, replaced ministers of energy and changed policies. "NIMTO"—Not in My Term of Office—became a Whitehall spoonerism describing deliberate inactivity. Compared to other oilfields, North Sea costs were high—$5 a barrel compared to $2 in Indonesia. New drilling in the North Sea declined however from 224 wells in 1990 to 116 in 1993, but at the same time, by using the new technology (especially horizontal drilling) to develop mature fields, the oil companies transformed uncommercial into commercial fields, and increased production sixfold, from 669 million barrels in 1991 to 920 million in 1993. On that basis, Norman Lamont, on the eve of his dismissal as chancellor in May 1993, again changed the tax rates, reducing incentives to reinvest profits and search for new fields. Although North Sea oil made up 7 percent of the world's production, Lamont was dubbed the "Driller Killer" because oil companies received no relief for unsuccessful exploration. Earning profits, the companies complained, was harder than ever. Gordon Brown, the new chancellor following the election of Tony Blair's Labour government in May 1997, continued Conservative policies, levying supplementary taxes, ignoring John Browne's warning that, as had happened in Venezuela, British policies would hasten a decline in discovery and production. In 2002, Brown's taxes had, according to the oil companies, stopped the development of 315 discoveries.

The oil industry in the North Sea may have been winding down, but Enterprise owned other assets. Based in opulent offices in Trafalgar Square, Graham Hearne was surrounded by expensive paintings, which attracted some criticism of supposed decadence. To predators, Enterprise appeared mollycoddled, aimless and weak, especially in the wake of Hearne's failed bids for Lasmo in 1994 and 1998. After Lasmo was bought by ENI, Enterprise's shareholders wanted their money. "It's a one-way bet," Watts told his board, suggesting that Shell buy a 25 percent stake from Norwich Union for £1.5 billion. His directors objected that that was too much, and that Shell should rely on organic growth. "We need bigger reserves," countered Watts. "We must buy them." Following further arguments, Watts was allowed to approach Hearne and Sam Laidlaw, the chief executive. Enterprise had just rejected a bid from ENI.

When Watts and Walter van de Vijver arrived at Hearne's corporate apartment in Whitehall Court, they got stuck in the elevator. Once released, Watts made little attempt to seduce his prey. Pulling out a legal yellow notepad, he said, "This is what the lawyers told me to say." The offer, snapped Hearne, was "derisory." In repeated meetings, Watts and van de Vijver increased the offers. In their opinion, "It was a game with Hearne's enormous ego" and his personal demands. The truth is impossible to ascertain. Clearly there was bad blood, and a climate of mutual suspicion. "He wanted to keep the paintings on the wall," recalled Watts. "He wanted the company apartment," Hearne spat back, saying that he feared the paintings would mysteriously disappear after a sale. Despite his fears that Hearne might still sell to ENI, Watts abandoned Shell's valuation of Enterprise's shares at 600 pence and agreed to pay 725 pence per share, making a total of $7 billion, a hefty premium that attracted public criticism. In his defense, Watts would say that Shell recovered the full price within four years, thanks partly to an inherited stake in the lucrative Tahiti oilfield in the Gulf of Mexico.

Watts proved less savvy by simultaneously selling Shell's oilfields in Rajasthan in India to Cairn Energy for £4 million. Two years later, Cairn found one billion barrels of oil, and the wells were valued at over $1 billion. Watts hoped that the loss would be more than recouped by spending $14 billion on refineries, gas stations and Pennzoil, making Shell the world's biggest producer of lubricants. Instead, he received more criticism. Compared to John Browne and Lee Raymond, neither Wall Street nor the City of London was enamored of Shell's chairman. "They've bought flowerpots, not even a greenhouse," was one comment. Some distrusted Watts's three-year strategy to maintain the annual growth of profits at 5 percent, his commitment to reduce costs, and his intention to focus on developments in Nigeria, China and Brazil — the first $1 billion investment by a foreign company in Brazil since privatization began in 1997. Ignoring the company's strengthening balance sheet and Watts's claim of "robust profitability," the mood within Shell in The Hague grew somber. Dislike of Watts hastened the transfer of departments, including gas and power, from London to Holland. The Dutch were entrenching their advantage, yet Walter van de Vijver was fretting. Shell's oil and gas production was not, as Watts had predicted, increasing, and its additional new reserves were questionable. Failure to meet the published targets, complained van de Vijver, endangered his inheritance of the chairmanship. In hindsight, Shell's senior directors had reached a perilous crossroads.

At the time, Watts was unfazed by van de Vijver's criticism. Russia, he hoped, would partly satisfy Shell's hunger for new reserves. Until the Sakhalin 2 contract had been signed in 1994, he had been denied access to key people in Moscow, but thereafter he felt on stronger ground to expand Shell's investments, visit ministers and establish a personal relationship with Vladimir Putin when the Russian president visited Britain just before Christmas 2001. Watts was invited to Chequers, the prime minister's country home, to meet Putin. "Don't steal my time," warned John Browne, another eager supplicant. Sitting by a roaring fire in the Elizabethan mansion in Buckinghamshire, Watts proudly brought Sakhalin "to the fore of Putin's mind." "We must capture the market," he told him. "It's worth $45 billion to Russia over its lifetime. So don't miss it, Mr. President." Watts believed that his message had been accepted as Putin stood up and said in English, "Mr. Watts, I wish you and your family a blessed Christmas."

Buoyed by that apparent warmth, Watts believed that the Russians would approve Shell's expansion to Zapolyarnoye, a gas and oil field, and Salym, a group of oilfields in west Siberia. Insensitive to the changing political mood but aware that "the Russians are ruthless people to negotiate with," during his next visit to Moscow he successfully harried Mikhail Kasyanov, the prime minister, out of a cabinet meeting to sign a letter of support for Shell's interests to Dmitri Medvedev, the head of Gazprom who would become Russia's president. To seduce Medvedev, Watts had secured the royal box in a London theater for a performance of *My Fair Lady*. Shortly after, Shell was awarded the license for Salym, but Moscow's latent antagonism toward Western oil companies threatened its revocation.

The doubts among Moscow's politicians about the sale of the country's mineral wealth had not disappeared. Unlike BP, which had constructed a network of advisers in Moscow, Watts relied for information on Andy Calitz, the local manager of Shell's operation in Sakhalin. Neither Watts nor Calitz had anticipated one unusual consequence of Moody-Stuart's reforms. Because of a historic anomaly, Shell's operation in Russia was reporting to Holland through the company's office in Dubai. To Russian officials, humiliated by the disappearance of the Soviet Union and the slow recovery from financial collapse, the link to Dubai confirmed their suspicion that Shell regarded Russians as primitive natives. Watts and Shell's directors were unaware that changing circumstances in the Kremlin, and Russia's increasing oil production, were propelling the corporation toward a self-made calamity.

Chapter Thirteen

The Shooting Star

————◁○▷————

BUYING AMOCO AND ARCO had transformed the slightly built, soft-spoken John Browne's reputation. Voted Britain's "most admired chief executive" in 2001, an accolade that would unprecedentedly be repeated for the next two years, Browne became entranced by the spotlight, and felt invincible on his pedestal. "The best is yet to come," he predicted. Employees and audiences hung on his attitude rather than challenging his opinions. The focus was upon him as the embodiment of BP. With the benefit of cost-cutting and the temporary oil price increases, BP's shares rose 11 percent in the year after the mergers. The glory was only tarnished by a handful of skeptical BP directors. "He's going for sizzle, not substance," complained one associate, carping about "St. John the Divine." The visionary CEO would brook no interference. But Browne, the minority suspected, was perpetually seeking the next challenge and forgetting to micromanage the details of the business. Delivering targets was delegated, and anyone mentioning problems was banished. Those who cited names of potential successors caused his particular resentment. Dominance was his style. Even when an executive made a presentation to the board, Browne would offer an introduction and a summary. In the glow of success, the doubts about his "sizzle" were rejected as churlish. No one dared steal his thunder. Undeterred, Browne evangelized about BP's image.

Stung by the ambiguous attitudes toward Bad Big Oil during the Arco bid, Browne pondered how to rebrand BP, consolidating the new acquisitions, and to rid the industry of the legacy of the *Exxon Valdez* and Brent Spar. At the same time he had become preoccupied by his latest passion: to save the planet from global warming. The energy companies, he believed, could lead the campaign to limit climate change. In a speech at Stanford University in May 1997, Browne had proclaimed BP as the first "green" oil major. To win over environmental sympathizers, he committed BP to reducing carbon dioxide emissions and supporting the 1997 Kyoto Protocol on climate change. BP, he said, would invest $1 billion in a solar business by 2010, as a contribution to the target that half of the world's energy needs could be met by renewables by 2050. Besides solar energy, he would later commit BP to invest $8 billion over 10 years in alternative energies, including a biofuels program, and would donate $500 million to a 10-year research program at the University of California at Berkeley to develop bioenergy. His speech, publicized as the first occasion on which an executive of a major oil company had pledged his opposition to climate change, was widely reported and welcomed. But not by Mark Moody-Stuart, who telephoned Browne to complain that climate change had been mentioned earlier in Shell's annual report. Graciously, Browne agreed to drop further claims to be "the first." By then, his speech had become a mere foretaste of a more fundamental initiative. After long discussions with the advertising agency Ogilvy & Mather, Browne had decided to rebrand BP as "Beyond Petroleum." The slogan was originally intended merely for internal use, but Browne seized on it to relaunch the corporation. The new BP, his mission statement would explain, intended to "reinvent itself as an energy company people can have faith in and inspire a campaign that gives voice to people's concerns, while providing evidence of BP's commitment, if not all the answers."

Browne's enthusiasm for rebranding had not been approved by all BP's directors. A minority feared that he was reacting against BP's original name, "British" Petroleum, and was preoccupied by the creation of a global corporation divorced from Britain, albeit one that would still rely on the British government's support. To those dissidents, Browne had also become excessively enamored of Tony Blair's zeal for rebranding everything—including Britain itself as "Cool Britannia." In reality, they argued, BP would remain an oil company, reacting in traditional ways to familiar challenges. Adamantly Browne insisted that BP would become environmentally friendly. The precedent was the company's reaction in 1997 to Greenpeace activists climbing onto a rig in the North Sea to disturb seismic survey vessels. To avoid a repeat of the Brent Spar disaster, BP officials had soft-pedaled their initial public reaction, offering the protestors tea and helping them return to land. After minimizing the problem, BP began a High Court suit to recover £1.4 million of its costs from Greenpeace that would eventually lapse. The company had, however, placated Greenpeace by expressing sympathy with its opposition to fossil fuels and its support for solar power. But skeptics in BP's boardroom were not convinced. "How can you prove BP is 'Beyond Petroleum'?" Browne was asked. "We're only a petroleum company. It's a simplistic and provocative slogan. The press will be cynical. They'll murder it. And so will Exxon." Browne was unmoved. No one, he knew, would dare to formalize a challenge to him. He was thrilled by Ogilvy & Mather's stunning logo, using Helios, the sun god of ancient Greece, with a vibrant burst of green, white and yellow, BP's traditional colors, shaped like a sunflower. Browne's intellectual strength and dogmatism smothered the dissenters. This was personal. "Beyond Petroleum" captured the spirit of the times, emphasizing Browne's insightful understanding of the future.

Browne was gratified that BP's rebranding, costing $200 million, was hailed as a triumph, amplifying his reputation as a trailblazer. Once again he had calculated the odds correctly. Overnight he was anointed "Sun King" by the media, increasing the reverence in which he was held and his seeming invulnerability. "There's nowhere to run and hide from me," he declared. To secure his destiny, Browne set about removing potential successors. Rodney Chase was delegated to "squeeze out

Peter Backhouse," among the most capable oil experts of the younger generation. No one on the board, including Peter Sutherland, the former attorney general of Ireland, banker and international functionary, could challenge Browne, and 28,000 gas stations were rebranded as environmentally friendly "bp."

Inevitably, some critics emerged, especially Lee Raymond, regarded by some as a heretic for championing the denial of global warming. In 1997 Raymond had defied the growing sentiment in favor of protecting the environment at the World Petroleum Congress. The following year, he dismissed renewables as "fashionable" and "scientifically unfounded scare scenarios." The API's chairman cited a petition signed in 1998 by 17,000 scientists questioning the evidence for global warming. Evidence emerged that the headquarters of the organizer of the petition, the Oregon Institute for Science and Medicine, was located in a tin shed, and that among the scientist signatories were "Ginger Spice" and a character from *MASH*. Undaunted, Raymond castigated "Beyond Petroleum" as "greenwash." "You'd better deliver," he sniped at Browne, irritated by his self-promotion. He, better than Browne, understood the fatal flaw of American governments. The oil industry's hopes for improvement under President George W. Bush after 1997 had been dashed. Recoiling from accusations of "blood for oil," Bush had distanced himself from the industry in the aftermath of the terrorist attacks of September 11, 2001. In the majors' eyes, Bush's only redeeming feature was that, unlike Jimmy Carter, he refrained from openly castigating the oil industry. In Raymond's opinion, Washington was congenitally deaf to the industry's predicament, and equally incapable of solving any big problems or of grappling with global warming.

Raymond's anger evoked Browne's disdain, even pleasure. More disappointing was Greenpeace's cynicism. BP, suggested a spokesman, stood for "Burning the Planet." The corporation, said the protestor, was financing the reelection of US congressmen with the worst anti-environmental records. Politicians voting in support of drilling in the Arctic National Wildlife Refuge had received donations from BP, including $7,000 for Senator Frank Murkowski, the chairman of the Senate Energy and Natural Resources Committee, and $11,000 for Representative Don Young, chairman of the House Committee on Resources. Two Republican senators, Trent Lott and Don Nickles, had received finance from BP toward their 1997 campaigns despite opposing Kyoto and other policies to combat global warming. Senator Nickles had voted for opening up the Arctic National Wildlife Refuge.

Browne ignored the criticism. Amid the euphoria, he presented his next ambition: BP's merger with Shell to create the world's biggest oil company. Browne knew the obstacles. His plan was not formally approved by BP's directors, and even informally it attracted limited support. Its opponents recognized Browne's indifference to the financial and regulatory hurdles, and to Shell's directors' disdain toward himself. In the Shell directors' opinion, he was a brilliant strategist, but was self-obsessed, the antithesis of Shell's management by consensus. To them Browne was a magpie, prone to stealing Shell's ideas. First, knowing that Shell was preparing to announce a sustainability program, Browne had jumped ahead to publicize BP's own program, despite the lack of adequate preparation to implement his plans. He had performed a similar wile by announcing BP's diversity program during a speech in Germany. Again, he had anticipated Shell's initiative without commissioning adequate research. Undeterred by the carping from Shell's directors, Browne was confident about his cause. Whatever the price of oil, all the majors were threatened by gradual decline. BP's union with Shell would forestall that threat. The alternative was the seeds of self-destruction.

Browne was determined to leave nothing to chance. If BP was to become the world's biggest oil corporation, both it and he had to be endorsed in the United States. An inspiration was his regular visits over the previous three years to Silicon Valley. As a nonexecutive director of Intel, a notable star of the dotcom boom, Browne had frequently discussed the prospects for growth with Andrew Grove, Intel's founder. On his return to London from the latest such meeting, he appeared motivated but depressed. Intel, he told his advisers, was planning 30 percent annual growth: "Why can't we do the same? I want growth from everyone." With a mixture of bravado and self-belief, Browne decided to commit BP to an annual growth rate of 5.5 to 7 percent, higher than Shell and ExxonMobil, the benchmarks against which he sought to be judged. He was prepared to bet the house that, despite sluggish oil prices, the higher volume would generate higher profits. Peter Sutherland and the board could not object, and Dick Olver, the head of exploration and production, reluctantly confirmed his ability to deliver Browne's target after persuading his boss not to aim publicly for 10 percent. Success would depend upon increasing the existing 30–35 percent recovery rate to above 40 percent. Since the production facilities existed, the additional oil would be pure profit. On July 11, 2000, Browne went for broke. In a dramatic presentation to analysts and the media, he announced that BP's oil and natural gas production over the next three years would grow by 5.5 percent to 7 percent every year, mostly in the Gulf of Mexico and Angola. That growth would generate an additional 10 percent in profits. "We are now ready to move from a phase of retrenchment to a phase of expansion," he said. His announcement was hailed in Wall Street and the City of London as proof of his genius, and BP's share price reflected his glory.

In 2001, BP was ranked fourth by *Fortune* magazine among American oil companies. Browne's targets and rebranding, he hoped, would improve that ranking, but "Beyond Petroleum" had generated problems. Surveys showed that the American public was puzzled by the word "Beyond." Investors were asking, "What is BP?" Was the exploration and production of oil BP's prime objective, or was it an oil company that did not want to be an oil company? Browne was undaunted. During 2001 he commissioned a "reputation renewal program" to refocus BP's employees and the American public on the importance of the company's size, scope and values. A "war room" was established in Houston to wage the equivalent of a political campaign to rebrand BP as a global player blessed by a visionary leader. Once again Browne was a pioneer. Although ExxonMobil and Shell had spent millions in self-promotion, neither had rebranded itself so aggressively as a friend of the environment. BP's competitors, Browne smiled, were taken by surprise.

Within the 200-page "Reputation Manual" dubbed "the cookbook" was an exhaustive list of BP's activities, investments, financial donations and scholarships to be cited during an unforgiving series of events, conferences and media appearances to promote BP as both America's biggest oil and gas producer and a company with deep American roots that was fulfilling its ethical and environmental responsibilities. Browne, the Reputation Team knew, expected to be at the center of the campaign, symbolizing BP's "humanization, personalization and identification." Thousands of opinion formers across the USA were targeted to attend the "engagement opportunities" at which BP's ambassadors would deliver the message. Invitations were also sought for Browne to attend important international meetings including the Aspen Institute and the World Economic Forum at Davos, and flattering profiles were arranged to appear in America's most prestigious newspapers and magazines, especially *Fortune*. BP logos and flags were attached to newly planted trees and balloons, and were brandished at a back-to-school parade by 200 BP employees and their families in Chicago. Browne spoke at Stanford, Harvard, chambers of commerce and on *The Charlie Rose Show* about corporate governance and BP's reinvention of itself to support the environment. Before each event, he required intensive briefings about the issues and the people he would encounter. After every "Big Moment," his advisers would spend hours analyzing the benefits and deficits.

Early results confirmed the program's success, and in 2002 the Reputation Team anticipated building on its achievements. *Fortune* magazine's survey in the 2003 Petroleum section would elevate BP from fourth place to first. But by then the message from London was negative.

At the end of December 2001, Enron's fraud had been exposed. Thousands of Houston's citizens had lost their jobs and savings because of their misplaced trust in the corporation. In a speech in Houston in August 2002, eight months after Enron's bankruptcy, Browne lamented that "Public trust in companies has deteriorated in recent months due to investigations into companies' financial record-keeping and other operational procedures." An opinion poll, he mentioned, showed that 84 percent of those questioned did not believe the statements issued by corporations. "I find this deeply concerning. The importance of trust," he emphasized, was linked to "the crucial responsibility of leadership." He added, "Trust comes not from size, or from words, but from practical action." Corporations, he believed, should show "practical commitment" to being fair and impartial. So BP was stopping funding political parties. To demonstrate BP's "practical commitment" to being fair and impartial, in March 2002 he had withdrawn the company's annual $1 million of donations to politicians and political parties, although the corporation continued to provide facilities and material assistance worth many millions of dollars. Browne's authority was barely enhanced by that pledge. In Britain he had taken a slew of Labour politicians and activists onto BP's payroll, including Baroness Smith, the widow of former party leader John Smith, as a director of BP's Scottish Advisory Board, and Anji Hunter, Tony Blair's former gatekeeper, as BP's director of communications. Neither woman knew much about oil or finance, but both enjoyed privileged access to Blair's inner sanctum.

Despite appearances, there were fundamental problems at the heart of BP. To speed up growth, Browne ordered that the business units should be linked by computer into a "virtual team network," to encourage internal competition and beat financial targets. Trusting as ever the advice he received from McKinsey, he added new criteria to the system of cascading contracts to enable his staff to monitor whether the 5.5 percent growth target would be met. The corollary was his self-immunization from those questioning his ambitions. In 2001 Browne's sense of infallibility was bolstered by his elevation to the peerage, as Baron Browne of Madingley, by Tony Blair. Dressed in ermine, he swore loyalty to the Queen and took his seat in the House of Lords. A subsequent letter signed "Lord Browne" to employees in Amoco refineries in America evoked the riposte from one recipient, "I only believe in one Lord." None of those employees had ever seen Browne. Only his closest advisers had noticed the significant behavioral change following the death of his mother Paula in 2000.

Browne had lived with his mother in Chelsea since his father's death. She often accompanied him to official functions, and was his partner at dinner parties. Following her death, Browne began for the first time to enjoy an unfettered billionaire's lifestyle. Dazzled by his fame, the respect he received from the City and Wall Street, and his popularity among grateful shareholders, his fellow directors tolerated his vanity. Challenging him about his preference for delivering speeches and attending A-list parties, enjoying unlimited expenses and the private jet available to the chief executive of the world's second-largest corporation by revenue, rather than wading through dirt in BP's refineries and distant oilfields, would have been churlish. Browne was trusted to ensure the reliability of the company's industrial processes using the management methods perfected by McKinsey. After all, unlike some of his predecessors he had actually trained as an engineer, although he no longer showed much interest in the details of exploration, production or refining. Nevertheless, the board of directors, with limited engineering expertise, relied upon Browne to integrate Amoco and Arco into BP. Distracted by the glorification of Browne, they were unaware that the early results, as Rodney Chase admitted, "felt more brutal than even we expected."

Until their first meeting in autumn 1998 in the cafeteria at BP's headquarters, Amoco's executives had believed Browne's pitch of a merger of equals. Their first surprise was Britannic House, a squat building rather than an imposing tower. They were equally surprised by the agenda proposed by McKinsey to impose BP's culture, structure and system upon what was privately called an "amateur oil company." Amoco executives had urged Browne to retain their centralized management. Browne insisted on the opposite. "We have to get through the harvesting of low-hanging fruit," he said, recalling BP's botched acquisition of Sohio. Culture clashes during integration could not be ignored. "I learned that you had to have clarity with an acquisition. You can't just let these things work themselves out," Browne, rarely troubled by the condition of the human spirit, explained. The two companies were to be consolidated within a single world market. Integration, in Browne's opinion, meant Amoco adopting BP's organization and jargon. He appeared impervious to the sentiment that Americans, Russians and other nationalities could not be easily assimilated into BP's culture.

Ever since their dismissal of Sohio as "flabby and costly," Browne and Chase had become antagonistic toward multilayered organizations with lengthy chains of command. Browne had championed his direct access to the asset performers. Amoco executives who spoke of "aspirations" discovered that BP talked about "hard targets." Browne accepted that McKinsey's plan of breaking BP down into four business units was the key to making a global company work. Each unit, he said, would have "operational freedom" to allow individuals to see their own achievements: "It's like they're running their own businesses." Each unit director would report quarterly to Browne during a one-and-a-half-day review. Success was rewarded with a bonus; failure earned dismissal. McKinsey's vocabulary of "peer group challenge" and "peer group assist," remarked Doug Ford, who was among the senior Amoco executives selected to remain, "takes getting used to. We weren't prepared for this." Doug Ford, a genial and talented engineer, attempted to adapt to Browne's "targets" culture, and Browne appeared pleased to receive obedience from the Americans rather than being challenged. Ford, responsible for refining at Amoco, and appointed to manage all BP's refineries, rarely contradicted Browne, and concealed his dislike of Rodney Chase, who appeared to delight in making others' lives uncomfortable. "Anyone showing weakness in front of Chase is fucked," said one BP executive. "He's an uncharismatic, complexed bully." Chase's mantra "More for less," which demanded that 100 percent of a task be completed at a cost of only 90 percent of the previous resources, was embraced despite its Orwellian overtones. As Tony Hayward would later describe the penny-pinching philosophy, "We have a management style that has made a virtue out of doing more for less."

In the first months after the Amoco and Arco mergers, Browne regularly commuted across the Atlantic to consolidate BP's American empire. Each visit confirmed the difficulty of managing the company's assets from London. "America is too serious a place to be used as a branch office," he admitted, anticipating the obstacles of integrating two cultures but impatient to micromanage the solution. Curiously for a man steeped in research papers and intensive discussions, Browne resisted adopting ExxonMobil's formula for success. Unlike BP and Chevron, Exxon imposed a military-style monoculture on its recruits. Every graduate was channeled through every experience in the industry, and conditioned to refer every issue to superiors and then to await their considered response. Cultural assimilation during the merger with Mobil had been relatively unproblematic, because Exxon's deep-rooted culture allowed no doubts about the guidelines. By contrast, BP's culture, with its mavericks shooting from the hip, was based on Browne's personal interaction with the business units. Unlike the managers of Exxon and Shell, where every decision was endlessly discussed and performance was minutely scrutinized, Browne gloried in the swift challenge: "Perform or you're out." While Exxon's and Shell's engineers constantly checked the technical performance and safety of each refinery, offshore platform and wellhead, Browne had dispensed with the layers of scrutiny as cumbersome and costly, in favor of personal responsibility.

Delegation was particularly pertinent in America, which was the heart of BP's empire, although it was ruled from London. One inherent confusion was ignored. While BP's organization was based upon a top-down culture, Browne directed an entourage of selected personal assistants known as "turtles"—those privileged to carry his burden on their backs—to cut across the units and deliver the imperial message. They reflected the reality of Browne's personal management, which was focused on profits rather than technical excellence, the fundamental values at Exxon and Shell. The casualty of Browne's focusing of authority in London was governance, an alien word among his inner circle. Browne believed that BP's strength was to centralize authority upon himself. Unlike Chevron, Shell and ExxonMobil, where the machine was entrenched and self-perpetuating, Browne made himself indispensable. After all, he reasoned, BP's resurrection was entirely his success. He could identify the purchase of Castrol in India and Verba in Germany, the development of alternative energies and his new negotiations with Mikhail Fridman to establish a major foothold in Russia as his most recent successes. But those prizes became irrelevant in early 2002 once the 5.5 percent growth target was missed. Three times in eight weeks, Browne's prediction was revised downward, finally to 2.9 percent. BP's share price fell to its lowest in four years. Within the company's headquarters, some spoke about Browne paying the price for his over-optimism; others mentioned panic. "In 2002, we at BP started stubbing our toes," observed Doug Ford, the Amoco director appointed to BP's main board. The financial results in the Gulf of Mexico were poor, there were operational problems in Alaska, and the Sidanco losses in Russia had not been recovered. Browne's forecasts of record production were unrealized, and his aura of invincibility was dented. Some blamed complacency; others said Browne was presiding over a culture of fear, which prevented his subordinates delivering bad news. His mantra of "more for less" to boost BP's share price was a poisoned chalice. "We always said BP would have its moment of truth," said one Shell director. The first casualty was talk about merging with Shell, an unlisted item on the agenda when BP's directors gathered for their regular board meeting in Seville in September 2002.

Peter Sutherland, BP's chairman, a physically big man with an engaging intellect, had been particularly irked by Browne's continuing quest to take over Shell. Ever since he was appointed chairman in 1997, Sutherland, like all BP's directors, had been impotent before Browne's juggernaut, rarely able to engage him about BP's problems or his quest for Shell. Browne's success in delivering dividends and rising share prices had secured seemingly invulnerable popularity in the financial markets. Although some close to Browne had whispered that the deal had been mooted with Shell's British directors — a risible idea, considering Phil Watts's disdain for BP's technological weaknesses — Peter Sutherland waited before declaring his opposition. The proposed merger, he believed, was delusional. The regulatory, financial and human problems were insuperable. Browne's ambition to merge with Shell, Sutherland suspected, had encouraged the continuing buyback of BP's shares and Browne's refusal to discuss a successor.

Since Peter Backhouse's premature departure, several other potential successors had also left the company. Much of BP's experience had been, according to the gossip by the watercooler, "individually strangled" by Rodney Chase. Browne, Sutherland decided, should be judged on his own terms. He confided to a fellow director that he had spotted Browne's limitations early on, but had been too weak politically to contradict the market sentiment. The City would have been aghast if Sutherland had successfully ousted Browne earlier, yet he deprecated Browne's fatal vanity. Only "yes men" were tolerated by Browne; dissenters were withered by his intellectual power. "He's become princely," said the director. "He only deigns to go to triple-A occasions, and fails to arrive at others. He will only speak to presidents, prime ministers and kings. This does not help the company's performance." Another director with nearly 20 years' experience agreed: "He's interested in the dynamics of the industry but not in what to do after you buy your latest toy."

Browne's allies suspected that Sutherland was encouraging dissent. The more extreme implausibly conjectured that Sutherland, acting as an agent of the American government, was stirring up dissatisfaction to prevent BP's merger with Shell. Most agreed, however, that Browne's conceit had damaged his credibility. "His fast-moving, high-wire act has crashed," admitted one of his allies. Knowing Sutherland's boast "I never move away from a fight," Browne suspected that the Seville board meeting would be bloody, but he rejected the notion of defeat. Unfortunately for him, he was trapped. He would be unable to avoid a fight. The first substantive item for discussion was the failure to hit the 5- to 7-percent growth target set in July 2000. An arranged confession by Dick Olver that Browne's objectives were overoptimistic provoked what eyewitnesses would call "a blazing row between Peter and John." At the end of the day, spurning the traditional dinner, Browne flew back to London in a huff. Inevitably, the news leaked.

Lord Browne squirmed. The supertanker was holed below the

waterline. The same newspaper columnists whom he had wooed now questioned whether he was losing his magic and, reflecting the opinions of Browne's enemies, whether he was a fallen god. Rather than confront his critics, on Thursday, October 31, 2002, Browne avoided journalists at BP's annual champagne party at the British Museum. The universal idolization had ended abruptly. Browne's swagger had become a stagger. "No one should stay chief executive for more than seven years," a director was told by Peter Sutherland. "Corporations need renewal." Sutherland suggested that the board should begin considering the succession, but Browne's considerable achievements prevented the stumble becoming terminal.

At the beginning of December 2002, to counter the suggestion that he was out of touch with his own company, Browne ordered an inquiry to discover why BP had missed its targets. To many, this appeared a mere gesture. In Blairite Britain, the public had become cynical about targets, with their Stalinist overtones, and inconsequential inquiries that appeared designed merely to distract attention from failure. Externally, Browne's credibility had been injured, but within the corporation he was undiminished, even if he was compelled to retrench. The most noticeable casualty was "Beyond Petroleum." Internal criticism had eviscerated the operation. Rodney Chase supervised the death. Executives at ExxonMobil and Shell were delighted. In response to BP's campaign they had increased their own environmental campaigns—Shell by an additional $30 million—and now BP had slackened the pace.

Detractors whispered that Browne's diligence was slipping. Options to drill for oil off West Africa, the Gulf of Mexico and the Caspian had been allowed to pass. BP's opportunities to expand in OPEC countries were waived. Browne's prediction of world recession during 2003 if oil rose to $30 a barrel lacked credibility. And yet, in the most unexpected circumstances, he was about to reestablish his supremacy, especially over Lee Raymond.

Chapter Fourteen

The Twister

———◀◉▶———

MIKHAIL FRIDMAN and his partners were not Browne's natural allies. Tough wheeler-dealers ignorant about the production of oil could not fit easily into a partnership with a multinational corporation. Browne's choices, however, were limited. Like BP's rivals, Browne's search for assets in Russia was constrained by President Putin's new conditions, but encouraged by his safeguards. During 2001, Browne considered a deal with Yukos before concluding that Mikhail Khodorkovsky, although keen, would not cede control, and the company was too expensive. "The discussions are terrible," a banker confessed. "Khodorkovsky and Lebedev don't make sense. Let's go with Fridman, the devil we know." Astutely, Fridman had bought oilfields around Sidanco's properties, including Samotlor, one of the biggest and oldest reserves in western Siberia. Combining TNK and Sidanco to create Russia's third-largest oil corporation would reverse BP's declining reserves by adding four billion barrels, nearly 30 percent of the total. In summer 2002, Browne finally decided to trust Fridman. "This would be the biggest deal in history," he smiled. The quest was also a challenge for Fridman. Putin was again changing the rules.

Vladimir Putin, like all KGB officers, enjoyed fostering uncertainty by mixing authority and enigma. His election as president in 2000 signaled to Fridman and other oligarchs the revival of a familiar atmosphere experienced before Gorbachev's introduction of perestroika. Portraying themselves as hard, wise and untouchable, the trusted *siloviki* who entered the Kremlin with Putin were all former KGB officers, technocrats and friends employed by Anatoly Sobchak, the mayor of Saint Petersburg whose Soviet outlook had become loathed during Yeltsin's era.

Fridman, an outsider uninterested in politics and lacking the finesse to cultivate relationships in the Kremlin, had astutely forged a partnership in 1993 with Peter Aven, first as an adviser and then as a 13 percent shareholder in the Alfa Bank. In 1991 and 1992, as the minister of foreign economic relations in Moscow, Aven had enjoyed a good relationship with Putin, his subordinate in Saint Petersburg. Eight years later, Aven was Fridman's invaluable passport into Putin's closed circle. Two men, Aven established, formulated Russia's energy policies. Sergei Priodka, a foreign affairs aide, guarded access to Putin. The second and more important was Igor Sechin. Born in comparative poverty, Sechin had been spotted by Western intelligence services while working during the 1980s as an interpreter for Russia's military and political advisers in Mozambique and Angola, supporting the socialists Samora Machel in Mozambique and Augustino Neto in Angola during the bitter strife in the former Portuguese colonies. Sechin would confide to the very few who penetrated his steely façade that his fluent Spanish and Portuguese were self-taught. "I never made mistakes in my essays or in the translation tests," he explained. On his return to Russia, he was initially employed as a translator in the mayor's office in Saint Petersburg and then promoted at Putin's suggestion. Sechin, the president believed, was discreet, diligent, loyal and "not interested in money." The workaholic grandfather of two could be trusted to exclusively pursue the interests of the state and to eschew self-enrichment through corruption. His priority, he had been told in 2000 by Putin, was to repair Russia's oilfields.

In a doctoral thesis written in Saint Petersburg in the mid-1990s, Putin had developed an argument that Russia's prosperity and restoration as a superpower should be built by reasserting sovereignty over its natural resources. In a speech in June 2000, he had criticized the Russian oil ministry for tardiness and failing to invest. More foreign expertise, Putin believed, could solve the dearth of management and finance. With the help of Western oil companies, Russia's production had already increased over the previous two years by 15 percent, providing about 10 percent of the world's supply. Although unsure about the precise terms of future agreements with Western oil companies, Putin was certain, after reading reports about the improvements executed by Lukoil and Yukos, that the oligarchs' ownership of the oilfields should be tolerated so long as the state's interests were unharmed. Sechin was to be the executor of that policy. His learning curve was steep. Uneducated in capitalist economics and competition, he read voraciously, listened to invited experts, encouraged debate and asked questions, harshly castigating anyone who revealed ignorance. He had been angered by Russia's painful experiences during the 1980s and 1990s, and his approach was colored by instinctive insecurities and suspicions about foreigners, mirroring Putin's own xenophobia. Western capitalists, Sechin believed, were intent on exploiting Russia, and the oligarchs were equally objectionable. Both groups, Sechin acknowledged, were unfortunately indispensable to fulfill the president's agenda. As he became more familiar with energy and commerce, visitors spotted a hint of humor. The agenda was still flexible, although he shared Putin's bewilderment about why Russian oil was priced lower than Brent. After all, he repeated, "Crude is crude." Unsophisticated about trade, he could not understand that his attempt to establish a futures market in Saint Petersburg would fail after traders discovered his attempt to control prices. Resistant to capitalism and uneasy about the Gorbachev and Yeltsin legacies, Sechin balanced Russia's integration into the world's market economy with single-mindedly protecting Russia's oil interests. As for Russia's gas industry, Sechin noted, that was the president's preoccupation.

Gazprom, the owner of about one sixth of the world's proven natural gas reserves and the source of 20 percent of Russia's taxes, was infected by corrosive corruption. Since August 1999 there had been a struggle over the company's fate between Viktor Chernomyrdin, the former prime minister who had become the company's chairman, and the oligarch Boris Berezovsky. Gazprom's American shareholders suspected that some Russian directors had sold shares in the company to relatives below market value, and had awarded contracts, especially to build pipelines, to accomplices. Some Gazprom assets, the shareholders found, had been sold for artificially low amounts, especially to Itera, an American company established in 1992 in Jacksonville, Florida, although Gazprom denied any links with Itera. Itera's 12 employees in Jacksonville, ostensibly trading food with Turkmenistan, actually employed about 7,000 people in 24 countries. As Itera expanded, Gazprom shrank. Following Berezovsky's complaints about this racket, Putin replaced Chernomyrdin with Dmitri Medvedev and Alexei Miller, both of whom had been Putin's subordinates in the foreign relations committee in Mayor Sobchak's office. By refusing Itera access to the Gazprom pipelines, Miller squeezed its directors in America to return the contracts to Gazprom for a pittance. Putin's attitude toward Western investors would be made evident by his rejection of the American investors' demand for compensation. Complaints of this nature went unreported as Putin eliminated Russia's independent media. The beneficiaries were Gazprom's new executives.

On July 28, 2000, the oligarchs were summoned to the Kremlin. Mikhail Fridman, Viktor Vekselberg and a handful of others gathered in a committee room to await the president. In advance, all had sought to discover the purpose of the unusual gathering. "Messages" from the Kremlin explained that after Chernomyrdin's removal, Putin wanted to set the parameters of the oligarchs' influence. When Putin appeared, he told the oligarchs that they could retain their billions, but they were requested to withdraw from politics. Mikhail Fridman's bank could keep its shareholding in TNK, but would be expected to invest in Russia, and not deposit its profits in foreign banks. Effectively, Putin had deactivated the oligarchs. Without political influence, they joined the ranks of the powerless rich. Berezovsky and Mikhail Khodorkovsky, who were not present at the meeting, decided to ignore the president's edict.

During 2001, the mood in Moscow changed. The increasing price of oil to $32 a barrel stiffened Putin's inclination to limit Western companies to difficult projects, especially offshore, and to decline their latest entreaties for PSA agreements. He accepted Khodorkovsky's claim that PSAs provided an incentive for the Western oil majors to increase costs and to siphon money toward corrupt bureaucrats. By contrast, standard contracts with the oil majors would encourage high profits and the payment of more taxes to the government. Putin, however, was equally hesitant about agreeing any private deals. To Washington's irritation, he appeared unsure about the mechanics of simultaneously increasing production and limiting Western operators' profits from Russian oil.

Reports from the Kremlin described a plan proposed by Russia's minister of trade German Gref, a champion of the free market, to increase production by compelling Russia's oil companies to improve their operations in existing fields. Under the headline "Gref Says It's Time to Squeeze Big Oil," the *Moscow Times* described Gref's proposal to intensify production in the existing fields, and block the development of new fields and new pipelines. The Western oil majors and Washington were partially sympathetic to the idea. Russia, many believed, possessed three times more oil and gas than was admitted in the official estimates, but that additional oil was worthless unless the old Soviet pipeline system linking the oilfields and refineries with military bases was replaced by pipelines from oilfields to ports to supply Europe and the USA. Those arguments were strengthened after the terrorist attacks on September 11, 2001. Strikes in Venezuela's oilfields and Iraq's threat to suspend exports increased the world's dependency on Russian oil. Putin obliged, and was hailed by the White House as a reliable ally against terrorism and a dependable source of oil compared to the Middle East. One landmark had been passed. In October 2001, crude oil had finally arrived at Novorossiysk on the Black Sea through the $2.6 billion pipeline from Chevron's oilfields in Tengiz.

Even though Moscow controlled the pipeline, Washington was relieved
to have Moscow as an ally against Saudi Arabia. Ali al-Naimi's visit to
Moscow in November, requesting a small cut in production to reverse
the oil glut, had failed to secure even a token half of 1 percent (160,000
barrels a day) response from the Kremlin. A follow-up visit by Mexico's
energy minister seeking Russian support for OPEC actually encouraged Russia to increase production to strengthen its finances. During
2002 Russia produced 8.5 million barrels a day, and briefly overtook
Saudi Arabia as the world's largest producer. Putin's dilemma was how
to reach his target of 10.5 million barrels a day in 2010, and 12 million
in 2012, without Western help. Despite his opposition to private pipelines, in April 2003 he did approve Khodorkovsky's plan to construct a
$4.5 billion pipeline from western Siberia to Murmansk, financed by a
consortium including Lukoil, TNK and American investors. Completion was set for 2007. Putin's unreconciled quandary about Western
investment coincided with John Browne and Mikhail Fridman finalizing their negotiations.

For more than a year Fridman had occasionally played hardball
with Browne. Although he had never properly visited an oilfield, not
least because Nizhnyvartovsk, TNK's operational headquarters, was in
the midst of the inhospitable Siberian tundra, where temperatures fell
to –50°C, he feigned uncertainty about whether BP was his ideal partner. Lukoil, Yukos and Rosneft, he said, appeared to prosper by hiring
Western specialists and admitting only minority foreign shareholders.
The attraction of a Western investor for Fridman was not only the
technical expertise, but also the money. Unfortunately, he acknowledged, BP was the only candidate able to offer cash, technology and a
measure of protection from the Kremlin. One obstacle was Browne's
dislike of Fridman, whom he regarded as arrogant and unpleasant.
Another obstacle for Browne was the price.

Ralph Alexander was detailed to value TNK's assets. The five refineries, 2,100 gas stations, six prospective ventures across Russia and 9.4
billion barrels of proven reserves in nine major areas were attractive,
not least because Russian engineers had recovered only 25 percent of

the reserves. BP's experts could easily extract 30 percent, and could expect ultimately to pump out 61 percent. The surveys commissioned by Alexander, however, were discouraging. American reservoir engineers were unaccustomed to the poor levels of maintenance they found at the dilapidated sites. The Russians, they reported, had cracked the reservoirs by injecting water at excessively high pressures to force out the oil. Even the major attraction, the Samotlor oilfield in western Siberia, one of the world's biggest, had been thrashed during the Soviet era, and was judged by BP's engineers to be beyond redemption. Similarly, TNK's refineries were rickety and mismanaged, precisely the kind of sites BP was selling elsewhere as unprofitable, environmental nightmares. Investing in TNK contradicted BP's policy of selling off old fields in favor of "elephants" at new sites. On those criteria, the conservative engineers reported, TNK's value was "zero." Fridman was unimpressed. "Alexander's sabotaging the deal," he scoffed, claiming that BP's experts clearly did not understand how to unlock the reserves' potential. "These fields are worth billions of dollars," he said. Samotlor, he would later smile, did eventually produce 140 million barrels a year.

In summer 2002, Browne still had his doubts about Fridman, but he admitted to a banker, "I can't ignore the historic opportunity." Russia had improved since 1999, and while there was always risk in oil, a joint company, he thought, would be too important to lose its shirt. Despite his concerns, he decided to trust Fridman. Putin, Browne believed, wanted Western investment, and TNK could be used as a gateway into the Kremlin to unlock other ventures. The partners could manage the political risk through their contacts with the Kremlin, and look for new opportunities while BP managed the business. Despite some BP directors opposing the deal, Browne appointed Rodney Chase to negotiate. "Rodney doesn't like or trust the Russians," Browne said. "He'll make sure the deal is okay."

The three Alfa-Access-Renova (AAR) partners—Fridman, Vekselberg and Blavatnik—flew to London. According to Fridman, Chase offered to pay $1 billion in cash, emphasising "in cash." The partners laughed. "Cash?" roared Fridman. "You think we're wild Russians. $1 billion in cash? You must do it our way or lose the deal." Chase has no subsequent recollection of the exchange, but clearly his doubts about TNK were not assuaged by his meeting with the proposed partners. He regarded Fridman and Vekselberg as thugs who had won the battles to assert control over their companies by using threats and muscle. German Khan, a key partner in the Alfa Bank since 1992 and worth over $10 billion, was described as "a gorilla," and like Chase was an enforcer—some described him as "Chase without the charm." Others regarded him as witty, and willing to "throw a grenade into a room just to see what opportunities would arise out of the chaos." Khan, born in Kiev, had been a close friend of Fridman's since they had met at university in 1982, and shared his ambition to build an empire. Unfortunately for Chase, walking away from the deal was not an option. BP needed the Russian oil.

The agreement with AAR was eventually priced at $6.8 billion—$2.6 billion in cash and $4.2 billion in BP shares, to be paid in three annual stages. A further $1.1 billion would be paid for a share of Slavneft, a Siberian oil company. Under the agreement, TNK-BP, with 113,000 employees and estimated reserves of 3.2 billion barrels of oil, would produce 1.2 million barrels a day in western Siberia and the Ural Mountains around the Volga. Russia's third-largest oil and gas company would be registered in Cyprus and the British Virgin Islands. Chase's final chore—an exhaustive undertaking—was his cross-examination of and agreement with the Russians about corporate governance, management and shareholders' rights. The initial agreement stipulated that BP would buy 50 percent of TNK. To those querying the unusual shareholding, Browne emphasized his astuteness in not insisting on 51 percent. In reality, Fridman refused to become a minority investor, and would accept only an equal partnership.

No deal could be finalized without Putin's blessing. Jealousy among Kremlin officials, and surviving anti-Semitic attitudes, could risk the stability of any business, especially oil. The oligarchs had identified those Kremlin officials requiring favors and those requiring reassurance about Western companies profiting from Russian energy. Still uncertain about their strategy, and unable to anticipate rising oil prices, Sechin and Putin were open to persuasion. "Never assume that the Kremlin has a detailed plan, only long-term ambitions," Fridman's adviser Peter Aven noted. Since Fridman barely knew the president, Aven drove out on a Saturday to Putin's dacha at Nova Ogmarc to explain the agreement. "Blair has already called," said Putin at the beginning of their 30-minute conversation. At Browne's request, the British prime minister had encouraged the president to agree to the largest Western investment in Russia. Cautious but positive and well briefed, Putin was persuaded that BP's investment would confirm to the global market that Russia's energy sector was a bankable asset, and that foreign investment in the country was safe. Aware of the president's priorities, Aven stressed that the same would apply to gas as oil. Putin only queried the equal shares. "Who will control the business?" he asked. "How can you successfully manage a company with 50–50 ownership?" "It won't be a problem," replied Aven. One prediction proved correct. After BP's purchase, Gazprom effortlessly raised $1 billion at good rates.

One week after the initial agreement was signed in February 2003, the two sides began negotiating about integrating TNK-BP's governance and structure. Browne chose Robert Dudley to represent BP. Dudley, a graduate of the University of Illinois, had worked in Moscow for Amoco between 1994 and 1997, but spoke little Russian. Clean-cut and slightly prim, he did not empathize with Russian-Jewish businessmen. Neither Tony Hayward nor Rodney Chase appeared keen on Dudley's appointment, but Browne was insistent that BP needed an engineer with political savvy in Moscow. Browne judged Dudley to be more than a mere manager. Loyal to BP's interests during the merger negotiations with Amoco, he was promoted as a rising star after 1999. Browne had been impressed by his skill while serving as an assistant or turtle, and told him to assert BP's domination of the partnership. "We'll go through these guys like water," the Russians would claim that they heard Dudley quip during the early negotiations, although Dudley would deny the suggestion. To persuade BP's shareholders that Browne was not repeating the Sidanco disaster, Dudley's brief was to affirm the sanctity of BP's rights in the shareholders' agreement. The Russians, he suspected, were not reading the document properly. In their credo, *ponyatiyno*, an understanding rather than the written agreement, counted. That misjudgment suited Dudley.

In the midst of the negotiations, Browne felt he wanted reassurance of the president's support. In April 2003, Browne, Chase, Fridman and Vekselberg met Putin for an hour in the Kremlin. Browne, even shorter than Putin, silkily offered the assurances the president sought about the 50–50 structure. Fridman also wanted assurances. The president was due to visit Britain in June, and through intermediaries both Fridman and Browne urged him to witness the signing ceremony of the final agreement. After the Kremlin meeting, Tony Blair also asked Putin to do this. Putin consented, in the expectation that when the occasion arose, Blair would return the favor. On June 26, 2003, Putin watched Browne and Fridman formally sign an agreement at Lancaster House, a 19th-century mansion in St James's. Since the final terms had not been precisely formulated, the ritual was symbolic, but Fridman was grateful nevertheless. "Today is a special day," he said, clutching the gray fountain pen with which the deal had been signed as a memento, "not only for the Russian oil industry but for the Russian economy as a whole." Browne, anointed in July 2003 by *Fortune* magazine as the world's most powerful businessman outside the USA, shared the enthusiasm, mentioning "my friend Michael." Browne had reason to be happy. BP's fortunes had been restored by his latest coup, one of the greatest of the era. Only envious rivals suggested that the company was dependent on the whim of Russia's president. For just $3 billion, BP had obtained instant access to vast reserves without risking, like Exxon and Shell, billions of dollars off Sakhalin. Conoco had bought 20 percent of Lukoil, but that agreement had soured after Lukoil's plan to merge with ConocoPhillips was resisted as undesirable to American interests, a snub interpreted in the Kremlin as a signal that the West's offers of partnership were flawed. BP, Sechin said, was exceptional.

The terms of the contract were finally completed on August 29,

2003. Among Dudley's first casualties was an experienced BP operator responsible for supervising the Sidanco oilfields. He was replaced by Larry McVeigh, a Texan who spoke no Russian and had not previously worked in the country. German Khan was unamused. Angered by Dudley's refusal to share control, he complained about the American's "colonial attitude" and his treatment of Russia as just another Third World natural-resource producer. "It's like two puppies fighting for dominance," observed a Russian participant as Khan began throwing ashtrays at walls. "German no longer carries guns into meetings," Fridman smiled about his partner's mellowing manners. "It's only knives." Strangely, in those early days neither Browne nor Fridman appeared perturbed by the hostility among some of the partners. The contest between Browne, playing chess, and the Russians, boxing like a heavyweight, was dismissed by both sides as "business, not personal."

The following month, TNK-BP's directors hosted a celebration for Moscow's major players at the Armoury, the museum in the Kremlin filled with priceless treasures owned by the Tsars. Beneath big photographs of Fridman and Browne in London with Putin in the background, Fridman offered his interpretation of the situation: "The sound you hear after the BP-TNK deal is the door slamming shut on any more deals." Fridman knew that the mood in Moscow was changing. Uninterested in politics, he had intentionally presented the deal with BP to the president as unthreatening to Russia's interests. Khodorkovsky's agenda had never attracted Fridman. The news in July of Platon Lebedev's arrest suggested that Khodorkovsky, his partner, should avoid irritating the president. After all, Vladimir Guzinsky, a media mogul, had agreed in 2000 to sell his television station and leave the country. He was followed by Boris Berezovsky, fearful for his life, fleeing to England and selling his Sibneft shares to Roman Abramovich. Khodorkovsky, to Fridman's surprise, was ignoring Putin's orders to stay out of politics.

Khodorkovsky's ambitions outclassed Fridman's. Yukos produced 20 percent of Russia's oil, and if Khodorkovsky's aspiration of merging Yukos with Sibneft materialized, the new $40 billion company would rank fourth in the world, producing 2.16 million barrels of oil a day, or 2 percent of the world's supply. To Khodorkovsky's delight, the marriage had been blessed by Mikhail Kasyanov, the prime minister, as "a flagship for the Russian economy." Abramovich would own 25 percent of the new company, while Khodorkovsky would exchange his 44 percent stake in Yukos for 61 percent of the merged company. Initially, the merger aroused less of Putin's interest than Khodorkovsky's plan to build a pipeline to China. Pipelines and the sale of oil to foreigners were regarded by the Kremlin as crucial to Russia's interests.

Both the Chinese and the Japanese governments were competing for the Kremlin's approval of a new pipeline to their countries. Rising oil prices and the competition between China and Japan enhanced Russia's strength. Putin was aggravated by Khodorkovsky's campaign for a pipeline to Daqing, the center of China's oil industry. The alternative, proposed by Junichiro Koizumi, the Japanese prime minister, during a visit to Moscow in January 2003, was a 2,500-mile pipeline to carry a million barrels a day from eastern Siberia to Nakhodka, a Russian port facing Japan. The Japanese hoped the Kremlin would also approve Exxon's construction of a 1,450-mile pipeline from Sakhalin to supply Japan. The two pipelines would provide about 25 percent of Japan's requirements. Tempted by the Japanese offer to pay $5 billion toward the pipeline, some in the Kremlin argued that a pipeline to a Pacific port would allow Russia to choose its customers. That faction appeared to have been outwitted by Khodorkovsky. President Hu Jintao of China believed that his agreement with Khodorkovsky in May 2003 for the shorter $2.8 billion pipeline to China, to be completed by the end of 2004, had been endorsed by the Kremlin. But the Chinese president was unaware of Putin's irritation over Khodorkovsky's unsated ambitions and callow presumptions.

The Kremlin was only hazily aware of Khodorkovsky's campaign to seduce American oil majors. American executives had been hired by Yukos, Khodorkovsky had pledged donations to American foundations and charities, he sponsored Formula One motor racing, and in 2001 he had hosted a dinner in Moscow for George Bush senior and the

Carlyle group. He even sought to finance American politicians. In conversation, the oligarch encouraged the belief that Russia's oil industry was open to foreign investment, not least by persuading Condoleezza Rice, the national security adviser and former director of Chevron, to approve American finance for Yukos's pipeline from Siberia to Murmansk to transport oil for the US. Although his American friends were less enamored by the deal to divert Yukos's oil to China, they were entranced by the sight of an oligarch playing both sides and irritating the Kremlin. To Western applause, Khodorkovsky's undisguised agenda was to leverage political power. Despite the criminal origins of his fortune, he had bought the support of politicians in the Duma to become a self-appointed scourge of corruption. Challenging the *siloviki*, he used his supporters to block legislation contrary to his interests. In Russian politics, Khodorkovsky knew, there was no place for a "loyal opposition." The Kremlin's reaction was predictable. Insecure and vindictive, Putin did not want merely to defeat, but to utterly crush his enemies. Khodorkovsky ignored those who warned him of the danger, and to finance his ambitions he sought American investment.

The temptation had been first offered to Exxon on a bitterly cold evening in Oslo in January 2002, when an American oil engineer employed by Khodorkovsky had suggested to an Exxon executive that the two companies try to do something together. Four months later an Exxon group arrived in Moscow to discuss an exploration project in Siberia with Yukos. After several days, Hubert Thouvenot, a French oil engineer employed by Khodorkovsky, confessed that the discussions were a ruse. "We're using these exploration talks," he told Tim Cejka, Exxon's president of exploration in charge of negotiations, "as a test whether you sincerely want to invest in Yukos. We want to see whether you're ready to do something more serious. Khodorkovsky is going to America to see government officials and would like to meet Lee Raymond." Inevitably, Raymond's reaction was negative. "Lee doesn't want to meet an unknown Russian," Tillerson responded. "Rex," said the imploring voice from Moscow, "this is something big. You've got to meet him."

Khodorkovsky and Thouvenot met Raymond in Dallas in June 2002. "All of Yukos is for sale," Khodorkovsky told Raymond, "but we must proceed bit by bit." Raymond was excited. He had made his name merging Exxon and Mobil in 1999, creating the world's biggest oil company. Buying Yukos would double Exxon's oil reserves, the size of which was the key to an oil company's survival. Disturbed by BP's success in Russia, and encouraged that Putin had apparently approved Shell's $1 billion investment in Salym, Raymond was spurred to forge close links with Yukos. He was unfazed by Khodorkovsky's high profile.

The negotiations, codenamed "Rugby" and confined by Raymond to six executives, involved few bankers: in a historic shift within ExxonMobil, Raymond declared that bankers added little value. To test the water, ExxonMobil intended to start negotiating the purchase of 10 percent of Yukos, but with Khodorkovsky's encouragement, Raymond's initial offer was to pay cash for a 15 percent stake in the corporation, with the intention, if the Kremlin approved, to buy another 25 percent after three years. Unsurprisingly, Raymond was fearful of the deal slipping away. Seeking reassurance, he asked Khodorkovsky for exclusive rights to negotiate, and in early 2003 Khodorkovsky finally agreed to sign a letter promising Exxon four months' exclusivity. Raymond's suspicions, however, had been justified. During his return flight to Moscow from Dallas in June 2002, Khodorkovsky had told an aide, "We'll negotiate with these guys. Now call Chevron and we'll offer them the same."

Chevron's interest in Yukos was deftly orchestrated by Khodorkovsky. Peter Robertson, Chevron's vice chairman, was invited to a riotous fiftieth birthday party in Moscow on March 27, 2003. Yuri Golubev, Khodorkovsky's partner, was seated next to Robertson in order to dangle the prospect of a merger. Robertson was visibly electrified. Chevron was in the doldrums, with fast-depleting oil reserves. Eager to catch up after Total and Exxon had done well in West Africa and the Middle East, Robertson and David O'Reilly, Chevron's chairman, were desperately hunting for a buy. Two months later, Sam Laidlaw, a scrupulous, efficient English technocrat, formerly of Enterprise Oil, arrived in Moscow. Laidlaw had been recruited by O'Reilly to finalize Chevron's merger with Texaco. He had saved $2 billion in synergies, and O'Reilly was hungry for more growth. In spring 2003, the invasion of Iraq was imminent, and O'Reilly shared the belief that the Middle East would finally be opened up to the oil majors. With teams searching for opportunities in Asia, Africa and Latin America, O'Reilly asked Laidlaw to secure assets in Russia. In his first meeting with Khodorkovsky, Laidlaw offered to buy 20 percent of Yukos, not with cash but with Chevron shares.

The Chevron team in Moscow felt well placed. With 12 years' experience in Kazakhstan and a strong base in Moscow, Chevron's managers spoke about their good relations with the Kremlin. Their confidence was ill-judged. Chevron's alignment with the Kazakh government against Moscow offended Putin, who refused to meet O'Reilly. Frustrated by John Browne's smooth diplomacy in using Tony Blair to engage Putin and finalize his deal, O'Reilly urged former US secretary of state George Shultz and Condoleezza Rice to advance Chevron's case. But, like the diplomats at the American embassy and the Kremlinologists hired by Chevron, they were unable to offer insight or access to Sechin or Putin. "It's too early to see Putin," O'Reilly told Khodorkovsky, to conceal Chevron's exclusion. "We've got to keep these negotiations confidential." In essence, Chevron's executives were offering Khodorkovsky shares for nearly the same valuation in a merged company as Raymond was proffering in cash.

The parallel negotiations by two compartmentalized Yukos teams were intended to be top secret, but Raymond had heard about Chevron's involvement within a week. He appeared unconcerned. Exxon's bid, he believed, would smother Chevron's. On Khodorkovsky's initiative, the two corporations were competing for different deals. Exxon was seeking a straight purchase, while Chevron wanted a merger. Khodorkovsky encouraged each to believe it was the favorite. In reality, the race was confused. Exxon's bid of cash and shares was financially superior, but Yukos found negotiating with Exxon difficult. Tillerson, complained the Yukos negotiator, "self-importantly spent the day constantly repeating himself." The Yukos team preferred dealing with Laidlaw, but in their lust for cash, personalities became irrelevant, especially as Raymond's ambitions increased.

Nearly 50 Exxon experts had been ordered to abandon other negotiations and number-crunch the Yukos bid. After exhaustive due diligence, they reported Yukos's value as $45 billion. In the wake of BP's coup with TNK, Raymond changed the terms. There was talk of Yukos and Sibneft merging. If that was finalized, he wanted Exxon to immediately purchase 25 percent of the shares. Combined with its stake in Sakhalin 1, this would mean ExxonMobil would overtake BP and become the biggest Western investor in Russia. Once the minority stake in Yukos-Sibneft was consolidated, he wanted to gradually increase Exxon's ownership to 75 percent of the giant.

Naturally, to cement the deal Raymond needed Putin's approval. Exxon's chairman always dealt with heads of state. He had always dismissed those decrying the competition in the 1990s among governments to make deals with the Russians. He did not sympathize with the critics of Clinton's and Gore's aggressive bid for Caspian oil, or overtly request protection under the "shield of government" from Washington to secure a level playing field. Nor was he particularly interested in persuading the Kremlin to adopt the rule of law. Although Exxon's adherence to the sanctity of contracts was inviolable, Raymond saw no reason not to exploit laws to advance the company's interests. Like John Browne, he paid scant attention to those fearful about the security of the West's energy supplies. In his opinion there was no reason to approach Russia with particular sensitivity. The country was just another source for an uninterrupted flow of oil at reasonable prices. Ignoring Khodorkovsky's relations with the Kremlin, Raymond abided by ExxonMobil's usual practice of standing aside from internal politics and being guided by Daniel Yergin's observation in *Commanding Heights*, his latest book, published in 1998, that the Big Four oil majors would dominate the world's supplies, casting aside the powerless oil-producing countries. In June 2003 Raymond met Putin, and was reassured that the principle of Exxon's investment was welcome. In the same month, President Bush met Putin at a summit in Saint Petersburg and declared his trust in Russia. Encouraged by Putin's imminent return visit to Camp David to discuss energy cooperation, Rex Tillerson announced that Exxon would offer $11 billion for a 25 percent stake in Yukos, and would submit a new application to develop oil and gas in an area called Sakhalin 3.

By contrast, O'Reilly was tardy and entangled. Having initially bid for 20 percent of Yukos, he also threw caution to the wind as the prize became more attractive. "Their eyes are getting bigger and they want more," reported Khodorkovsky's negotiator. O'Reilly was warned, "Khodorkovsky's valuation is greedy because he knows you need oil." He acknowledged the high price, but like Raymond he was desperate to match John Browne's deal. He suspected that Raymond and Tillerson had ordered their staff to whisper it around Moscow that "Exxon is much better known to the White House, and a deal with Exxon will improve Russia's relations with America." In his telephone conversations with Khodorkovsky, O'Reilly struggled to pose as Lee Raymond's equal.

In early July 2003, Raymond returned to Moscow to participate in a business conference with Khodorkovsky. "Raymond thinks it's all sewn up with Khodorkovsky," ExxonMobil's rival reported from Moscow to O'Reilly at Chevron's headquarters in San Ramon, California. Exxon's chairman had arrived in frenzied times. Khodorkovsky was due to formally announce Yukos's merger with Sibneft, but days earlier, on July 2, Platon Lebedev, a major shareholder in Yukos, had been arrested on suspicion of fraud. Khodorkovsky recognized the warning shot, but chose to ignore it. Raymond followed suit, disregarding a local battle. Russia's internal politics, he decided, were unimportant, and he and Tillerson continued to discuss with Yukos's staff Exxon's investment and arrangements for Khodorkovsky's visit to America.

Raymond had good reason to be confident. Mikhail Kasyanov, the prime minister, had agreed that ExxonMobil would be allowed to buy a stake in Yukos, and Putin, he believed, was keen to do business. He had heard that James Mulva, the chief executive of ConocoPhillips, had recently met Putin to ask for his approval to purchase 20 percent of Lukoil for $7 billion. Mulva had also mentioned that ConocoPhillips was part of a consortium planning to develop production of 13 billion barrels of oil in Kashagan, and that the company intended to collaborate with Gazprom and extract natural gas from beneath the Barents Sea. Mulva had emerged from the meeting grinning, encouraging Raymond's belief that Russia's need for financial and technical support to extract the giant reserves in Shtokman, Kovytka and Sakhalin would favor Exxon. Commuting from his office in Irving, Texas, around the globe on his private jet with his wife and bodyguards, Raymond was too remote from Putin's prejudices and anxieties to understand the confusion and anger within the Kremlin.

Putin and the *siloviki* resented Russia's humiliation by Washington's boast of victory in the Cold War. The Americans and Europeans spoke about friendly alliances with Russia, but both had reneged on their agreement with Gorbachev to limit their military expansion in return for Russia's agreement to Germany's reunification. The former communist states, even those bordering Russia, had joined the European Union and NATO, and, encouraged by Clinton and Bush, had treated Russia's oil reserves around the Caspian Sea as booty. The swagger of some visiting Western oil executives was intolerable. Restoring Russia's prestige and power, Putin believed, depended upon hard bargaining over the country's commitment to supply Europe with energy. Sechin, by then also the chairman of Rosneft, was calculating how to extract the maximum income from oil exports. Beyond the Kremlin, few realized that a big NOT FOR SALE sign was being erected across the country. A new fear had gripped the capital. The president and his entourage would not tolerate critics. Journalists, bankers and businessmen were being gunned down in broad daylight.

By then, unknown to Raymond, Khodorkovsky had been barred from speaking directly to Putin, and had to rely on intermediaries to find out what deal would be acceptable to the Kremlin. The breach had occurred following a televised encounter on February 19 between

Putin and 20 Russian businessmen in the Kremlin. From the outset, Putin was annoyed that Khodorkovsky was wearing a turtleneck
sweater, although he had worn a shirt and bow tie for a recent meeting
with President Bush. The president's irritation only grew during the
program. Khodorkovsky complained about a license awarded to Rosneft, the government's oil company, in northern Russia despite higher
offers. "Your bureaucracy is made up of bribe-takers and thieves," he
brazenly told Putin. "That was corrupt." Visibly angry, Putin replied
sharply, "We might see how you got your money."

The turmoil between the Kremlin and Khodorkovsky had not
been reported to Dallas. Neither Tillerson nor Raymond made the
connection with the news on July 26, five days after Khodorkovsky
had accused Putin of orchestrating theft, tax evasion and murder, that
a Moscow court had issued an arrest warrant for Leonid Nevzlin, a
major Yukos shareholder living in Israel, on charges of murder (he
would be found guilty, in absentia, of five killings). Tillerson was preoccupied by his suspicion that Chevron, to prevent itself being squeezed
out, was leaking poison about Exxon to the Kremlin. Over the following weeks, while Tillerson continued to negotiate with Yuri Golubev, meeting once in Luton to "bullshit each other like there was no
tomorrow about a supposed deal-breaker," Khodorkovsky's ambitions
had become unacceptable to the Kremlin. Security officers, bankers
and ministers close to Putin individually told Khodorkovsky and his
close adviser to "stop causing more trouble," especially after he bought
a newspaper and radio station. Khodorkovsky understood the possible
consequences. "The Exxon deal might not be doable," he admitted to
a senior aide. "But keep Exxon in the game to bargain with Chevron."
Khodorkovsky suspected that O'Reilly's cashless offer to merge would
be more acceptable to Putin. The oligarch might not be allowed to
walk away with any money.

In early September, Putin's suspicions about Khodorkovsky's
ambitions hardened, but neither Tillerson nor O'Reilly would have
gleaned from Putin's warm welcome to an oil conference in Saint
Petersburg that their ambitions toward Yukos were jeopardized. Rather,

the Exxon team had become concerned about Russian corruption. In conversations with Yukos executives, the corporation's representatives explained their policy never to pay bribes. The Russians believed, however, that in certain circumstances there might be exceptions. In Africa and the Middle East, for example, oil companies had at times been required to pay local "consultants" for "services" to construct roads or buildings. The companies would have maintained that such payments were not bribes, but were merely a condition of doing business in those countries. Corruption in Russia was equally endemic, and in special circumstances a payment might be made to the single person who mattered—the person at the top. Payments further down the food chain, to ministers and bureaucrats not empowered to make decisions, were pointless. "You do one deal with the top and the rest would take care of itself," was the memorable phrase uttered to a senior member of Khodorkovsky's team.

Finalizing the deal with the "top" was Raymond's intention when he met Putin at the Waldorf Astoria in New York on September 25. Unconcerned by any friction between the president and Khodorkovsky, Exxon's chairman sought approval for purchasing Russia's largest privately owned oil company. Ignoring Russia's identity crisis and unaware of Khodorkovsky's deception about his negotiations, Raymond plowed ahead for 30 minutes, dealing with Putin as he would any other Third World ruler. He did not anticipate the consequences of this conversation for Khodorkovsky, and for the West's reliance on Russia as a secure supplier of energy. One week later he arrived in Moscow to finalize the deal. David O'Reilly was also in Moscow, convinced that he too was close to agreement with Khodorkovsky. Chevron's outstanding issue was whether Khodorkovsky would own 42.6 percent or 37 percent of the merged company. O'Reilly believed they were heading toward a 60–40 split, with the new company having headquarters in both Moscow and San Ramon. Khodorkovsky wanted Exxon's cash, but sensed that Chevron's offer was more politically acceptable—although, cut off from Putin, he was unsure. Relying on Roman Abramovich and his two lieutenants, Eugene Tannenbaum and Eugene Shvidler, for news from the Kremlin, Khodorkovsky was warned that his profile was becoming too high. During those three weeks, in his cause to build a civil society, Khodorkovsky had openly challenged Putin for power. Putin's warning that oligarchs could not buy political power was ignored. "Khodorkovsky," Shvidler reported to a Yukos executive, "has screwed things up. He's on a path of no return. Yukos is going down because Rosneft wants it." Putin, encouraged by Igor Sechin, agreed that Yukos's merger with Sibneft should not proceed. Instead, their assets should be diverted to Rosneft, which was controlled by Sechin.

As he sat on the podium of the World Economic Forum on October 3, Raymond must have been heartened that Yukos's shares had risen in anticipation of his deal, and by the announcement that Yukos's merger with Sibneft had been finalized. The only jarring note was Khodorkovsky's agitation on the stage. He had received a telephone call from his wife informing him that police had surrounded his home and were searching a nearby school building, seizing equipment provided by Khodorkovsky. The oligarch left the hall just before Putin arrived. The president announced the Russian government's intention to reduce its interference in business and encourage foreign investment. Those close to Raymond and Tillerson did not notice either man seeming alarmed by the contrast between Khodorkovsky's plight and the president's assurances. Nor did they comment about Khodorkovsky's response to the Kremlin after the police raid. Describing the "intimidation" as "shameful to our country," he said, "What we are seeing is the most repressive and aggressive part of the bureaucracy in its dying throes. We got over communism in 1996. Now the last sickness is the absolute power of the bureaucratic system."

If Exxon ignored local Russian politics, the Europeans did not. On October 16 Pascal Lamy, the European commissioner for energy, arrived in Moscow. Gazprom supplied 40 percent of Germany's homes with energy, and was seeking a greater share of the European market for natural gas. But its slogans "Gazprom is a reliable partner" and "What is good for a strong Gazprom is good for the world," did not appeal to Europeans accustomed to Russian subservience. Lamy delivered a condescending ultimatum to ministers and officials about Russia's threat to Europe's energy market. Putin and his aides decided to draw the line.

Khodorkovsky's arrest on October 25 surprised Europeans and Americans. On that Saturday evening, both Raymond and O'Reilly possessed "term sheets" from Yukos suggesting near agreement for their respective deals. With or without Khodorkovsky, Raymond still wanted his prize. Two weeks later, he suspected the worst. Officially, Yukos's merger with Sibneft had been "suspended." Raymond and Tillerson agreed that the Yukos deal was off, although many suggested that Khodorkovsky would be released within weeks. And then came the twist. Igor Shuvalov telephoned, summoning an Exxon representative to Old Square. "The president is concerned," said Shuvalov, "that certain people in his entourage want to exploit the Yukos situation, and we would like you to continue negotiations with Khodorkovsky." "The problem is that he's in jail," replied Exxon's employee. "We will arrange it," said Shuvalov, adding that Exxon should improve its position by co-opting a Russian partner. Raymond was reluctant to participate, but since Chevron, after receiving the same message, had recruited Gazprom, ExxonMobil signed Subneftigaz, a smaller Russian company, as a partner. Tillerson delivered Exxon's tough conditions for any deal to German Gref at the Ministry of the Economy. Yukos, said Tillerson, should be given "a clean slate and there would be no tax increases." Gref raised his eyes in disbelief. The negotiations were restarted through Anton Drel, Khodorkovsky's Russian lawyer. Tillerson flew regularly from Texas to Moscow, Luton Airport north of London or Geneva to meet Golubev. He was searching for an angle: was Khodorkovsky a personal issue for Putin, and what part of Yukos was for sale? Since Chevron had departed the scene, clearly scared by the political turbulence, he suspected that with Yukos's share price still high, the Kremlin might be persuaded to sell a stake. By December he accepted the inevitable truth.

In Dallas, assured that his conversation with Putin had not triggered Khodorkovsky's arrest, Raymond offered an unruffled assessment of history. "People need to take a deep breath," he said. Risk was always a factor in the oil business. "Back in the 1910s and 1920s," he continued, "everyone went to Latin America and talked about risks. Indonesia was meant to be terribly politically risky. We went to Saudi Arabia and we went to Iran and got nationalized twice. We got thrown out of Libya. My point is that's the nature of the business." Taking the long view, the perils in Russia left him undisturbed. Exxon would never compromise by accepting lower returns on its investments. There would come a time when Russia, like all producer countries, would pay the price to attract Exxon and the other oil majors. "We bring capital, we bring technology and we bring people," he said unemotionally. Concessions were unacceptable. In the short term, Raymond's judgment was spotted by traders as a chance for serious profits.

Chapter Fifteen

The Gamble

—◂◇▸—

A FTER 30 YEARS in the business, Andy Hall was confident about his unique gamble. Since the 1973 crisis he had traded oil from troughs to peak prices and down again. He believed that 2003 was the beginning of another watershed.

The previous year, the EIA had estimated that the world enjoyed a healthy excess of production of about eight million barrels a day, confirming John Browne's assurance, "Supply will always meet demand." In 2003, the chief executives of the oil majors spoke confidently about oil continuing to trade between $25 and $30, the average price since March 1991. Hall disagreed. Prices, he was sure, were about to increase. The market suggested the opposite. Buoyed by the EIA estimate, "spot" prices on Nymex were higher than prices for oil to be delivered five years in the future. Convinced of ample supplies, investors could buy oil for delivery in 2008 at 15 percent less than the current prices. Technically, the market was in backwardation. Hall believed that the market was wrong. "It's a layup," he said. But before placing an enormous bet against the market, he immersed himself once again in the recent history of oil.

The Big Four majors, Hall believed, were losing control of supplies and prices. In 2001, Shell's economists correctly forecast that oil prices would fall below $20 by the end of the year, and with herd instinct, BP's and Exxon's economists had echoed that prediction. But careful study of oil statistics showed that overall production was static, and in some places was actually falling. Unable to increase production in the North Sea, Alaska, the USA and other mature areas that were now beyond their peak, the Western industrialized countries had increased their dependence on OPEC countries and Russia. Silently accepting their decline, each of the Big Four chief executives preferred to buy oil supplies cheaply from the producers, boast about huge profits and return capital to their shareholders, rather than drill for oil in risky areas. That was in the public domain, but a new truism, Hall noticed, had infiltrated the market. For generations, sages had preached that knowledge was power. In the new millennium, access to unlimited information was easier than ever before, but Hall's rivals lacked the skill to profitably use that benefit. Unlike him, they were ignoring glaring inconsistencies.

The West's principal intelligence agencies for oil, the EIA and the International Energy Agency (IEA) in Paris, were using "murky data" and guesswork. In 2003, the IEA, dependent on OECD forecasts of economic activity, estimated that $16 trillion would need to be invested over the next 30 years to maintain existing supplies of energy, and even more as demand for oil surged from 77 million barrels a day in 2002 to 120 million in 2030. Low oil prices, complained the IEA, were limiting investment. Within its report there was a passing reference to China's increasing demand, but ignorance about its extent. Data showed that the inventories of oil had fallen since 1998, but the consequences were ignored. That vagueness fed Hall's interest. In his opinion, there would be shortages.

Oil prices at $20 to $30 a barrel were encouraging the heads of the major oil corporations to misinterpret the market. In unison, they ignored the size of China's growing demand and the Kremlin's use of oil to increase its political influence. None anticipated that the cushion to cope with increased demand was diminishing.

Most traders and oil producers had foreseen a glut of Iraqi oil after

Saddam Hussein's defeat in 2003, but contrary to expectations, prices rose after the invasion of Iraq. OPEC countries, anxious for more revenue, had increased their production by 900,000 barrels a day (to 25.4 million barrels a day), but shortly after the Allied victory prices unexpectedly fell to $23. Based on his conviction that oil prices would permanently hover between $20 and $30, John Browne sold BP's Forties field in the North Sea to Apache, a company based in Houston, for $630 million. In Browne's opinion, the North Sea's decline, low oil prices, the eventual cost of abandonment and an oil major's overheads made Forties peripheral and offended his "more for less" philosophy. Outwardly, he blamed repeated huge tax increases levied by Gordon Brown, which would hasten the North Sea's decline. Unspoken was BP's deficit of reliable in-house expertise in the North Sea, a corrosive consequence of Browne's outsourcing. The Forties sale signaled the end of the oil majors' investment in the North Sea. Events over the following five years would challenge their assumptions. BP had estimated the field possessed reserves of 150 million barrels of oil. Apache would invest $2 billion, and five years later the amount of recoverable oil had increased to 200 million barrels; the corporation had similar results from old wells in Utah that it had bought from Chevron and Exxon. But even Apache was cautious. Advised after buying Forties that prices would fall, it sold production for the following two years at $26, a decision that would cost it about $650 million.

Andy Hall knew few details about the technical problems in the North Sea, the Gulf of Mexico or elsewhere around the globe. The glory of pinpointing a reservoir through thick salt and extracting oil from miles beneath the seabed aroused no curiosity in him. He was interested solely in the statistics of supply and demand, which would enable him to calculate whether to bet long or short. Naturally, his rivals' bets were significant. In 2003 he was amused that Morgan Stanley, after brokering the purchase of oil wells in the Gulf of Mexico on behalf of Apache for $1.5 billion, had invested in the well and then presold the oil over the following four years for around $35. Hall intended to bet that Morgan Stanley was wrong. "Be long or be wrong," some believed was his motto: his rivals thought he always betted the same way. "It's so dangerous," retorted Hall. "I'd be bankrupt if I did that." But he hated going short, which meant risking a total loss. The beauty in 2003 was the safety of going long.

John Browne, Lee Raymond and Phil Watts all believed that regardless of the turmoil in the Middle East, the supply of oil from Russia and other non-OPEC countries was safe. "Nations need not be overconcerned about energy security," Browne had said in July 2002. Relying on BP research that the world's energy consumption had grown in three consecutive years by less than 0.5 percent, Browne's confidence was bolstered by the diversity of supply, neutralizing OPEC, and by the 2.4 percent decline of consumption in America. But his reasoning, like that of Raymond and Watts, was flawed. None of their experts had accurately assessed all the consequences of President Putin's seizure of Yukos in November 2003. Nor did they link it with the decision by Gennadi Timchenko, Putin's key adviser about oil during the Saint Petersburg era, to move Gunvor, a Russian oil trader, from London to Switzerland. Some said the move was executed to exploit an easier climate, beyond regulatory reach. Others assumed that Gunvor was involved in transfer pricing by reselling Russian crude oil and refined products bought for artificially low prices. A few in Moscow felt that Timchenko was doing the opposite, paying top prices for Russian crude to get market share and earning his profits from refined products. In more prosaic language, Albert Helmig, a former vice chairman of Nymex, who had served on several of its regulatory committees, explained that in Switzerland Timchenko could benefit from "better vocabulary texture." Few in Europe and America spotted a more important development: the Kremlin was positioning itself for an anticipated increase of oil prices.

Mikhail Khodorkovsky's arrest had flummoxed the Chinese. Beijing's agreement with Yukos to build a pipeline and supply crude to Daqing, China's oil capital in the northeast, was canceled after six months by Putin. "When we faced them at first," said Yang Bojiang, a Chinese observer, "I thought they were sincere. Now, I think they are probably playing a game." Despite the arguments advanced by the Chinese to the Kremlin that Japan's economy was fading, Putin and his advisers were tempted by Tokyo's offer of additional money and opportunities. Putin did not however ignore China's urgent need for more oil. The Chinese economy, Beijing's emissaries explained, was expanding, and electricity produced from the country's coal-powered stations had become unreliable. As a substitute, households across China were buying diesel generators. To feed the increased domestic and industrial demand, China would need an extra million barrels of oil every day. Reports in New York and London about China's increased demand were common, but Andy Hall knew that no one could accurately predict the amount. Even he did not anticipate that China's demand would increase during the following year by 2.6 million barrels a day. Global demand for oil in 2004 would increase to 82.4 million barrels a day, and in 2005 would hit 84 million. "How did we miss that?" American experts would ask three years later, contemplating the consequences of a decade of underinvestment. Hall could claim to be the exception. He had been "mulling the idea for around two years." The best oil traders spent more than half their day reading—everything on the Internet and in trade journals about politics, geology, economics and commodities. Unlike traders who feared competing against the intelligence fed to BP's, Shell's and Vitol's traders, Hall trusted his own sources, and everything he read suggested it was "a perfect opportunity to place the bet."

Hubbert's Peak (2001), a book by Kenneth Deffeyes, a retired petroleum engineer, had been inspirational. Deffeyes described the prediction in 1956 by M. King Hubbert, a Shell geophysicist, that oil production in the "Lower 48," the United States excluding Alaska and Hawaii, would peak between 1966 and 1971, and would decline thereafter. Hubbert's prediction was based upon an equation using accurate production and reserve figures. The image he drew—a bell-shaped curve showing the rise and fall of oil production—was unusually accurate. In 1970, oil production in the Lower 48 did peak at nine million barrels a day, and then declined. Hall was impressed by that accuracy, and by a wave of other predictions, based upon Hubbert's assessment, published under the umbrella of the "peak oil theory."

Like-minded petroleum experts and environmentalists grasped at peak oil to support their campaign against the excessive use of fossil fuel, and in particular against what President George W. Bush would call America's "addiction to oil." Emboldened by his new fame, Hubbert had predicted in the early 1970s that production in the North Sea would peak in 1985 at 2.5 million barrels a day, and global crude production would peak in 2000 at 12.5 billion barrels a year. Neither Hall nor other "oil peakists" would be disillusioned when Hubbert proved mistaken. In 2000, the North Sea was still producing over two million barrels a day, and in 2007 global crude production reached 31 billion barrels. Like religious fundamentalists, orthodox peak oil theorists, alias "potheads" or "oil peakists," preferred to ignore any evidence that contradicted their beliefs.

Hubbert's gospel had been embraced by Colin Campbell, a retired geologist who had worked for most of the major oil companies, including Amoco, Fina, BP, Texaco, Shell, Chevron and Exxon. "The peak of oil discovery," wrote Campbell in June 1996, "was passed in the 1960s, and the world started using more than was found in new fields in 1981. The gap between discovery and production has widened since. Many countries, including some important producers, have already passed their peak, suggesting that the world peak of production is now imminent." He was convinced that the reserves reported by the oil majors and producers were inaccurate, and concluded, "It is now evident that the world faces the dawn of the Second Half of the Age of Oil, when this critical commodity, which plays such a fundamental part in the modern economy, heads into decline due to natural depletion...Petroleum Man will be virtually extinct this century, and *Homo sapiens* faces a major challenge in adapting to his loss. Peak oil is by all means an important subject."

In Campbell's scenario, oil production would peak in 2000 with 24 billion barrels and the ultimate recovery of 1,750 billion barrels. In 1998, he reassessed, or in his words "updated," his forecasts. Once the oil-price slump had passed, world production had risen to 65.74 million barrels a day, and "proven" reserves were assessed by the IEA and EIA to be about 1,830 billion barrels, sufficient for the 21st century. Campbell challenged that estimate of "proven" reserves, pronouncing that oil production was flat, and would begin to decline in 2010. His argument was enhanced by an allegedly irrefutable "fact": no major oil reserves had been discovered since the 1960s, and since 1985 consumption had been higher than all the combined new discoveries. Every area in the world, he claimed, including Greenland and Tibet, had been surveyed, and no more great discoveries could be anticipated. Therefore, he wrote, since the depleted reserves were not being replaced, "we conclude that the decline will begin before 2010." In 2000, Campbell wrote: "The myth of spare capacity is setting the stage for another oil shock." In his alarming scenario, he asserted that at the current rate of consumption, the new deepwater discoveries of oil would be depleted in just four years, not least because Norway's reserves were not 49 billion barrels, but 28 billion. Firmly, he predicted that by 2009 the Middle East would be "close to its depletion," and that oil supplies would peak between 2005 and 2010.

By 2003, both Hubbert's and Campbell's methodologies were proved to be flawed. Hubbert's original forecast of production in America was based upon data compiled over decades, but new technology and prices challenged its accuracy. While America's "proven" reserves in 1991 were 24.7 billion barrels, the total reserves, including tar oil, oil shale and other condensates, were estimated in 2003 to be 204 billion barrels, and some experts calculated that new technology could increase the ultimate recovery to between 263 billion and 368 billion barrels; and that figure did not include an estimated 180 billion barrels of oil in the Canadian tar sands. Price and technology would determine the amounts: producing a barrel of oil from bitumen cost about $35. "Let me know when we reach peak technology, then we can talk about peak oil," Paul Siegele, Chevron's vice president in charge of strategic planning, would say.

Studies in 2005 would show that even Hubbert's original estimate of America's peak was mistaken. Between 1970 and 2005 he had underestimated production in the Lower 48 by 15 billion barrels. Failing to take technological advances into account, his mistake in the North Sea was more profound. Production did not peak at 2.5 million barrels a day in 1985, but rose to 2.7 million in 1999, and only fell to 2.09 million in 2003. Hubbert's—and after his death in 1989, Campbell's—calculations had been based on inaccurate estimates of the reserves reported by the oil majors and the national oil companies. Despite Hubbert's tarnished reputation and the error of his own previous predictions, Campbell issued a revised forecast in 2003 that the peak would be reached between 2010 and 2015. Maximum production, he said, would be 87 million barrels a day, including 7.2 million barrels from the four major deepwater areas. He disputed the forecast issued by Guy Caruso, the head of the EIA, that the producers would be able to supply 116 million barrels a day in 2020. The gap between Campbell's and Caruso's estimates was enormous. Campbell believed in 2003 that the world's maximum annual production would peak in 2004 at 28 billion barrels, while the EIA reckoned it would eventually rise to 43 billion barrels. The EIA reported that the oil price would remain at $20, while Campbell was certain there would be rises and "shocks."

The huge disparity sprang from differing interpretations of statistics. Campbell relied on information provided by OPEC and non-OPEC producers who, for political and financial reasons since their expulsion of the Seven Sisters, had deliberately published inaccurate information and underestimated reserves in order to resist pressure for extra production. Russia's production and reserves were unreliable on both counts. Similarly, forecasts of supply were always mistaken, because no one could accurately foresee the influence of prices. Campbell's insistence that no new major discoveries of oil had been announced since the 1960s was also questionable. Although the discovery of new oilfields had declined since 24 were pinpointed during the 1970s (there had been 46 during the previous 20 years), 26 new supergiants had been found in the last 20 years, and four giant fields since

2000—in Cantarell, off Mexico; Tengiz and Kashagan in Kazakh-stan; and Azadegan in Iran. While the bald statistic that 20 percent of oil production came from just 14 fields appeared alarming, the rate of replacement, said Campbell's critics, was underestimated.

Using contradictory data, the two sides clashed over a core issue. Peak oil's advocates required a "final" estimate of the ultimately recoverable reserves. Campbell estimated that the world had consumed about a trillion barrels of oil, and that a further trillion barrels remained, which at current levels of consumption would last about 34 years. The EIA, BP, and the optimists calculated that there were at least three trillion barrels still to be recovered. ExxonMobil disagreed. Based on its "global geoscience toolkit" and an online database with over 100 terabytes of data, five times more than the entire Library of Congress, Exxon calculated that 15 trillion barrels of energy equivalent remained to be extracted. The combination of Exxon's exploration costs falling from $2.75 to find a barrel of energy in the 1980s to 44 cents in 2004, and the power of its computer comparisons between the tectonic history of the earth's plates and the rock distribution in specific locations, magnified the disagreement between it and Campbell about the amount of oil that could be recovered in the Arctic and other remote areas.

Drilling in deep water was akin to putting a man on the moon. In the 1960s and 1970s, Hubbert could not imagine successful drilling six miles beneath the sea's surface. By the end of 2003, over 1,800 wells had been drilled in deep water in 70 areas, with 990 in the Gulf of Mexico, 383 in Brazil, 111 in Angola and 84 off Nigeria. By then, 47 billion barrels of oil had been found, but most experts believed that at least 100 billion, and possibly 700 billion, barrels of undiscovered oil remained in the Gulf of Mexico. Peak oil, according to Hubbert and Campbell, meant that after 50 percent of the oil had been produced from a major discovery, any further discoveries in the same area would be small and immaterial. The statistics appeared to support their argument. The number of offshore wells in Brazil had declined since 1987, and had almost halved in the Gulf of Mexico, from 109 in 2001 to 56 in 2003. The contrary opinion was equally plausible. Low oil prices during those years deterred expensive exploration, but once prices improved, new possibilities arose. Undiscovered reserves were still likely to be found in the vast, uncharted deepwater areas. Explorers using 3D and 4D seismic, horizontal drills, multilateral wells and smart infill drilling were likely to increase production from mature wells and revive dry ones to extract over 50 percent of the oil, as BP had accomplished at Thunder Horse. The world consumed about 30 billion barrels every year. Contrary to Campbell's scenario, the problem was not how much oil was in the ground, but how much the producers would spend to extract it.

Guy Caruso, the head of the EIA, was the man in the middle of the two sides' increasingly sterile arguments. Appointed by the George W. Bush administration in 2002 after serving for 12 years as an energy analyst at the CIA, Caruso had won star status by correctly forecasting the 1973 crisis, but he had also been involved in the mistaken CIA forecast in 1977 that Russia would become a net oil importer. Twenty-six years later, he acknowledged his misunderstanding, and disparaged the pessimism among the peak oil advocates, who relied on conservative, erratic and inconsistent data when so much depended upon price and technology. Russia had an abundance of oil—an estimated 250 to 300 billion barrels of reserves, of which 100 to 150 billion were recoverable—but extraction was fraught. The Russian slump in the early 1990s had been reversed with Western technology. The country contributed between 10 and 15 percent of the world's production. Caruso predicted in 2003 that the world could rely on non-OPEC producers increasing their output from 40 million barrels a day to 54 million by 2025. Only production data, he said, was truly reliable. Annual production in 2021, predicted the EIA, would be between 48.5 billion and 78 billion barrels. Caruso's forecast depended on the national oil producers expanding production; his optimism was shared by Peter Davies, BP's chief economist. Pessimism in the industry, said Davies, an authoritative spokesman against peak oil, stemmed from weak demand causing less oil to be produced. Scientifically, he felt, there was no proof that the world was approaching "peak production."

Caruso's and Davies's optimism was rejected by Peter Wells, a former BP engineer and peak oil believer, as "highly unlikely." While OPEC producers, Wells argued, had failed to invest, and their spare capacity had fallen from 10 million barrels a day in 1987 to 1.5 million in 2004, the non-OPEC producers had also failed to replace depleted reserves. The result, he predicted, would be an oil-supply crisis that "can be expected to be acute after 2005," causing actual shortages in 2007. Wells was supported by Jean H. Laherrère, a colleague of Campbell's, who projected hitting the peak in 2010 at 33 billion barrels a year; by the IEA's experts, financed by the OECD, who dated the peak as starting in 2012; and by Chris Skrebowski, the editor of *Petroleum Review*. Skrebowski reckoned that Venezuela's oilfields were depleted, only clever tricks would recover oil from the fast-declining North Sea, and Russia could produce no new reserves, especially offshore. He predicted the peak starting in 2010 or 2011. That group questioned whether Caruso was clever enough to understand their arguments, and whether, as a Republican appointee, he withheld arguments that conflicted with the administration's policies.

Besides the irreconcilable optimists, wishful thinkers, pessimists and alarmists who had provoked Hall to "get thinking," he also noted the indisputable upheaval. Since 2001 Venezuela's President Hugo Chávez had been gradually expelling the oil majors, initially by increasing their royalties to 30 percent; in Mexico, Vicente Fox's stumbling failure to reform Pemex suggested a fall in production by another major supplier; China's demands were increasing beyond anyone's guess; the global economy was in the early stage of a boom; and in Washington there was disarray. The election of George W. Bush had been interpreted as an advantage for Big Oil, but like his predecessors, Bush had failed to formulate a coherent energy policy. Paralyzed and ignorant, the White House was suspicious of the CIA's annual estimates of the state of the oil industry, but Bush rarely invited the presidents of the Big Four oil companies to the White House to hear their opinions. He could not change the laws governing the building of refineries although the newest had been built 30 years earlier, and he was unable to persuade Congress to permit drilling for oil off the coasts of Florida and California, or in Alaska's wilderness. Fearful that oil supplies were peaking, energy secretaries spoke in terrified tones about their own weakness, dithered about strategy, toyed with dependence on Russia, and instinctively accused the producers of blackmail. Having studied all those patterns, Hall was sure that oil would eventually be priced far above $20. The uncertainty was whether his superiors would support his gamble against the whole industry. Fortuitously, he was asked by Bob Rubin, President Clinton's former treasury secretary who had been appointed as a board member of Citigroup, to offer a presentation on oil to the bank.

Hall owed his continued employment in the Citigroup to Rubin. In 1998, Salomon-Phibro had been bought by the Travelers Group, which later that year merged with Citicorp to form the world's biggest bank. Based by then in a squat office building in Westport, Connecticut, an hour by train from Manhattan, Hall could have been in Siberia as far as Sandy Weill, his ultimate boss and the joint chairman of Citigroup, was concerned. Indeed, if Weill had had his way, Hall's operation would have been closed down after Citicorp bought Salomon Brothers. He was unimpressed by its lackluster profits even in the good times. "Cents rather than dollars," he said, unconvinced about a bank actively trading commodities, and especially oil. Traders, in Weill's opinion, were an aberration. Jamie Dimon, the president of Citi, had until his departure in 1998 agreed. Initially, Weill limited Phibro's exposure to risk; then, still dissatisfied, he ordered Hall to liquidate the operation. Hall had found a buyer at the time that Bob Rubin joined Citigroup. Unlike Weill, Rubin had been involved in Goldman Sachs's development of J. Aron, and he understood the profitable advantages of trading commodities. Rubin also acknowledged Hall's reputation. He had been around for longer than others, and although he lacked the conviction and the finance to copy Morgan Stanley's and Goldman Sachs's operations, in recent years he had earned profits as a trader. Rubin ordered Hall to stop the sale of Phibro and address a group of Citi executives.

Hall's theme was the world running out of oil. "I've dug into the facts," he told his audience, "and saw that supply will not meet demand." He intended, he said, to bet against the oil majors and his rival traders. "It's a three- to four-year bet," he explained. "It's not a question of 'if' but 'when.'" Oil prices, he thought, would head toward $100 a barrel. "Everyone else," he said, "is behind the curve." That included Morgan Stanley, which, he explained, believed that the network of tankers, storage and billions of dollars of hedged oil contracts provided invaluable intelligence about the future. "Advanced intelligence is bogus," said Hall. "Customers don't tell the brokers their intentions. Those close to oil can lose more than the consumers. The flow of funds is more important than customers." Morgan Stanley, he could have added, had risked much more than Phibro. To earn $2 billion profits, the bank owned assets worth $25 billion. The risk involved in that 10 percent return included being left with the liabilities if the investors crashed. He concluded with the critical challenge: "If my thesis is right, the bet will perform the same as buying oil at 'spot' but with less risk. So if I buy at $30 and the price goes to $120, I'll earn 400 percent profit. But if I wait until the price hits $120 and then rises to $200, my profit will not be as interesting."

Citi's chief economist disagreed. "As prices rise," he argued, "so will supply to meet demand." Shortages, he explained, could not last forever, because rising prices would drive investment in high-cost production, and demand would eventually decrease. He echoed Peter Davies in believing that the pessimists were guided by weak demand, which meant that less oil was being produced, rather than that peak production was imminent. Hall's reply was scathing. "It's nothing to do with price. It's geology. The world's running out of oil." Economics, he argued, would be turned on its head. "No one will want the oil when the prices get too high."

Hall's argument, carefully based on Hubbert and his disciples, won Rubin's support. "No one paid any attention to the economist," said Hall on his return to Connecticut. "I hate going to New York." Hall had mastered the commodity and the market, read the literature, spoken to the right people, and was ready to stake about $1 billion in margin calls. He would buy oil two to five years in advance. Since the market was in backwardation, oil was cheaper in the future than on the instant "spot" market. Holding 90 percent of Phibro's "book," or positions, he wanted to bet against the world, particularly the banks, rather than the oil majors.

Speculating about oil prices had been transformed by the major players. The growing popularity of "swaps" and over-the-counter (OTC) contracts, developed during the 1990s to reduce and spread the risk assumed by energy providers, merchants and traders, had become frustrated by the limitations of trading and pricing through report-ers and the existing exchanges. Brokers crying out on the floor of the Nymex exchange in New York represented traditional traders buying oil for immediate delivery and speculating about future prices. After 1983, the "spot" trade had extended into "swaps" in the OTC market. "Swaps" were perfect for two traders to agree a deal specifying a range of variables including the price, location and specification of products, which was unavailable on Nymex. Not only was the agreement cus-tomized to suit the two traders, but there was no commitment actually to deliver the oil or natural gas, merely an agreement to pay the differ-ence in price on the day the swap expired. In effect, "swaps" were trades in price, not in oil itself, and were used by both speculators and trad-ers to protect or hedge their positions, reducing or spreading the risk. The final price would be calculated against Nymex's. In 1990, traders demanded that regulatory authorities acknowledge that the trade was legitimate and their agreements were enforceable. Responding to those demands for unhindered growth, the CFTC agreed in 1991 that com-plex derivatives could be bought and sold using borrowed money or on margin calls. Under pressure from the traders and bankers, in 1993 Wendy Gramm, the head of the CFTC, announced that the regulator would not exercise its authority over the "spot" trade or over swaps or "forwards" on the OTC market, the equivalent of Nymex's "futures."

As the trade expanded, the OTC traders speculated on the price of oil, natural gas and electricity free of any controls, and simultane-ously expressed dissatisfaction with Nymex. "Nymex has too many shenanigans on the floor," complained one of the leading bankers pioneering the OTC trade. "No one's policing the brokers. They are greedy and rigging the market." The traders believed Nymex's brokers were deceiving their clients about the prices on the floor for their own profit. Equally unacceptable was the iron grip imposed by Enron. Electronic trading of energy swaps was only possible through Enron-on-line, a facility that allowed Enron's own traders to see every bid and deal made by rival traders and major energy corporations, and to adjust their own prices accordingly. The expanding "voice-brokering" community trading swaps acknowledged its own inefficiency. Trading by telephone calls and telexes was time-consuming, and Enron-on-line proved the advantage of electronic trading.

John Shapiro of Morgan Stanley wanted a neutral exchange and a genuine market to trade customized "swaps" electronically 24 hours a day, a facility Nymex's directors refused to provide. After securing the agreement of Tim O'Neill at Goldman Sachs, Morgan Stanley sought the agreement of BP and Shell to establish the Intercontinental Exchange (ICE), an electronic exchange or bazaar to be used by major oil producers and traders on their own terms. The oil companies' response was positive. Like the banks, they scorned Nymex's floor traders as compromised, and Platts reporters as "kids two years out of college who can be easily outwitted by those knowing the business and can't be relied on to get today's price." Nine other major market-makers agreed in 2000 to share ownership of ICE, trading "swaps" using a computer in Atlanta. Their deals, the founders agreed, would be settled by using the average of Nymex's prices on the day. Motivated by their desire to be free of Nymex, the founders successfully argued that since only large corporations were participating, they did not require regulation or government protection from misconduct. Subscribers trading through ICE would be able to avoid US regulators.

Jeff Sprecher, ICE's first chief executive, was keen to establish the new market. In 2001 he bought the limping International Petroleum Exchange in London, which traded futures in Brent and natural gas delivered in Europe, and merged it with ICE. ICE's subscribers were now further removed from American regulators. By trading WTI in London, the newly named "ICE Futures," an exclusively electronic exchange, was nominally supervised by the British Financial Services Authority, an understaffed, underfunded and inept regulator. That encouraged aspiring speculators to trade oil. Rapidly, ICE's footprint spread across the world, attracting over 1,000 OTC spot traders and 440 futures traders using 9,300 active screens. The explosion was noted by the regulators in Washington and London, and was ignored. Asian and the Middle Eastern traders were using ICE by hedging oil derivatives (CFDs), although the identities and honesty of those influencing prices beyond the regulators' knowledge was unknown. The ignorance was compounded by the exceptional complications introduced into oil trading by investment fund managers, bankers and speculators.

The new complexities did not deter Andy Hall. As he began taking long positions, he believed that prices could only go in one direction if the market became the epicenter of craziness. Over several months he bought call options on Nymex and "swaps" from the banks, not least from Goldman Sachs, which owned a stash of Mexican oil. "We're easing our way in," he told his staff, "to prevent the market moving against us. At the end of several months, he had invested the first $1 billion for the delivery of oil over the next three to five years.

Hall's rivals were unaware of his plans. "Andy," said one of his awe-struck competitors, "doesn't speak to people. He distills the salient points out of the essence from reading, which with hindsight is brilliantly obvious, but hidden by specious and wrong 'facts.'" No one within that small community doubted that Hall embodied "the perfect combination of immense intellect and off-the-scale commercial brilliance." None of his admirers was aware that the basis of Hall's mammoth bet was an old, controversial and, to some, discredited theory.

The slow increase in the demand for oil reinforced Hall's conviction that supplies were peaking and prices would rise. The number of identifiable speculators was limited. The principal players on ICE, especially the banks—Morgan Stanley, Goldman Sachs, Lehman's, J.P. Morgan and Barclays Capital—rightly resisted being tarnished as price fixers or a cartel. But beyond New York, in the Middle East, Russia and Asia, were anonymous traders blessed with privileged information and anxious to ride a rising market, not least by occasionally introducing erroneous facts into the media to enhance their profits.

Their paper chase was noticed by some in Congress. Oil prices at the end of 2002 had unexpectedly risen from $18 to $40 a barrel. On February 12, 2003, despite nervousness about war, terrorism in Saudi Arabia and environmental extremism, Anne-Louise Hittle, a senior director of CERA, predicted that the price was likely to fall. She believed OPEC would resume setting prices, and huge supplies of oil would flow from Iraq. Staff employed by the Congressional Permanent Subcommittee on Investigations were unconvinced. "Gas prices: how are they really set?" was the committee's question. The world, the investigators puzzled, was not short of oil, yet storage tanks were at record lows, and prices were at a 12-year high. The committee collected evidence that traders in London and New York were manipulating crude prices. But the situation remained inscrutable. The impenetrability of ICE, the investigators admitted, made supervision impossible. "There are defects in the market," the senior investigator noted, but could offer no solution to the puzzle.

The beneficiary would be Andy Hall. The confirmation of his judgment was the news in January 2004 that Shell had grossly overstated its oil reserves. Reading reports about the exaggeration, Hall smiled: "The fiasco confirms my suspicion that the IOCs can't replace their reserves."

Chapter Sixteen

The Downfall

———◄○►———

IN 2000, SHELL'S executives were surprised when the Securities and Exchange Commission (SEC), the American regulator, published on its website new guidelines to the oil companies about reporting the assessment and the value of their reserves. Until then, Shell and all the oil companies had defined their reserves according to the SEC's Rule 4-10, which required "reasonable certainty" that the oil would be produced. There had been no prior discussion about the SEC making any changes to rules that had originally been commissioned by Congress in October 1973, after the Arab–Israeli war. Mindful of America's vulnerability to oil scarcity, President Nixon had launched "Project Independence" to plan for energy security. At a time when the US imported 30 percent of its oil, it made strategic sense to know just how much oil America could produce domestically. The SEC was the agency that applied rules, finally approved in 1978, that assessed how much oil existed at the bottom of vertical wells in Texas, Louisiana and Oklahoma, the principal locations of America's reserves. The definitions were written to protect competing mineral rights within 40-acre sites, and ignored seismic data. Since then, judging reserves had become more complex than had been conceived in 1973. Experts could give differing but justifiable estimates dependent on their interpretation of

3D seismic images, the extent of horizontal drilling and the depth of the wells beneath oceans. Even the definition of oil had changed to include Canada's tar sands, Venezuela's oil in the Orinoco and liquids extracted from coal. The SEC was also operating in a different environment. By 2001, America's share of the world's proven reserves had fallen from over 65 percent to 17 percent, and the country was importing 60 percent of its consumption. Over 80 percent of the world's oil and natural gas was controlled by non-American national producers. Just 10 percent of the world's reserves and 40 percent of the production of oil and gas, deposited mostly outside the US, was controlled by the major oil companies. Although the majority of these were non-American companies operating in foreign countries, they were quoted on the New York Stock Exchange and governed by the SEC rules.

Walter van de Vijver, the director of Shell's exploration and production, sought to understand the background of the revised rules. Ron Winfrey and James Murphy, both former petroleum engineers based at the SEC's office in Fort Worth, Texas, had noticed before 2000 that during the previous 20 years all the oil companies had practically doubled their reserves. They questioned whether the increase represented real growth, or was merely a maneuver to maintain the companies' share prices. Only one fifth of the oil Shell was selling was drawn from its reserves; the remainder was resold after purchase for the full market price. Shareholders, Winfrey and Murphy believed, were being misled. In an attempt to remove "confusion" from the companies' filings because of improved technology and their operations in uncertain environments like Kazakhstan, the two officials announced that since universal classification had become "difficult, if not impossible," the rules were "in urgent need of modernisation." Rather than the "reasonable certainty" defined in Rule 4-10, the SEC would in future restrict the oil companies to announcing reserves only when there was "absolute certainty" of production.

The SEC believed there were good reasons for doubting some bookings. Shell had booked reserves in 1999 after drilling only two exploration wells at the Ormen Lange field off Norway. While no one doubted that at least 109 million barrels of oil and gas could eventually be produced there, Shell's partners—BP, Exxon and Statoil—had not booked the reserves, because complicated technical problems remained unresolved. Studies showed that 8,000 years earlier a giant slide of sediment, equivalent to the size of Iceland, had been dumped onto the seabed. Settlers on Norway's coast had been drowned by colossal waves, which had reached as far as England. To check that another disaster would not be triggered by drilling through the sediment, the consortium had commissioned a $1 billion study that would not be completed before mid-2003. After the safety issues had been dealt with, the partners would still need to lay the longest pipeline in the world, 747 miles across the seabed to England, and would finally have to confirm that the gas could be marketed. Only then, five years after Shell, would the other companies, abiding by the SEC's rules, book the reserves. On that basis, Shell's booking in 1999 was premature. Winfrey and Murphy noted that the oil majors had also booked reserves in Kashagan, an unstable region, where the start of production had been repeatedly delayed from 2005 until 2014. To officials in Texas, the notion of relying on the verbal promise of an African or Kazakh dictator to formally book reserves influencing the price of shares in New York was risible.

The two officials wrote to the major oil companies stipulating that reserves could only be booked under the new conditions. They could no longer be booked in anticipation of the renewal of a license to produce oil in the future, but only if the company was in possession of completed legal documents. The sale of natural gas had to be contractually watertight, and signed government approvals in "frontier areas" would be required before any reserves could be booked. The SEC's new regulations challenged the oil companies to radically reduce their values. "Prove it to us or take it out," Murphy would tell a meeting of petroleum engineers. The industry was alarmed by the new rules' apparent arbitrariness. No one had anticipated back in 1973 that the SEC would become a global regulator. Too many officials in America's regulatory agencies, fed by distrust and even envy of big business, had become a dangerous cancer. Lawyers at ExxonMobil, BP, Shell and other companies protested. It was folly, they argued, to demand that oil companies prematurely sign a contract for the sale of natural gas from the North Sea, because prices would inevitably rise, and the eventual sale of the gas in Britain was certain. Similarly, valuing their oil and natural gas reserves according to the price on the last day of the year was distorting. ExxonMobil's booking and valuing of its reserves, protested the company's lawyers, had always been based upon the "conservative, disciplined" management of the entire business. Politely, the SEC officials were warned that their interference with Exxon's "certainty and consistency" was unlawful. Rule 4-10 could only be changed after formal consultation. Winfrey and Murphy were unmoved. Discomfiting Big Oil, they knew, would attract public applause, as the oil companies understood only too well. Nevertheless, Winfrey and Murphy were not immune to persuasion. By deft legal arguments, stubborn lawyers employed by ExxonMobil, BP, Chevron and Total either accommodated the SEC's new strictures or successfully opposed them. Only Walter van de Vijver remained alarmed.

Van de Vijver had become head of Shell's exploration and production after Mark Moody-Stuart had reorganized the 35 upstream businesses into five regional centers. The five new technology managers had been asked to reexamine their reserves. Their reports unnerved van de Vijver. "I have a gut feeling there are problems in Oman, Brunei and Nigeria," he told his deputy. "Production is not rising and I want to know why." In Oman, he knew, Shell's relations with the government had been fractured after Philip Watts's optimistic promises to produce more oil had not materialized, and production had actually fallen. Irritated, Sultan Qaboos bin Said, the ruler, had invited other oil companies to compete with Shell. In Brunei, the government was empowered to declare its own reserves regardless of accuracy. The situation at the Gorgon field in Western Australia was ominous. In October 2000, Shell's local expert had recommended that the reserve be debooked, but Anton Bylondrecht, a reservoir engineer appointed in 1999 as the group reserves auditor, ignored the suggestion after Shell's group reserves coordinator declared that debooking was "too big to swallow." At Sakhalin the reserves had been booked although the development was delayed by problems; and in Nigeria, to avoid blame for any wrongdoing, Shell had sought to satisfy successive bankrupt dictators' demands and overbook the reserves, although the regimes were unwilling to pay for the development of the oilfields. Nigeria's anticipated reserves, van de Vijver suspected, could not be exploited, although the company expected that new discoveries would compensate for any losses. But even that was uncertain. The Shell team in Nigeria was constantly asking for money to finance development, and, considering that the company should be self-financed from profits, van de Vijver was puzzled. He asked questions but received either unsatisfactory answers or assurances that "Everything's fine." To some extent he was reaping the whirlwind of the changes that had encouraged engineers to refuse uncomfortable jobs in the oilfields, preferring to PowerPoint their engineering from Houston, London or The Hague. Frustrated, he dug deeper, and found only ignorance or embarrassment. He suspected the worst. Under pressure to maintain targets, he believed that Watts had exaggerated the reserves in order to enhance his bid for the chairmanship. Van de Vijver feared that his own promotion would be jeopardized by any shortfall. His salvation would be to discover the truth. Ordering a "top-down investigation," he dispatched Frank "the Tank" Coopman, Shell's chief financial controller, to Nigeria.

On February 11, 2002, van de Vijver distributed a Note for Information to the committee of managing directors, the most senior group of executives, warning that Shell's reserves were possibly overstated by 2.3 billion barrels of oil equivalent (boe). According to the SEC's latest guidelines, he explained, one billion were overbooked and a further 1.3 billion in Oman, Abu Dhabi and Nigeria were dubious because there was no "reasonable certainty" that the licenses would be renewed. Unmentioned was Anton Bylondrecht's assessment. He would subsequently admit that some of Shell's bookings in January 2002 had been "too lenient," and that greater honesty on his part "would probably have cost me my job." Van de Vijver could rely on the fact that neither Price Waterhouse nor KPMG, the company's auditors, had expressed any doubts about Bylondrecht's reports. Both firms would assert that certifying oil reserves was not their task, yet Shell's executives would assume that both had passed the reserves statements. In any event, the implications of van de Vijver's report should have been clearly understood by all Shell's managing directors. They all knew that Shell applied its own interpretation of the rules. Some would say that estimating the reserves was a "black art" that had plunged Texaco into costly controversy for "stretching," but Shell's directors basked in the public perception that their company could be trusted. All the managers simply assumed that the reserve numbers were immune to challenge by the SEC. None questioned van de Vijver's reaction to the SEC's changes, or queried the fact that Anton Bylondrecht, now part-time and officially retired, was the company's sole auditor. None knew that, by contrast, Exxon employed eight full-time reservoir engineers, advised by a former SEC lawyer, to manage its bookings. Hindered by isolation and fragmentation, the directors assumed that Watts would resolve any problem and ignored van de Vijver's warnings.

Over the following months, the problems festered. Van de Vijver received hazy reports about the reserves, while his relationship with Watts deteriorated. In the chairman's opinion, van de Vijver lacked the ability to energetically manage Shell's global portfolio. For his part, van de Vijver repeatedly complained that his prospective inheritance was in a far worse state than he had imagined, and considerably worse than was portrayed by Watts to the board of directors. Watts, he felt, had been "aggressive" and "premature" in his bookings between 1997 and 2001. Believing that he was being misled about the reserves, van de Vijver became frustrated by Watts's directive that the SEC's new guidelines were van de Vijver's problem, but that legitimate juggling could resolve the issue. Van de Vijver decided to mention his unease again to the conference of managing directors. Besides his interest in protecting his own reputation, he was prompted by an instinctive suspicion of Watts. The Dutchman disliked the replacement of Shell's technical vigor by entrepreneurial priorities. He also disliked Watts's disdain for the internal politics in The Hague's headquarters, and blamed the

British for giving the country managers an incentive to exaggerate their reserves. By failing to obey the SEC's new guidelines, he believed, Shell was losing the plot.

Van de Vijver was due to make his second presentation to the managing directors on July 22, 2002. Seven weeks before then, Watts sent him a memo urging that while "leaving no stone unturned," he should ensure that, if possible, the existing reserves were maintained. In his assertive manner, Watts placed the onus on van de Vijver: "It's Walter's plan," he said, fixing the responsibility. He would later appear to contradict his own memo by claiming that he had commissioned van de Vijver's review six months previously. "Do we need to debook?" he had asked van de Vijver. The reply, according to Watts, was "No." Van de Vijver had obliged his chairman, who "didn't want to hear the bad news." Conveniently, van de Vijver agreed to play for time at the July presentation by suggesting that any problems would be resolved. He could do little else, because the facts remained elusive. None of the managing directors urged him to hasten his review. All assumed that reserves were "Walter's problem," and the solution was simple.

During the summer of 2002, van de Vijver became noticeably agitated. On September 2, he wrote a memo blaming "aggressive booking in 1997–2000" for giving outsiders a false impression of Shell's reserves, which according to the SEC's new guidelines, he concluded, had been overstated by at least 20 percent. "We are struggling on all key criteria," he wrote. While Shell's credibility remained "high," there were "dilemmas," and "the market can only be 'fooled' if . . . positive trends" could be shown. Out of Watts's hearing, he described the "inflation" of the statistics to several senior Dutch directors, and protected himself by recording in a note marked "strictly confidential" that the information about the problem had been "obviously 'transmitted' in a careful fashion so as not to compromise/undermine the previous leadership." In other words, the message was obscured. "The severity and magnitude of the EP [Exploration and Production] legacy issues," he added, "may therefore not have been fully appreciated."

Finally Watts realized there was a problem, but he blamed van de

Vijver. The Dutchman, he was certain, was diverting attention from his own unsatisfactory performance by repeating his doubts about the reserves—a view with which van de Vijver strongly disagreed. Nevertheless, to placate van de Vijver, Watts invited him for dinner. The atmosphere was not harmonious. The chairman did not conceal his irritation that van de Vijver was disparaging his inheritance and failing to control his department. He also scorned the aspersions about his honesty, and the idea that he had personally booked the reserves. They disagreed over whether to limit any announcement to merely noncompliance, or to disclose the true extent of the overbooking. Remarkably, the decision could not be made by Watts. Although he was the company's chairman, he was employed by Shell, and was unable to issue direct orders to van de Vijver, who was employed by Royal Dutch. Frustration, mutual suspicion and allegations of deception meant that the meal ended acrimoniously. Watts blamed the argument on van de Vijver's depression over his father's recent death and problems with the funeral. Van de Vijver resented that suggestion. On October 22, 2002, he wrote to Watts, "I must admit that I have become sick and tired about arguing the hard facts and also cannot perform miracles given where we are today." He continued, "If I was interpreting the disclosure requirements literally... we would have a real problem." In effect, van de Vijver accused Watts of a cover-up toward the SEC and Shell's directors, especially about the Nigerian reserves, which were said to be overstated by 220 million barrels—which outraged Watts. With conflicting reports about the reserves' accuracy, relations between the two men over the following months became poisonous.

On August 25, 2003, van de Vijver wrote another memorandum implicitly blaming Watts for Shell's "shrinking opportunity" to find new oil, which had been "exacerbated by... too aggressive reserves bookings in the past." Watts ignored the allegations. None of those whom he respected doubted his integrity. Just because he had adopted less conservative guidelines to reporting the reserves did not amount to fiddling the books. "Walter," he commented, "is playing his own game." Disunited and squabbling, Watts and van de Vijver failed, unlike the rival oil companies, to vigorously resist the SEC's new guidelines. Instead, Watts mismanaged Shell's official response. The company's audit of its reserves was repeatedly delayed, and in April 2003, undirected by van de Vijver, the formal reply to the SEC composed by Cravath, Swain, Shell's lawyers in New York, was tepid. "We think your guidelines go counter to the spirit of 4-10," Cravath's wrote, and nothing more. Shell was meekly unwilling to challenge the American government, and its protests to the SEC officials, unlike those from other oil companies, failed to elicit any concessions. Throughout those weeks, Watts behaved in a self-destructive assertive manner, while van de Vijver, having failed to meet and micromanage the problem with Anton Bylondrecht, the part-time reserves auditor, relied on Frank Coopman, Shell's financial controller.

Coopman had been identified as a troublemaker by the company's British executives. Mischievously, he ignored Judy Boynton, the newly appointed chief financial officer. The American Boynton had been nominated by John Hofmeister, Shell's human resources director, in an effort to break the incestuous culture of internal appointments, an initiative that was unwelcome to many of the Dutch. Van de Vijver showed his disrespect for Boynton by failing to mention the reserves issue to her. Coopman showed the same insubordination. "He's poison in the soup," Hofmeister told Boynton, but, reluctant to exert her power, she failed to remove him. Her inaction encouraged Coopman to assume measurable control of the reserves investigation, leaving Boynton unaware of the problem.

In May 2003, Coopman delivered an unpleasant report to van de Vijver about Shell's operations in Nigeria. Pipelines and equipment from wellheads were being stolen on an industrial scale. Repairs and replacements costing vast sums were not covered by production profits. Shell's operation in Nigeria was running at a loss, prompting demands for money from Holland. Van de Vijver dispatched Coopman back to Nigeria to compose a definitive report. Coopman's new discoveries were startling. Shell's reserves in Nigeria, he revealed, were massively overstated. Instead of the original assessment that the overstatement was 220 million barrels, the true amount was 1.2 billion. On his return to Holland, Coopman did not reveal his findings to van de Vijver, but instead consulted Michel Braundjes, a company lawyer. Alarmed, Braundjes sent Coopman's report to Cravath's in New York.

Unaware that Shell's most sensitive issue had been disclosed to outsiders, van de Vijver was disturbed by Coopman's perfunctory report in September 2003 that the reserves in Nigeria had been "overstated" and were "unsatisfactory," especially after the SEC had further tightened the rules in the wake of the Enron fraud. Even van de Vijver was still unsure about the precise facts. In that atmosphere of discontent, Watts was due to write his annual review of van de Vijver's performance. With production failing in Nigeria and the Gulf of Mexico, and the new estimate of costs at Sakhalin expected to exceed the $9.6 billion budget, Watts's report was outspokenly critical. Van de Vijver regarded the written reprimand as a step too far. The business plan was, he believed, flawed because of Watts's exaggerated assertions. No extra production existed to replace the depleted reserves, and he resented taking the blame for that humiliation. "I am becoming sick and tired," he wrote to Watts on November 9, "about lying about the extent of our reserves issues and the downward revisions that need to be done because of far too aggressive-optimistic bookings." Shell's directors, he added, were concerned that the market would continue to be "fooled." In reply, Watts expressed his surprise at van de Vijver's allegation of "lying." The problem, Watts convinced himself, had been addressed by the management conference in mid-2003, and everyone had been reassured. Van de Vijver's written allegation, he said, "blew up out of the blue," but he decided not to mention its contents to anyone. "I had to scream at him to stop ignoring the problems," van de Vijver would later say. "The Shell board were also at fault for trying to shove the problem aside as a row between two men." Critical reports about the overstated reserves were submitted to Shell's audit committee, and managing directors were briefed about it, yet those same directors would subsequently claim to have been unaware of the full extent of the problem until December 2003. The culture of secrecy and the lack of mutual candor within Shell was fracturing relationships as events spun beyond the control of the company's chiefs.

The advice received from Cravath's by Frank Coopman was devastating. Shell, the lawyers in New York warned, was required to substantially downgrade its reserves and to advise the SEC about a "material matter" affecting the company before it revealed the serious breach of rules to the stock market. On December 2, Coopman told van de Vijver about the lawyers' advice. Failure to comply with the SEC's regulations, he warned him, would expose the company to prosecution. "This is absolute dynamite," van de Vijver replied, shocked by the unintended consequence of his campaign, adding, "Not at all what I expected and needs to be destroyed." His belief that the genie could be pushed back into the bottle, an unlawful act, showed naïveté at best, and at worst raised questions about his motives. After reflection, van de Vijver recorded in an internal memorandum his intention not to be blamed for "a watershed reputational disaster…and I do want to stick to some very firm criteria: the problem was created in the 90s and foremost in 1997–00 and any clean-up must reflect that…I will not accept cover-up stories that it was okay then but not okay with the better understanding of SEC rules now and that it took us two and a half years to come to the right answer." The "clean-up," he ordered, should reflect that the problem was created by Watts.

The implications of Coopman's activities could no longer be ignored. To protect Shell's relationship with Nigeria, the world's seventh-largest oil exporter, which shipped 40 percent of its production to the USA, Shell's accountants had wrongly categorized 60 percent of Nigeria's oil as "proven reserves." Van de Vijver suggested keeping the true level of the Nigerian reserves "confidential in view of host country sensitivities." Shell, he knew, would not want embarrassing revelations to damage its relationship with other members of OPEC, or to undermine the continuing negotiations with the Nigerian government for a $385 million bonus payment. He was too late. Two days later, Shell's bureaucrats finally understood that containment was impossible. On December 5, chief financial officer Judy Boynton was told for the first time about the problem. "I'm gobsmacked," she spluttered. "I've never come across such duplicity."

Shell's board of managers was due to meet for Christmas dinner in The Hague on Monday, December 8. That morning, van de Vijver submitted a 42-page report to the board describing the crisis. His report was not the first time they had been made aware of the situation. Ever since February, he had warned that one billion barrels of the reserves were "no longer fully aligned" with the SEC rules, and that an additional 1.3 billion were at risk because of expiring licenses. Within the corporation, officials had even drafted an "external storyline" and an "investor relations script" to explain the problem. An added complication on that day was that Watts was in Moscow, where he was waiting to meet Putin to discuss a draft decree threatening to revoke Shell's license in Salym. A telephone call from Jeroen van der Veer warned him that difficulties had arisen that would need to be discussed on his return later that day. This timetable was jeopardized by Putin's decision at the last moment to meet Watts at his dacha outside Moscow rather than in the Kremlin. As he was being driven at speed into the countryside, Watts was overtaken by the president's motorcade. On arrival, he and John Barry, the head of Shell's Moscow office, were filmed being received by Putin and a group of ministers. "They're revoking the Salym licence," Watts told Putin. "I'm sure you're committed to the rule of law and that you won't think this is impertinent of me, but if that happens we will fight it in the courts." Putin, Watts believed, understood his threat before the interpreter had finished, but after the translation was completed, he turned to his ministers and said: "I'm sure that a solution to this matter will be found amicably." To Watts's relief, Putin provided an escort to speed him to his private jet.

During the flight, Watts could reflect that Shell's investment in Salym was safe, and that Sakhalin 2, despite internal criticism of tardiness and excessive cost, proved that Shell's engineers could carry out a long-term plan. Indeed, a laudatory newspaper article would subsequently appear under the headline, "Oil Giant's Patience Pays Off in Russia." Watts was proud of his record. Although BP appeared to have outperformed Shell, he believed that his competitor's investments were risky. Shell's future was assured by its global brand in gas stations, enhanced by the purchase of the American retail group Pennzoil for $1.8 billion, making Shell the biggest gasoline retailer in the US, with 14 percent of the market. Although profits had fallen, he had consolidated the corporation's interests in chemicals, natural gas, tar sands and oil wells. He had even cut over 5,000 jobs in the previous two years. He expected to enjoy Christmas and the New Year in the sunshine in Oman.

The whispered conversations during the pre-dinner reception in The Hague ruined Watts's airborne reverie. Jeroen van der Veer was unusually animated. Until Judy Boynton had arrived in his office that very morning, Shell's director responsible for chemicals had assumed that van de Vijver was coping with the problem of the overstated reserves. No one, he would explain, had previously explained to him the differences between proven and unproven reserves. Well aware of the unsustainable relationship between Watts and van de Vijver, van der Veer had assembled a committee to resolve the issue. Watts was told about that initiative, and given graphic accounts of van de Vijver's accusations of lies and concealment, as he joined the 50 guests heading for the chairman's dinner in the city hall. Protesting his innocence, he said: "This is all news to me. I had no idea that any figure reported to me was wrong." An argument erupted among Shell's directors. Each claimed to have worked within his own silo, unaware of events in the rest of the business. All had previously chosen to ignore the intense animosity between the two most senior directors, and some would subsequently claim that even at that late stage, Watts had not informed the conference of managing directors or the audit committee about the full extent of the problem, although Watts would presumably have regarded this as van de Vijver's responsibility, rather than his own. Jeroen van der Veer, although admitting receipt of documents after July 2002 showing that the reserves did not match SEC definitions, would say, "I did not know about incorrect bookings. I did not appreciate the severity and the magnitude of the problem." Amid those unconvincing denials, the directors agreed to urgently consider a review, code-named "Project Rockford," reestimating the reserves in nearly 300 oilfields, 90 percent of Shell's assets, and disclosing the report to the SEC and the stock exchange. Controversy was inevitable, but the only director to raise the unpleasant question of whether there had been deliberate deception—although there was no evidence of this—was Larry Ricciardi, a New York lawyer and nonexecutive director, who expressed his outrage, hinting that many would suspect that deviousness rather than arrogance lay at the heart of the misreporting. None of the British or Dutch directors agreed. Without concrete evidence of dishonesty, they were unprepared to formally voice any misgivings, or to anticipate the consequences for Shell of the fragmented relationship between Watts and van de Vijver.

Compiling "Rockford's" conclusions about the overbooking of the reserves ruined Watts's Christmas holiday. Ostensibly celebrating with the Sultan of Oman, he commuted back to Holland to implement Cravath's advice that Shell needed to debook 2.3 billion barrels of reserves in accordance with the SEC's requirement of "absolute certainty." Uncurious about Cravath's lack of forensic accounting, and oblivious to the doubts about himself among his fellow directors, Watts obeyed his lawyers and auditors to hastily complete a self-destructive operation. Disturbingly, the *Wall Street Journal* had hinted at Shell's predicament. The leaks of the critical reports by Anton Bylondrecht describing the unsatisfactory reserves in Oman and Nigeria were clearly intended to destroy Watts and other managers.

On January 9, 2004, Watts was due to address analysts and the media. Before his appearance, he agreed that Shell should publicly issue a catastrophic admission. A statement was prepared that between 1996 and 2002, 3.9 billion barrels of oil equivalent, or 20 percent of Shell's oil and gas reserves, had been wrongly booked and would be recategorized. Watts had not anticipated the deafening echo as the news ricocheted around the world. Fearing that the British media would enjoy the "blood sport to tear me apart," Watts decided to duck the challenge. Shell's spokesman was ordered to explain to the skeptical audience that under the rules it was impossible for Watts to appear publicly in the "closed period" before the financial results were published on February 5. "I'm going into the bunker," Watts told John Hofmeister. "Are you sure that's wise?" asked the American. "Hitler went into his bunker and never came out alive." Watts ignored the advice. No other voice within Shell warned him of the danger. Having crushed too many people with an elbow in the eye or a kick in the shin, Watts lacked friends at that critical moment. He was also incapable of self-criticism. "We're going to be crucified," he was told after the meeting. The damning headlines in the media were greeted by his fellow directors with stony faces. As the criticism and questions fell relentlessly on the corporation's directors, Watts belatedly understood the catastrophe that had befallen him. "I'm the hero," van de Vijver crowed. "Philip's the goat."

On January 16, Watts addressed the staff on the corporation's website: "It is important to bear in mind that this recategorization was the result of our own internal processes... Based on those reviews, I believe that the individuals concerned worked in good faith to the interpretations in use when the bookings were made following proper processes, and that there is no evidence of misconduct." There was no consensus about whether Watts was the perpetrator, victim or scapegoat, but all agreed he was isolated. Shell's spokesmen quietly advised journalists to consider BP's booked reserves, which during 2000–02 were double its own growth history and double Exxon's. BP insisted it was not revising its reserves.

The world's second-largest oil company could still have saved itself if Larry Ricciardi had not demanded an independent inquiry. "We need a forensic investigation to discover the truth," he said. No oil company had ever exposed itself to such self-examination. Exxon after *Valdez* had resisted every inquiry, but Shell's directors, unable to explain that estimating oil reserves had never been an exact science, or that the crisis had erupted because of van de Vijver's schemes and ambitions, succumbed to the American's demand for a full investigation. In the post-Enron era, any resistance to an American nonexecutive would have been futile. Davis Polk & Wardwell, a New York law firm, was appointed to conduct a swift review. At that moment Polk was also advising other oil companies how to resist the SEC's demands. "We've unleashed an unnecessary witch hunt," mourned a British director.

The mood at the company's annual strategy review in London on January 25, 2004, was acrimonious. Not one person accepted blame, and the directors were divided by their suspicions. Jeroen van der Veer felt neither sympathy for Watts nor trust of van de Vijver, whose stupidity had encouraged Cravath's "reporting" to the SEC. Van der Veer was certain that Watts could not survive, and expected Polk to deliver the fateful blow. Meanwhile van de Vijver, encouraged by his forceful wife Bernadette, expected the events to deliver the crown. His ambitions were fueled by a simultaneous outburst of detailed descriptions of Shell's secrets in the *New York Times* and the *Wall Street Journal*.

On February 5, Watts and van de Vijver finally appeared in public to present Shell's annual review. During the long presentation, Watts admitted his mistake in failing to appear on January 9, but refused to resign. "This thing has happened on my watch. I have the will and determination to see us through," he said. Both men suggested that the reserves problem had only emerged during 2003, rather than in 2002. Van de Vijver described his "shock," which had sparked the Rockford inquiry, while Watts stated that after the facts appeared "late last year," "It was a matter of all hands on deck. And I remember writing down the words 'Get the facts and do the right thing.'" There was, he added, "no evidence of any misconduct." By giving the impression that the reserves issue had only just arisen, both men unwittingly aroused suspicions about their own conduct. Watts's submission that "I thought we had booked in good faith and applied SEC guidelines retrospectively" lacked credibility. To the Polk investigators, the directors' statements had the flavor of a cover-up, although they resisted stating as much as a conclusion. The flaws reappeared at a shareholders' meeting the following day at the Saint Regis hotel in Manhattan. "I'm sorry I got it wrong," Watts sighed in a tired voice. His defiance had evaporated. Shell had consistently underperformed over the previous decade, and he offered no hope for the future.

Polk's interim report, completed with undue speed at the end of February, cast Watts as unreliable. Summoned to a special meeting in The Hague, Shell's directors read confirmation of van de Vijver's judgment that the reserves booked during Watts's era were "aggressive" and "premature," and had failed to comply with Shell's own guidelines, and by implication the SEC's. Watts's professional "success," reported Polk, could be attributed in part to his ability to "meet or exceed reserve expectations." Shell was advised to debook 4.47 billion barrels, or 23 percent of its reserves, in four stages between January 9 and May 24. Other reductions would be made in March. At the end of the meeting, the directors were angry. Watts's position, they agreed over dinner in a hotel, was untenable. At 8:30 the following morning, March 1, Watts was waiting in his office for the inevitable. Overnight, his resignation letter had been finalized. The demand for his signature was delivered by Ronald Oxburgh, the British nonexecutive chairman of Shell. "Why?" asked Watts. "Misjudgments," replied Oxburgh. With dignity, Watts signed. "Can I have the company plane to take me back to London?" he asked. "I'd prefer not to queue at the airport." "No problem," he was told. Only Ricciardi would challenge, albeit unsuccessfully, the financial terms of Watts's departure, which included the company's agreement to finance all his future legal battles.

Oxburgh next called on van de Vijver. The Dutchman had been expecting to be offered the chairmanship, but instead Oxburgh said, "We want your resignation." Van de Vijver's face drained. "But I discovered the truth," he protested. "I led the charge against Watts. This can't be happening. I don't believe this." Oxburgh remained impervious. In tears, van de Vijver telephoned his wife. "But I told the truth," he shouted. Shortly after, his wife appeared at the office and firmly shut the door while he slumped over his desk. At 10 o'clock, the directors nervously dispatched John Hofmeister to extract a signed resignation. Pushing hard to open the door, he was greeted by an angry Bernadette van de Vijver, who uttered grossly unfounded allegations. Hofmeister issued van de Vijver with an ultimatum: "You have until noon to resign. If you don't, the record will show that you've been removed for cause." In the aftermath, Chris Fay told Mark Moody-Stuart, "I warned you that Watts was a cover-up guy, that there was a geological mafia cover-up."

Jeroen van der Veer was appointed Watts's successor, although he had been party to the mismanagement since 2002. He was, it was decided, the least guilty of the group. Paid one quarter of John Browne's salary (£2.9 million in 2006), the cost-conscious Calvinist who cycled on the weekends to save gasoline appeared in a crumpled shirt with frayed cuffs and an old tie to utter a suitably morose mea culpa to the public. "It has been a very tough time. This is by far the most difficult period of my whole Shell career...We had all kinds of processes and procedures, audits, signing off letters of representation, external accountants, so there was a whole list of things to make sure that we did not drop the ball. There was a basic belief that nothing major could happen in the company. But it happened."

Shell's share price plummeted. Polk's final report was delivered four weeks later. Quoting e-mails and internal documents, it condemned Watts's conduct and invited an investigation by the Financial Services Authority (FSA) in London and the SEC. Shell's directors used Polk's conclusion as the basis for blaming Watts for masterminding a conspiracy. In a statement following Shell's $150 million settlement, the SEC attacked the company as "reckless" for having repeatedly ignored its warnings: "In each case Shell either rejected the warnings as immaterial or unduly pessimistic or attempted to 'manage' the potential exposure by, for example, delaying debooking of improperly recorded proved reserves until new, offsetting proved reserves bookings materialised." Watts believed those conclusions to be untrue. "They produced a report dripping with innuendo to cast me as a scapegoat," he complained. Refusing to be seen as the villain and condemned to lifelong disgrace, he retaliated. Hiring lawyers in Washington, he subpoenaed the SEC's correspondence with rival oil companies about their reserves in Kashagan, Nigeria and Norway, the same oilfields as Shell's. Among nearly one million documents was correspondence showing that in exactly those areas where Shell had debooked reserves, SEC officials had allowed ExxonMobil, Chevron and BP to keep the same reserves booked. Indeed, just before Shell debooked Kashagan in 2004, the same reserves were booked by ExxonMobil, despite the SEC's orders to debook them. Watts could only speculate about why ExxonMobil, a collaborator in Kashagan like Total, had both stayed silent about its confrontation with the SEC, and refused to offer any help to Shell, its beleaguered partner.

Lee Raymond and Rex Tillerson told analysts on March 11, 2004, that all the circumstances justified ExxonMobil's booking in Kashagan because, in their judgment, "all necessary approvals" would eventually be given. Faced with the company's insistence, Winfrey and Murphy had retreated. ExxonMobil was too big and too powerful in Texas to risk confrontation. "Our policy is the most conservative and rigorous and disciplined," said Raymond. "You may call that risk-averse. I call it prudent management." ExxonMobil simply resisted the SEC's pressure to adopt the new rules for booking reserves. Total had also been permitted to keep its reserves booked. Yet in ostensibly similar circumstances, Shell—and by implication Watts—was accused of securities fraud. Its rivals had stood silently on the sidelines watching Shell's self-inflicted evisceration for their similar conduct.

Watts was incandescent, but was alone in his battle to show that Winfrey and Murphy had picked on Shell in order to berate the whole industry, and had exploited van de Vijver's ambitions. The two officials had failed to comply with the statutory requirement to consult the industry before unilaterally changing the rules, and Watts was their trophy. For regulators, he knew, the process was not about the truth, but about winning. Neither SEC investigator understood the contradictions contained in their correspondence with all the oil companies. With hindsight, Watts realized that his timid obedience in debooking the Kashagan reserves as ordered by the SEC had been foolish. ExxonMobil had refused, and the SEC meekly acquiesced. If ExxonMobil had informed Shell of its stance, as exposed in the correspondence, Watts lamented, he would have been partly protected. Accustomed to danger in Nigeria, where guns could be hired for $20, he challenged the SEC to a fight in court. The public cross-examination of Winfrey and Murphy would prove, he promised, that the SEC officials were technically incorrect, even if they had acted in good faith. In support of his argument, CERA published two reports in 2005 and 2006 confirming that the 27-year-old system was "under stress," unreliable for shareholders and unsuited to world pricing, where up to $6 trillion would be invested by 2030 to increase production of oil and gas.

Watts's battle against the SEC began in London. The FSA, relying unquestioningly on the Polk report, had in May 2005 rejected Watts's defense. "The FSA thinks they've got my head on a plate," he told his lawyers. His submission, exposing flaws in the background research commissioned by Polk, persuaded the FSA to abandon its case. Fearing embarrassment after the collapse in London, the SEC terminated its own investigation in August 2006, and declared Watts to be innocent. Only the stigma remained, and the absence of support from his former colleagues at Shell.

By then, Jeroen van der Veer had capitalized on the disarray. "Don't waste a crisis," he was advised. The moment was right to merge Shell with Royal Dutch, but on different terms from those Moody-Stuart had proposed six years earlier. Some preparation had been undertaken in secret by Watts, despite a promise extracted by van Wachem not to consider any change. "I know the difference between 60 and 40," Watts had told van Wachem. Van Wachem had since retired, and in the aftermath of the reserves crisis, with a defeatist mood prevailing among the British directors, van der Veer could rely on their support against the Dutch conservatives. The only issue was whether the headquarters should be in The Hague or London, where the company would be quoted to avoid punitive Dutch taxes. "The queen will never agree to the company leaving Holland," said van der Veer, explaining the delicate problem he had faced in persuading Queen Beatrix to allow the company to retain the word "Royal" in its title only on condition that its headquarters remained in Holland. On that trifle, the British directors surrendered their power.

Once the news emerged toward the end of 2004, van Wachem

sought to persuade the Dutch not to sacrifice their legacy, but it was too late. Shell's headquarters were moved from London to Holland, where the corporation would be encumbered by Dutch labor laws and conservatism, and estranged from British government support. In the short term, it was easier to fire British employees.

The oil-producing nations were delighted by Shell's humiliation and confusion. Weakening the Big Four reinforced the power of the national oil companies. The traders were also satisfied. When Andy Hall saw that Shell had exaggerated its oil reserves, he realized that he now had convincing evidence for an inevitable shortage of oil and rising prices. "The trend is your friend," he smiled, certain of huge profits.

Chapter Seventeen

The Alarm

——◦——

TOWARD THE END OF 2003, there was an abrupt change in the oil market when prices unexpectedly passed $35 a barrel. "Prices are too high," admitted OPEC's spokesman, adding that its target price was between $22 and $28. The unexplained bubble, OPEC feared, would explode, and prices would collapse. Blaming "speculators" and "geopolitics," OPEC's statisticians initially cast doubt upon their own data; later they were surprised when the steady decline of oil delivered during the year was unexpectedly reversed, and were bewildered about the destination of the extra supplies. In early January 2004, the mystery was resolved. IEA researchers in Paris belatedly identified China as the "culprit." Consumption there had risen during 2003 by a remarkable 11.5 percent. The extra demand reflected not only the widespread purchase of new generators and industrial expansion, but China's inefficiency. The country required three times more oil and coal to manufacture the same item than the US and Europe. The equivalent of 16 million barrels of oil was being squandered every day. China's demand for oil had doubled in 10 years, and was accelerating. Every year, the country's energy growth was the same as Britain's total consumption. Between 2000 and 2004, the unexpected surge had consumed an additional 5.8 million barrels a day (out of the world's total consumption of

82.4 million barrels). Previously, there had been no fear of shortages: the oil producers had been confident of their ability to produce an extra 12 million barrels a day if required. But within months, the combination of China's demand and disruption in several oilfields meant that the excess capacity had fallen to one million barrels. The IEA presented an alarming scenario: "What if everyone in China swapped his or her bicycle for an automobile?" In 2003, China was importing 16 percent of America's total oil production, but if the Chinese owned as many cars as Americans by 2010, imports would rise to 40 percent of America's total, absorbing all the world's anticipated extra production.

China's unprecedented demand enhanced the oil peakists' argument. Daily production, Colin Campbell was convinced, could not increase beyond 87 million barrels. The dwindling safety net of excess capacity was revealed by the stark statistics. Fifty-two out of 99 oil-producing nations, it was said, had passed their peak, and a further 16 would hit their peak soon. In the short term, the world could rely on Russia and the Middle East to satisfy the extra demand, but soon even they would "go short," and, since the discovery of new oil was declining, an oil crisis was inevitable. By May 2004, problems in Venezuela, Iraq and Nigeria, where violence forced Shell to cut production by 10 percent, had pushed oil above $40. Reports described the market as being "on a razor edge," and the prospect of $50 oil as having "lost its shock value."

Reassurance was offered by Guy Caruso and the EIA. Despite the uncertainty and the decline of production in the North Sea and Alaska, the EIA and the US Geological Survey assumed that there would be an abundance of supply. Production, forecast the EIA, would increase to 107 million barrels a day in 2020, and to over 120 million by 2025. According to the EIA, the OPEC countries would increase their daily production from 29 million to 61 million barrels, while production in the non-OPEC countries would rise from 50 million to 64.6 million barrels. Oil over that period, the EIA estimated, would cost about $35 a barrel on average but would rise to between $49 and $54 in 2010. No serious shortages were anticipated before 2037, and even they would be soluble.

Caruso's optimism provoked ridicule from the oil peakists. The EIA, Campbell and his supporters believed, was in league with the Bush administration to underestimate future price rises and overestimate the reserves. One foundation of their argument was the seemingly authoritative death sentence pronounced by Matt Simmons, the Houston banker and energy consultant, on Saudi Arabia's prospects.

Simmons was completing his book *Twilight in the Desert*, an examination of hundreds of surveys by engineers working over the previous decades in Saudi Arabia. By analyzing their reports about the reserves in the world's biggest oil producer, Simmons claimed to have pierced the remarkable wall of secrecy imposed to prevent outsiders discovering whether the Saudi assertion of seemingly limitless oil reserves was true, or whether supplies from the Kingdom would begin to decline. Sensationally, Simmons concluded that Saudi Arabia was going to run out of oil soon. The giant Ghawar field, he wrote, was being filled with water to force the oil toward the surface, proving that the gusher would soon be dribbling. That assertion made Laney Littlejohn incandescent. "Matt doesn't hear very well," he commented. "Thirty percent of water is necessary in Ghawar to get pressure, and it's been like that since the 1970s." Littlejohn's message was ignored, as was his successor Carl Calabro's testimony that "Saudi Arabia says it has 75 years of reserves left," and Guy Caruso's that "Saudi Arabia is working towards 75 percent recovery." Sixty-five percent of the world's proven reserves—1.05 trillion barrels of oil—were said to be in the Middle East, but since the Arab oil states regarded the quantity of their reserves as a state secret, even that was uncertain.

Colin Campbell supported Simmons by describing the oil states' inconsistencies. In 1980, he said, Kuwait had reported reserves of 65 billion barrels. In 1985, having produced over 20 billion barrels, it restated its reserves as 90 billion barrels, and added another two billion in 1987, although no new discoveries of oil had been announced in those seven years. Similarly, Abu Dhabi increased its reserves from 31 billion barrels in 1987 to 92 billion in 1988. In the same two years, Iran increased its reserves from 49 billion barrels to 93 billion; Iraq increased its from 47 billion barrels to 100 billion; and Saudi Arabia would announce that its had increased from 170 billion barrels to 258 billion, again without revealing any new discoveries. The following year, Khalid al-Falih, an Aramco senior vice president, stated that Saudi Arabia possessed 260 billion barrels of provable oil reserves and assumed that a further 100 billion would soon be found, some offshore. The Kingdom, according to al-Falih, had 25 percent of the world's reserves, and had only used 25 percent of them. Production, he said, would soon rise to 12 million barrels a day. "Peaking won't happen any time soon," he concluded.

Campbell suspected that the announcements were phony. The OPEC producers were, he believed, jockeying to increase the amounts they were permitted to sell according to their quotas. But even if the new reserves were accepted, Campbell argued, the Middle East's maximum oil reserves would have been 696 billion barrels. By subtracting past production of 255 billion barrels, only 441 billion barrels remained. Considering that the world had so far consumed 944 billion barrels and, according to Campbell, 65 different authorities agreed that the world's total reserves were about 1,930 billion barrels, that meant the world had used 49 percent of its oil, and was about to hit the peak. If the OPEC countries' suspicious figures were discounted, then the peak had already been passed.

To reconcile the incompatible predictions, a public debate was organized in February 2004 in Washington between Matt Simmons and Mahmoud Abdul-Baqi, the vice president of Aramco's exploration. Simmons's challenge was piercing. Abdul-Baqi, he believed, could not prove the Kingdom's ability to fulfill the EIA's forecast that it would expand production from 9.95 million barrels a day in 2004 to 13.6 million by 2010, and 19.5 million by 2020.* "The sweep of easy conventional oil is ending," he said, and future production from Saudi Arabia's new fields would be "complex." If Aramco was to disprove

---

* At the end of 2004, America was producing 8.84 million barrels a day, and Russia 8.44 million.

his assessment, he challenged Abdul-Baqi, it should be transparent about its reserves and their rate of decline. Without that candor, his forecast of Saudi Arabia's imminent crisis was undeniable. Abdul-Baqi appeared undaunted. Saudi Arabia, he replied, could currently produce 10 million barrels a day, and could increase that to 15 million until 2054, considerably longer than the EIA's forecast. "We will continue to deliver for another 70 years at least," he pledged. Spurred by Chinese demand, the prospect of oil rising to $50, he continued, would encourage Saudi Arabia to immediately increase its production by 3.5 million barrels a day without any problems.

"Simmons is asking all the right questions," said Nansen Saleri, an Aramco reservoir manager, later that day, "but giving all the wrong answers." Simmons's specific technical descriptions, said Saleri, were not a full picture of daily operations in the oilfields but a mere snapshot, ignoring the ability of Aramco's engineers to solve the problems Simmons had identified. By the end of the day, one irrefutable truth united Simmons and Abdul-Baqi: Saudi Arabia would be unable to permanently produce unlimited quantities of oil. Simmons's doubts about Aramco's candor remained. He had persuaded even Saudi Arabia's sympathizers in the audience to question whether Aramco's "tired" reserves could cope.

Those firmly unconvinced by Simmons included the chief executives of the oil majors. Lee Raymond did not believe in oil shortages. Contrary to the alarmists' scenario, he knew the world had become less reliant on oil. Since the 1970s, conservation and oil substitution in the industrialized countries had reduced oil consumption per unit of GDP by 40 percent. China's surge of demand, ExxonMobil's planners believed, would prove to be temporary. Rising prices would eventually force the country to limit its consumption. The strategists also knew that the oil market was more complicated than the oil peakists portrayed. Energy was no longer a wholly free market determined by economic forces, but was increasingly political, controlled by governments and foreign policy. Talk about the decline of non-OPEC oil supplies was as unreliable as the uncertainty about the published data describing reserves and decline rates. Reliance on that data fed the pessimism. Production data was the only reliable guide, and that remained healthy.

Nevertheless, a conversation with the billionaire financier T. Boone Pickens during a round of golf at Augusta National in Georgia was retold with relish by Matt Simmons. "Why are you advertising against peak oil?" Lee Raymond was asked by Pickens, an oil peakist. "Because we have very bright people and they don't believe it," replied Raymond. "So why haven't you increased production?" asked Pickens, knowing that ExxonMobil's reserves had risen 26 percent, from 9.6 billion barrels in 1994 to 12.1 billion in 2004, but production had dropped 2 percent, from 909 million barrels to 893 million in 2003. Pickens pounced on Raymond's failure to convincingly explain ExxonMobil's' flatlining production over the past 10 years. To Pickens and Matt Simmons, Exxon's stagnation confirmed the end of unlimited oil. Raymond's reply encouraged Pickens's speculation in energy, reportedly earning him $1.5 billion in 2005 while promoting the value of wind farms. To prove his conviction, in 2005 Simmons would bet John Tierney, a *New York Times* columnist, $10,000 that the average daily price for crude in 2010 would be over $200.

Lee Raymond's refusal to be alarmed by the rising prices was shared by John Browne. Browne had disclosed in March 2004 that for investment decisions, BP was pricing oil at $20, although he believed that the price would rise. With a sense of satisfaction, he announced that BP's reserves had increased from 8.6 billion barrels in 1997 to 18.2 billion boe in 2004. The corporation's profits had also increased, by 35 percent in the second quarter and 43 percent in the third. Rather than invest all the profits to discover new oil, he pledged to return $33 billion in cash to shareholders over the following three years. Boosting BP's share values silenced the doubters. "I think," said Browne, "that it is not helpful for the world to believe that we are running out of oil, which we are not." His certainty relied on Peter Davies's assessment that the world's unexplored reserves probably amounted to five rather than three trillion barrels, and that high prices would be temporary.

To many, Browne continued to appear a genius. BP had sold its shares in PetroChina, earning $1 billion profit, and he had bought Arco's and Amoco's reserves for between $10 and $15 a barrel at a time when prices were hovering around $55. "The best is yet to come," he boasted. Only a handful judged that the "Sun King" was exaggerating.

CERA, the industry's authoritative consultants, also contradicted the oil peakists and predicted in early 2005 that oil production could increase to 101.5 million barrels a day by 2010. This contribution by Dan Yergin was condemned by Matt Simmons as "part of the complacency machine." "Dan Yergin," said Simmons, "is the 'Peace in our time' propagandist, the most dangerous man in our time." In that cacophony, criticism of Simmons was muted. Few could disagree with Colin Campbell's observation that the debate had unexpectedly changed. "Just months ago," he said, "the debate was whether $20 could be held. Now $50 is a floor rather than a ceiling." Even his critics agreed that the predictions of doom were causing prices to rise. "Simple economics says that this [surge in demand] can't continue," commented *Oil & Gas*. President Bush's vow to end America's "addiction to oil" and encourage the switch to renewables added to the atmosphere of peril. Even a CERA researcher spoke of a peak in 2020. The oil economists employed by Saudi Aramco at Box 8000 in Dhahran could only smile. Since 1999 the Saudis had been controlling production, hoping to influence prices. They had never anticipated that a combination of oil peakists and speculators could achieve their goal.

The momentum was strengthening Andy Hall's credibility. Arjun Murti, an analyst at Goldman Sachs in New York, had examined the latest data, which showed that energy demand in 2004 had outpaced the 10-year trend. Consumption had grown by 4.3 percent, the highest increase since 1984. Over the previous three years, Chinese imports of some refined products had risen by 65 percent. Murti was convinced that this was the prelude to a "super spike." The moment was propitious to make a headline. "In a few years," Murti announced on March 30, 2005, "oil will hit $105." Outrage and skepticism rippled across the industry. Few believed that Murti's bombshell was based on facts.

Executives of the oil majors "would not bet" on Murti because the economy "would not take it." Some suspected that Murti's bolt from the blue was a ploy to help Goldman Sachs's traders. In 2004, Goldman Sachs and Morgan Stanley had together earned an estimated $2.6 billion by trading energy commodities. The suspicion of self-interest was rebutted by Hank Paulson, the bank's chief executive, who was obliged to defend Murti's honesty until, toward the end of the year, his analysis had gained respectability. Even Guy Caruso was revising his opinion. The oil peakists' criticism of his excessive optimism, he realized, was partially justified. His forecast of higher production by the national companies was wrong, and his confidence in extra production had been dented by the "difficult political environment." "Peak oil is not the issue," he nevertheless insisted. "It's about investment. But I now realize that there isn't a silver bullet." Even if the oil majors were given access to exploit the difficult reserves to produce additional oil, there was no evidence of their willingness to invest an additional $1 trillion every year to produce the extra 40 million barrels a day that Caruso anticipated. In his opinion, the majors were willing their own decline. By buying back shares rather than investing in exploration they appeared to be perpetuating a downward spiral.

But reports from the Gulf of Mexico suggested the opposite. BP had found a new elephant on August 31, 2006, at Kaskida, and Chevron reported that a strike in Jack 2 suggested a field containing "billions of barrels of producible oil and gas." New technology was opening up new fields on the Abyssal Plain, beneath 10,000 feet of water and at the edge of the salt line, where explorers would drill 32,000 feet, or more than six miles, into the rock to find whether the oil had been trapped. Whatever was discovered would benefit from new subsea hubs, which increased every well's productive life by two years. Recovery, which used to be 25 percent, had risen to 35 percent, and would certainly one day reach 60 percent, and possibly 80 percent. Farther south, Petrobras was on the verge of confirming the discovery of a field at Tupi, 175 miles off the coast of Brazil. Trapped under a mile of salt, the oil was in a field 500 miles long and 125 miles wide, encouraging the geologists to anticipate that Brazil's oil wealth could equal Venezuela's, or even Saudi Arabia's. To extract that oil would require endless inventions. First, because the salt was like "plastic," and collapsed once contacted by a drill; and second, because the oil was acidic, metallic and under low pressure. Brazil would have to develop equipment to pump the oil from the reservoir, but if 200 billion barrels of extractable oil were discovered offshore, the oil peakists would need to revise their forecast of doom.

Chapter Eighteen

The Struggle

———◄◦►———

RESOLVING THE CONTRADICTIONS of peak oil depended on the chairmen of the oil majors. Although they owned just 6 percent of the world's reserves, the executives, including David O'Reilly, the 60-year-old chairman of Chevron, set the tone about whether to invest or to retrench across the globe. Competing against Exxon, Shell, BP and Total, O'Reilly faced the challenge of preventing Chevron from tipping into the second division. In the maelstrom of American and European politicians seeking security of energy supplies, Chevron became a litmus test of whether smaller oil companies were doomed. Hidden away at the company's headquarters in San Ramon, a nondescript backwater near San Francisco, O'Reilly and his predecessors had plotted during the lean years to restore Chevron's fortunes. Ranking fifth among the oil majors, Chevron had struggled for the past 25 years to assert its credibility and survive. Decline was inevitable if it excluded itself from the annual $1 trillion cost of finding new oil. Bankruptcy or bonanza was the perpetual risk. While O'Reilly emphasized operational excellence, he could not deploy the financial weight of Lee Raymond or Jeroen van der Veer, and he lacked John Browne's vision and commercial cunning. His inheritance, however, was blessed by a stroke of luck.

An unexpected bonus from Chevron's merger with Gulf in 1984 was the offshore oilfields in Angola. Chevron's geologists had not appreciated the work of Hollis Hedberg, Gulf's geologist. Sailing up and down Angola's coast, he had selected what he regarded as the ideal acreage. "He had the vision of finding the equivalent of the Gulf of Mexico off Africa," said an admiring Chevron executive, delighted that in the decade after the merger Chevron had exchanged its ranking with Texaco. In 1985, Texaco was bigger than Chevron, but by 1999, profiting from the Angolan discoveries and pumping 240,000 barrels a day in Kazakhstan, 10 times more than in 1993, Chevron was double Texaco's size, and ranked as America's third-largest oil company. In order for the company to grow still further, Ken Derr had wanted to buy either Amoco or Arco. Thwarted by Browne in both cases, the only remaining possibility was a merger with Texaco, an option he regarded with distaste. Texaco had a reputation as the meanest and the most disliked of the Seven Sisters. "If I were dying in a Texaco filling station," said a Shell man, "I'd ask to be dragged across the road." Derr would have echoed that sentiment. During Chevron's collaboration with Texaco in the Gulf of Mexico, Brazil and West Africa, he had observed how under Peter Bijur the company had become faded, defensive and steeped in a military culture. "We send two people to meetings," critics carped in San Ramon, "and they send ten."

There appeared, however, to be no alternative after Bijur suggested a merger. Shell had refused to bid for Chevron, and Total was on the verge of buying Elf, France's biggest oil company. While America was immersed in the Monica Lewinsky affair, France had become preoccupied by its own scandal. A group of Elf executives had stolen at least $259 million from the company over a four-year period; Loïk Le Floch-Prigent, Elf's former president, had been arrested for fraud and was subsequently imprisoned; and an attractive female employee had confessed that she had offered money and sexual favors to Roland Dumas, France's foreign minister. To clean up the mess, Total added Fina to its purchases, becoming the world's fourth-largest oil company. Chevron's survival, Derr decided, depended on a merger with

Texaco. Together, they would be the biggest operator off Angola and in the Caspian, and the second largest in Nigeria. But as a sign of his aversion to Bijur, Derr's offer in May 1999 was low—just $37.5 billion—and he rejected Bijur's request to remain as a senior executive. "It was a lowball, cheap proposal," snapped a Texaco executive. Relations between the two chairmen disintegrated, and in early June their discussions were abandoned. "What I was interested in doing was an opportunity for a marriage, as opposed to selling the company," commented Bijur after the breakdown. Three months later, Derr retired in favor of David O'Reilly. During that period, Texaco's share price fell by 9.6 percent, while its rival's value rose. Bijur's self-respect was the only obstacle to resuscitating the merger. That was resolved over a round of golf. O'Reilly agreed that Bijur could remain as a token vice chairman, but the purchase price would be reduced to $35.1 billion. In October 2000, the deal was approved. Chevron's executives, surveying their new empire, worth about $100 billion, with 8.6 billion barrels of oil reserves and the equivalent of 2.8 billion in gas reserves, were comforted by the restoration of "critical mass." Under O'Reilly's regime, financial management was improved and the focus on engineering was resurrected. The obliteration of Texaco's culture was merciless. "Quicker homogenizing" was the headline of the amalgamation philosophy. "You can't run two cultures," O'Reilly's executives declared. "Those unwilling to make sacrifices will have to leave."

In the race to survive, O'Reilly knew that markets would not judge his performance by the provision of cheap oil or by satisfying a social agenda. The management of money was as important as geology, technology and politics, with a corporation's executives required to mitigate risks and manage uncertainties. The long-term incentive to search for more oil was always tempered by the short-term cost. Securing guaranteed supplies depended on fostering the corporation's political and financial interests. Ever since the mergers in the late 1990s and the unexpected price collapse, the oil majors had focused on big projects to meet shareholders' demands for profits. The importance of earnings did not change as prices rose after 1999 and the majors increased their investment. Between 1999 and 2003, the eight biggest oil companies spent $242 billion in their attempts to find and develop oil and gas reserves, and had replaced 134 percent of their production at a price of $5.18 a barrel.

Despite that success, their influence had declined. In 2004, the five oil majors controlled just 15 percent of oil and gas production. Limited access to oil reserves was only part of the reason. More important was the majors' retreat from risk. To justify high investment to shareholders, David O'Reilly, like his rival chief executives, needed to satisfy their quarterly scrutiny of profits, which undermined the incentive to take risks. Within the oil majors' magic circle, conservative decisions based on short-term concerns about prices restricted developing oilfields' potential. Even proven elephant oilfields required at least four years to develop, and if progress or production faltered, lower profits jeopardized further investment. Ken Derr had balanced Chevron's fate in the Gulf of Mexico and Kazakhstan; the corporation was not in the league to speculate in Kashagan. Kashagan's estimated reserves had fallen from "jaw-dropping" initial figures of 25 to 100 billion barrels to between 38 and 60 billion barrels, and had then been further reduced by some experts to 13 billion, while the cost of exploration — $29 billion and rising — was beyond Chevron's resources. Confronted on one side by the national oil companies' sovereignty over their oil reserves, and on the other by the oil majors' financial muscle, Chevron and the second-division companies had good reason to fear decline.

To resist the squeeze, Chevron's executives bolstered their courage by expressing disdain of the bigger companies. BP and Shell were dismissed as "in decline" and "unwilling to take risks any more." The playing field, Chevron's executives said, had "leveled." Shell was mocked for possessing insufficient "acreage" and having "lost out." ExxonMobil's employees were disparaged for their willful ignorance of local customs. "Exxon don't have people in the country looking at the whites of the locals' eyes and knowing what's happening," was the opinion expressed by one senior executive. "They turn up for meetings in Kazakhstan in blazers and white shirts." Referring to the Gulf of Mexico, another believed that "Exxon is out of deep water. It's not a serious player in the Gulf." Irritated that "Exxon doesn't play well with the other children," one Chevron executive concluded, "The Exxon model will break."

These negative attitudes were born of frustration. Natural growth had become too difficult, especially in America. Uncertainty about US taxes and regulations had earlier persuaded Ken Derr to expand in foreign countries, but breaking out of the straitjacket was a gamble. The usual options, Derr and O'Reilly understood, were either to lead the pack or to invest with the rest. "There's always someone willing to pay more than us," O'Reilly said. The smaller oil companies, the Chevron executives reckoned, could rarely negotiate straightforward access to oilfields in Africa or Asia without agreeing to special financial arrangements. Despite America's statutory obligations requiring corporations to behave scrupulously, there were occasions when ignoring those virtues was critical to prosperity. The alternative to operating on the edge was an uncertain fate, possibly following Amoco and Arco into the abyss. Salvation depended on taking shortcuts.

The integration of Gulf in the late 1980s had temporarily paralyzed Chevron, and had also infected the corporation with an unexpected ethos. Gulf's executives had succeeded in Angola and Nigeria by paying handsome bribes to local politicians and officials. Chevron's executives did not reject Gulf's culture outright. Instead, the senior directors, while adhering to their wholesome scripture, acknowledged the need to adopt Gulf's methods in the Third World. In West Africa, Chevron was suspected of collaborating with Shell to bribe James Ibori, the governor of the Nigerian Delta state. Both corporations had hired houseboats from the governor, and in return deposited $2.3 million into the account of MER Engineering at Barclays bank in London. Global Witness, an organization dedicated to exposing corruption, accused Chevron of paying "signature bonuses" to politicians in return for securing leases to offshore oilfields. In 2002, at the end of the 27-year civil war in Angola that had cost half a million lives, Chevron, in partnership with ExxonMobil, developed four blocks, each producing around 100,000 barrels a day, and in 2007 it would sign a $5 billion natural gas deal, leading a partnership with Total, BP and ENI. The companies put money into the hands of the dictators with little care for the consequences, except for the shareholders. Other companies, including Elf, had paid corrupt politicians in Congo Brazzaville, Equatorial Guinea and Angola, in cash or by payment to a president's bank account in Washington, DC, but Chevron's stance was unusual. The corporation's senior executives indulged in questionable transactions while publicly extolling probity. "Our responsibility is to make the system as transparent as possible, but it's their oil and money," said one senior executive. Nigerian politicians, the most corrupt among all oil producers, were resistant to such transparency. "I can't make them more transparent than the government wants," one of Chevron's senior executives admitted. "We push the 'transparency,' but only up to a point and only in private." They did not push too hard. Knowing that China's, India's and Norway's state oil companies unhesitatingly paid bribes, and anxious to expand Chevron's oil acreage, the company continued operations in the delta long after BP and Shell had withdrawn because of the region's perils.

Uncontrolled violence in Nigeria had cut production by 25 percent, or 600,000 barrels a day. Despite the technical problems and political complications, Chevron planned to produce 250,000 barrels a day in the Agbami field, although the estimated costs had risen from $3 billion to about $5 billion. Reality would interfere with the timetable. Sucked into the struggle between the delta's tribes, Chevron was accused of providing helicopters for Nigerian troops to liberate 200 Chevron workers held by local criminals, and of ending the occupation of the Parabe offshore oil rig by hiring armed toughs. Protestors claimed that the soldiers ferried to the area by Chevron had ransacked their villages and tortured the inhabitants. Summoning the army, according to Chevron, was the only solution. In December 2008 the company was cleared of all charges of human rights violations in connection with the incident.

That was not the company's defense when it was accused of flouting UN sanctions by paying illegal surcharges through middlemen to profit from the Iraqi "oil for food" program. Over 2,000 companies paid an estimated $1.8 billion in bribes between 1996 and 2003 to Saddam Hussein's agents in exchange for oil. Chevron, in common with Glencore, Vitol and many Russian and Swiss traders, profited from that illegal trade. Vitol would plead guilty in November 2007 to grand larceny, pay a fine of $17.4 million and promise to abandon the cowboy era and legitimize itself because, with crude so expensive, it, like Glencore, needed banks to finance its trade; Oscar Wyatt of Coastal would be fined $11 million and sentenced to one year and a day's imprisonment; Chevron paid $30 million to settle corruption charges. Trafigura was also swept into the opprobrium. In August 2006, a tanker chartered by the trader had unloaded cargo at Abidjan, in Ivory Coast, and later dumped toxic waste into the ocean. There were claims that many people had died from the polluted water, and the company would pay a $200 million settlement to Ivory Coast. In September 2008, Trafigura agreed to pay $30 million in compensation to the victims.

In San Ramon and at Chevron's office in Houston, the former headquarters of Enron, these infractions were brusquely pushed aside. John Browne's "Beyond Petroleum" campaign was a template for Chevron's self-promotion as a socially conscious corporation. Inconsistency was not an issue. The oil majors, O'Reilly and vice chairman Peter Robertson believed, needed to reinvent themselves with new "value propositions," because their rivals were always willing to pay more. In public, employees were urged to "look at ourselves as invited guests" and "behave properly." Oil had always been about big politics, but Chevron's new leaders, publicly espousing Californian values, believed they were inventing a new craft by speaking about "making sense of local politics," especially in sub-Saharan Africa and Kazakhstan. "We don't buy people the fish," staff were told by Peter Robertson, "but we show them how to fish. We have responsibilities where we operate towards employees, communities and the environment. We offer project management for schools, water and services." Chevron's image-makers invented a "vision statement" about "people, partnership and performance," and pledged "to make the world a better place."

Funds were provided to fight malaria and AIDS, but overcoming the reluctance of engineers to work in Africa undermined the Californian virtues. Despite their "sensitivity training," they arrived at the oilfields on private planes and lived in isolated communities, often offshore, protected by security men. Unsurprisingly, the oil engineers appeared "flash" to local Africans.

As well as having to smooth local sensitivities to retain access to the oil, Chevron was faced with unprecedented hurdles. The company's fortunes were threatened by the technical superiority of state-funded corporations, especially Petrobras and Statoil of Norway. Statoil had become the world leader in drilling in the Arctic seas, deploying unique subsea production technology to convert natural gas into LNG on the seabed, while Petrobras had proven its expertise finding oil in deep water under a mile of salt. In 1992 the Brazilian company, founded in 1953 as a government monopoly, had been categorized as among the least efficient in the world, and was even prosecuted for impugning the national honor. Over the following 10 years, as the state reduced its stake to 32 percent, new managers transformed Petrobras into a world-class operator, drilling 200 miles off Brazil's coast to similar depths as in the Gulf of Mexico. Encouraged to plan over the long term, with global aspirations, access to domestic oilfields and unique technology, Petrobras and Statoil could, over the next 20 years, cripple Chevron and every other second-division player.

In March 2003, Chevron was directly challenged by two Chinese oil companies, Sinopec (the China Petroleum and Chemical Corporation) and CNOOC (the China National Offshore Oil Corporation), both of which had negotiated to buy stakes in Kashagan from BG, the British oil and natural gas corporation. After these attempted purchases were blocked by other shareholders exercising their right to buy the stakes, Sinopec successfully bid $2.2 billion for two fields in Angola, accompanied by promises of loans in the future. Soon after, Chinese planes and ships arrived with men and materials to build an airport and workers' accommodation as a preliminary to wrenching control of Angola's deep-sea oil wells from Chevron and the other Western companies when their licenses expired. The pattern repeated itself in Kazakhstan, where dozens of Chinese transport planes delivered men and equipment to build an oil pipeline to the Chinese border. In both countries, corrupt local politicians were common. Unlike Western corporations, which were subject to laws and scrutiny, the Chinese companies' largesse was sanctioned by their government. "China's doing now what we did one hundred years ago," admitted a Chevron executive. The Chinese even followed Chevron into the Sudan, despite the civil war.

But there was a limit to tolerating competition, especially in America itself. In March 2005, CNOOC indicated its intention to bid for Unocal, America's ninth-largest oil and natural gas company, operating in the Gulf of Mexico, Southeast Asia and Azerbaijan. With a market value of $14 billion, Unocal's principal attraction was 1.8 billion barrels of proven reserves, especially in Burma and Thailand. CNOOC's bid prompted counterproposals from ENI and Chevron. Chevron appeared to be vulnerable. At the end of an 11-year streak, the company had not sanctioned a single large-scale project during 2004, and had only replaced 18 percent of the previous year's production. Having made a $13.3 billion profit in 2004 and increased its cash deposits during 2005 by $1 billion every quarter, it had spent nearly $2 billion buying back shares, but had failed to restore its reserves. O'Reilly's solution in April 2005 was a $16.8 billion bid for Unocal. Financially the bid could only be justified because oil had risen to $65 a barrel, a price O'Reilly said was too high. Two months later, CNOOC offered $18.5 billion for Unocal, an apparently knockout bid.

The Chinese had not anticipated the xenophobic outcry their bid would trigger. Although Unocal was small and lacked any strategic importance, officials in the Bush administration were persuaded of the danger of foreign ownership. Ignoring the hypocrisy that American oil companies, supported by Washington, demanded the right to invest in Russia and along the Caspian on their own terms, O'Reilly urged politicians in Washington to block CNOOC's bid. The national oil companies, he said, were competing on unfair terms against the privately owned oil majors. CNOOC's higher bid for Unocal, claimed O'Reilly, was only feasible because the Chinese government had provided interest-free loans. State companies like Petrobras and CNOOC, Chevron's spokesmen complained, were untroubled by low profits, and their purpose was clear. CNPC, a third state-owned Chinese corporation, had bought oil assets in 30 countries including Sudan, Angola, Russia, Kazakhstan, Australia and Saudi Arabia, and planned to invest a further $18 billion by 2020. Sinopec had signed a 25-year contract worth $100 billion to buy LNG from Iran's reserves in South Pars, although America's embargo had rendered that agreement useless: without Western expertise, Iran's oil and gas industry was crippled. Chevron painted a scenario of oil shortages in the future and the national oil companies exploiting their advantages if Washington's politicians did not favor American corporations. CNOOC's bid was rejected, and O'Reilly improved Chevron's offer to $18.3 billion in cash and stock. On the first day of its ownership, Chevron obliterated Unocal's culture. The company's name disappeared and the employees who remained were absorbed into Chevron's organization.

Chevron's victory intensified competition between the national oil companies and the oil majors—with an unexpected twist. Higher oil prices gave an incentive to all the producers to earn more profits by increasing production; or, as the national companies discovered, they could enjoy the same income by producing less oil. Russia and the socialist countries, increasingly influenced by political rather than commercial criteria, challenged the oil majors to defy their command over access and prices.

Producing sufficient quantities of oil at Tengiz to make the field profitable had never been in doubt until rising prices encouraged President Nazarbayev to demand that the 1993 PSA agreement, which was tilted in Chevron's favor, be rewritten. Nazarbayev wanted a $2 billion low-cost loan, and more income—immediately. In theory, he could impose higher taxes, demand a bigger share of the royalties or insist that Chevron and ExxonMobil, the junior partner, produce more oil. The obstacle was the terms of the original contract, which stipulated that Kazakhstan would receive only 2 percent of the revenue for 10 years, about $120 million, until the oil companies' costs had been repaid. Under Chevron's plan, those costs would increase if the company decided that production should rise from 270,000 barrels a day to 430,000. That expansion, said Chevron, would cost $3.6 billion and would delay Nazarbayev's receipt of royalties. To persuade Chevron to pay royalties immediately, Nazarbayev fined the company $73 million for allegedly storing sulphur in a manner contravening Kazakhstan's environmental laws. Surprised by the existence of these laws, especially since its storage method was common in America and Canada, and fearing that Nazarbayev would also seek to rewrite the Kashagan agreement, Chevron announced on November 15, 2002, what was called "a lesson in oil politics." Expansion of production was halted until Nazarbayev agreed to withdraw his request to rewrite the 1993 agreement. "It's about time someone drew the line," said a Chevron executive, buttressed by ExxonMobil's traditional resistance to changing contracts. Nazarbayev surrendered. He would be less amenable in Kashagan.

The discovery of oil in Kashagan had been a relief for the majors. Earlier, Total's directors had despaired after drilling dry holes around the Caspian, and were on the verge of quitting the region, while ExxonMobil had accepted the loss of millions of dollars on unsuccessful trials. Both corporations had looked on enviously at BP's successful production of 100,000 barrels a day from the Azeri field, due to be pumped in 2005 through the Baku–Tbilisi–Ceyhan (BTC) pipeline, and hailed as an American triumph because it bypassed Russia, while Azeri natural gas flowed 1,100 miles from Deniz to Turkey through the South Caucasus pipeline. By contrast, the consortium in Kashagan, which included ExxonMobil and Total, originally estimated that $29 billion would have to be invested before production could begin. After five years, the estimated cost of developing the oilfields had risen to $136 billion and the production of oil had been delayed from 2005 to 2015 at the earliest. The blame was heaped on ENI, which had demanded the right to manage the project.

Paolo Scaroni, ENI's ebullient chief since 2005, had similar ambitions to David O'Reilly, but inherited a tradition of taking bigger risks. He committed $3 billion to produce oil in a violent offshore area in Nigeria, paid $6 billion to purchase Yukos assets, signed a long-term supply deal with Gazprom, and bid a record $902 million to search for oil in one block in Angola. "At the time it seemed crazy," he conceded, but ENI's ambitions demanded risks. Scaroni's predecessors had aggressively insisted on taking the lead in developing the Kashagan field, probably the most complicated project in the oil industry's history. ENI, the Italians insisted, possessed the technical expertise to develop the world's biggest discovery in 40 years. Capitalizing on the dislike between the executives of ExxonMobil and Shell, another major investor, ENI's bid to become the operator slipped through as a reluctant compromise. ENI's initial $10 billion investment required its engineers to work in temperatures ranging between 55°C in the summer and –40°C in winter. The principal contracts had not been awarded to specialists, but to inept Italian companies favored by ENI, and in every task the engineers blundered. They encountered innumerable difficulties, not least while navigating with special icebreakers through the shallow waters during the winter. The wind, tinged with

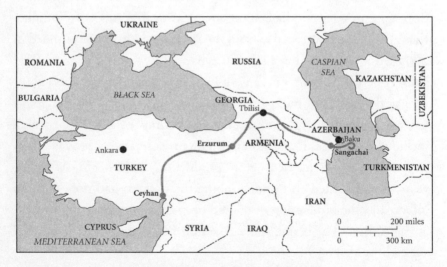

The Baku–Tbilisi–Ceyhan (BTC) pipeline

poisonous hydrogen sulphide, exacerbated the perils of extracting the crude oil, trapped under huge pressure and liable to explode. Drilling safely down to 20,000 feet proved to be a near-insuperable challenge, and $2 billion spent building accommodation on an artificial island was wasted after jet vapors of lethal hydrogen sulfide sprayed over the new buildings. The original plan to produce 100,000 barrels a day by 2005 was abandoned, and the ultimate ambition to produce 1.2 million barrels a day was indefinitely postponed.

Frustrated, Nazarbayev successfully demanded that the original contracts be rewritten, tilting the advantage away from the oil companies. Unusually, ExxonMobil's objections to this demand were defeated. Kazakhstan was given a bigger stake in the oilfields, the oil companies agreed to pay compensation for the delays, and ENI was stripped of operating control. Nazarbayev's victory was a blow to the oil companies. Without their expertise, Kazakhstan's oil would have remained unobtainable, yet rather than gratitude, the country's dictator had voiced anger. The relationship between the oil-producing nations and the private oil corporations, especially the smaller companies like Chevron and ENI, was jeopardized. Rising oil prices were creating an arrogant conviction among all the oil-producing nations about their permanent indispensability.

Chapter Nineteen

The Survivor

―◄◦►―

ENI'S INCOMPETENCE made Lee Raymond impatient. In his opinion, the free market in oil was diminishing. Instinctively, he asked himself of every proposition, "Is this serving the interests of Exxon's shareholders?" Exxon, he needed to remind outsiders, was not a government agency. The corporation's prosperity, he believed, depended on producing huge volumes of oil. He dismissed those who argued that Exxon's long-term survival depended on focusing more on the value of a project than on pure profit. In the hostile business environment of high taxes, overbearing regulations and growing nationalism, he would opt to stay out if the returns for Exxon's shareholders were not guaranteed. In an earlier era, Raymond could comfortably have identified himself as a professional imperialist. His virtuous cause was to enrich ExxonMobil's shareholders by any legitimate means. He made no pretense about himself or the corporation. The values of the chairman and of ExxonMobil were interchangeable: impervious to criticism, neither sought love, and both assumed admiration. Aloof and awkward in the limelight, Raymond cultivated an image of inflexibility and irrefutable success. "We are gigantic," his ambassadors would tell new recruits, listing ExxonMobil's unchallenged superiority as the biggest oil major. "Exxon owns 2 percent of the world's reserves.

We understand the molecular structure of oil better than our competitors, and how to get more oil per barrel than our competitors." Since
America's fate depended in part on Exxon's continued success, those
doubting that oil was a religion as much as a science were methodically
excommunicated from the ranks. While extolling the purity of exploration and production, Raymond mocked BP and his smaller rivals as
grubby for trading oil. ExxonMobil was in a race, he preached, not
only to rank first, but to last the longest.

After 1998, as his exploration engineers failed to find sufficient
large reserves, Raymond began limiting the investment budgets and
used the company's profits to buy back shares and pay high dividends
to sustain its rising stock market value. Even Exxon, the champion of
the long-term view, occasionally switched to short-term performance.
Raymond was reassured that besides steadily accumulating billions of
dollars in cash — about $40 billion would be deposited in the bank by
the end of 2008 — ExxonMobil's reserves of around 72 billion barrels of oil guaranteed its survival. Some would criticize his "inordinate
emphasis on reserves" as a mistake, embodying a complacency that
would threaten the corporation's ultimate fate, but they were ignored.
The chairman paraded his realism as a badge of pride.

In the hunt for more oil, the prime target for Raymond had been
the Middle East. Following the meeting at the Saudi Arabian embassy
in Washington in December 1998, he had been encouraged to extract
oil and natural gas in the Kingdom's "empty quarter," which required
expertise the Saudis lacked. But by 2000, Raymond was dissatisfied.
Exxon was being asked to undertake the exploration in a $600 billion
project, but was denied ownership of the oil and natural gas. During
a year's negotiations, the relationship between the corporation and the
Saudi rulers became acrimonious. "This is a bad deal," Crown Prince
Saud al Faisal, the foreign minister, reported Raymond saying. Exxon,
complained Raymond, was not a contractor providing services like
Schlumberger, but an oil major, and was only prepared to take a risk
in return for some ownership of the reserves, and profits. Nothing less
than 12 percent, and preferably up to 20 percent. Exxon's resources and

equipment were too costly to be employed merely to provide services, especially if there was no certainty that oil and natural gas would be found in the area. After Shell and Lukoil had been awarded the rights in the region, they would discover that Raymond's doubts were well-founded. That would be little satisfaction for Raymond.

Exxon's failure to reenter the Middle East at the end of the 1990s had coincided with problems in Russia. Ten years earlier, the corporation's executives had arrived as Cold War victors to take the spoils. Famed for remaining impervious to pressure, Harry Longwell, the tough and reliable head of Exxon International, was taking no prisoners. Neither Raymond nor Tillerson had appreciated the history of Russia's oil industry, and when they saw its dilapidated state, they were unimpressed. To some Russians, the two Exxon executives appeared unable to decide whether they were visiting an eastern country like China or India, or a former colony like Nigeria. Some concluded that neither man cared. Uninterested in obliging the Russians' request for money and assistance, Exxon's representatives insisted on buying stakes in virgin sites.

Avoiding ENI's and Total's loss-making investments, Raymond refused to develop Sakhalin 1 unless Exxon's shareholders profited. The risks were considerable. Amid drifting ice packs, snow, fog and earthquakes, a $1 billion platform would be constructed to extract two billion barrels of oil and 15 trillion cubic feet of natural gas. Tillerson approached the negotiations according to the Exxon rulebook. Since no Russian official or engineer properly understood the complications, and since only after 15 to 25 years would the project's profitability be certain, Exxon could only be protected by the sanctity of the contract. Its principles had been developed from the experience of the previous hundred years. Every agreement was influenced by different circumstances, but as an Exxon senior executive maintained, "There's nothing new in oil — only the players and the country are different. And you don't negotiate your principles. If something is unknown, you must negotiate it according to principles." During the negotiations in Moscow, Tillerson and Longwell expressed no sentiment. Their task was not to help Russia—although the country would benefit from Exxon's expertise—but, after crunching the detail, to impose a contract ensuring Exxon's profit, and to abide by it. They had no doubts that a familiar cycle—as in Venezuela and the Middle East—would recur: "Mega-projects are lumpy, so there's no easy way to show continual progress. There'll be the thrill of agony, victory and defeat."

Success at Sakhalin encouraged Raymond's optimism. Exxon's most experienced teams had assembled 40,000 tons of equipment to drill nearly two miles down and over six miles horizontally into the Pacific Ocean. Over 10 years, Exxon was committed to invest nearly $4 billion and employ 70 Russian design institutes to construct two thirds of the infrastructure in Russia itself. That commitment, Raymond reckoned, would surely persuade the Russian government to grant the company more licenses. But to Raymond's disappointment, despite US vice president Al Gore's representations, the Kremlin declined to oblige.

A $1.5 billion agreement to develop oil production in Timan-Pechora province in the far north was rescinded in August 1997 by Viktor Orlov, the natural resources minister, blaming "legal irregularities" and Exxon's "unacceptable conditions" for Rosneft's participation. The Exxon executives' zest for "business with integrity," combined with their passion for secrecy, was regarded by the Russians as dogma. Soon after, another agreement for Exxon's development of two more blocks in Sakhalin and the Nenets oilfield was canceled. "We don't need Russia," Raymond declared, surprised that the Russians' need for Exxon's expertise did not override their irritation about the inflexible contract. "If we drill a dry hole we can't go back and ask for a contribution to the cost," was his reasoning. The Exxon contract protected the company's investment. Exclusion from Russia did not make Raymond fear that other countries might also reject Exxon. He would not, as a Chevron vice president would describe it, attempt to "seduce and innovate by offering solutions to the local government's problems." The Kremlin would accept Exxon's way or no way. To his critics, who believed that Raymond's intransigence would "break the Exxon model," a senior executive replied, "What's happening in Russia is nothing new. The same happened in Venezuela and Africa. It's always the same cycle and balance of risk that you win some and lose some."

This flippancy was not merely for public consumption. Ever since the crash in 1998, ExxonMobil had reduced investment to increase its profits. In 2000, after cutting investment by 24 percent, the company's profits rose in the first quarter by 122 percent, to $12.5 billion. At the same time, Raymond's skepticism about Russia and the Middle East was balanced by his successful persuasion of George W. Bush and Dick Cheney to invest time in sub-Saharan Africa, in recognition of a new linchpin of US energy. West Africa supplied 14 percent of America's oil and gas, and ExxonMobil had a stake in about 10 percent of the area's estimated reserves of 77.4 billion barrels of oil. With ExxonMobil's encouragement, in July 2003 President Bush visited the continent to reinforce the corporation's stake, especially in Angola, a safe country for ExxonMobil's expansion.

The aversion to risk shared by the oil majors after 1998 had not deterred the next-largest 20 oil companies, including BG, Apache, Devon Energy and Anadarko, from exploring for new oil. Unlike the Big Four's economic experts, who as usual reached a consensus, the chief executives of the next tier of oil companies believed that oil prices would rise. Raymond's caution benefited Hess, a $50 billion company based in New York and Houston, which anticipated that supplies of oil would not match rising demand. Focused on exploration, Hess recruited experts from Exxon and other majors to find oil under 5,000 feet of salt in the deep waters off Brazil. This commitment was especially risky, as Shell, BP and Total had all drilled dry holes in Brazil's acreage, and given up. Hess's interpretation of the seismic data on lease BMS22 in the Santos Basin pinpointed the ideal spot to drill. The problem was financing the $100 million well. The only way the company could raise the money was if an oil major was prepared to take a stake in the license, owned 80 percent by Hess and the remainder by Petrobras.

ExxonMobil agreed to pay 100 percent of the exploration costs in return for 40 percent of the profits. Having "farmed in," ExxonMobil's experts in Houston violently disagreed with Grant Gilchrist, Hess's geologist, about his interpretation of the seismic data. In the normal Exxon manner, their assumption was that only their judgment could be correct. The disparity in size between the two corporations encouraged the ExxonMobil team's belligerence. While ExxonMobil needed to discover at least 1.4 billion barrels every year to replace production, Hess required 140 million barrels "to stay out of liquidation." "Our production doesn't even cover Exxon's needle," commented John O'Connor, Hess's director. Eventually, Exxon's team were persuaded by Gilchrist's version. Just as when they had reluctantly bowed to BP's analysis and found oil in 1997 in the Hoover field, 165 miles south of Galveston in the Gulf of Mexico, Exxon's bluff about its deepwater expertise was exposed in Brazil. The discovery of a large reservoir in January 2009 was announced as an ExxonMobil success. The reserve was in an area where Petrobras had already discovered Tupi, a field beneath a salt layer with between five and eight billion barrels of oil, and another find with possibly 100 billion barrels. In the short term, ExxonMobil's need for more oil could usually be met by investing in rivals' expertise, but over time, the Hess directors knew, Exxon's survival depended on more "consolidation," meaning mergers or takeovers. Only the oil giants possessed sufficient money to discover and exploit the billion-barrel offshore elephants that were essential to future supplies.

ExxonMobil's success at Sakhalin confirmed the reliability of the company's model. The revised estimate of the reserves in the area had soared to 60.5 billion barrels of oil and 200 trillion cubic feet of natural gas. Alexei Miller, Gazprom's chairman, was surprised. The fate of that enormous quantity of natural gas, he felt, should be his decision. Either the gas would be used in Russia, or exported through Gazprom's pipeline to Europe. Both options were rejected by Raymond. ExxonMobil would earn more by selling the gas to China or Japan, and Gazprom was contractually bound to a unanimous agreement with the corporation. Miller's efforts to take control were stillborn. But Raymond's intransigence had a cost. ExxonMobil lost the license to develop Sakhalin 3, and the Kremlin decided that development of Shtokman, an elephant oil and natural gas field 340 miles offshore in the Arctic, would not be licensed to an oil major as an "owner," but only as a contractor. Raymond excluded Exxon from the development. On the "win some lose some" equation, he was unconcerned that rejecting Russia's terms fettered the growth of energy supplies. Sakhalin had enhanced Exxon's reputation for operational excellence, and that skill was only available on Exxon's terms.

Financially, as oil prices began to rise after 1999, there seemed no reason to relax the corporation's autonomy. Exxon's size immunized it from negative news. By the fourth quarter of 2004, ExxonMobil's profits were $8.4 billion, $3.8 million every hour. When he was asked what the company would do with the money, Raymond replied, in a rare moment of humor, "First of all, we'll sort through it. And secondly, why in the world would we ever tell anybody in advance what we're going to do with it?" He did not believe such profits would last forever. At around $45 a barrel, oil prices were, he believed, only temporarily high. Looking back to the mid-1990s, he said, "We had a view that the price structure then could not last—that it was fundamentally unstable." The same thing, he believed, would happen in 2004. To calculate future exploration and production, Shell's strategists projected an oil price of $50 over the next 30 years. Exxon's price was $25, tenaciously low. "They're operating outside the rulebook," a Shell executive exclaimed. "You need to keep a long-term perspective in this industry," retorted Raymond. "You have to recognise that pressures come and go. It's swings and roundabouts." As a reward for keeping the faith, earning Exxon $36 billion in profits in 2005 while the share price had nearly doubled since 2003, Raymond's personal annual income was $69.7 million. Recognizing his importance to the industry, Dick Cheney, the US vice president and former chairman of Halliburton, declined invitations to parties in Washington on the night of President Bush's second inauguration in 2005 and flew to New York for a celebratory dinner with Raymond.

Much united Raymond and Cheney, not least their frustration at Congress's general ignorance. Although America was the world's largest energy producer and consumer, Washington's lawmakers, they both believed, never ceased pursuing contradictory policies toward energy, except when they united to attack the oil industry for profiteering and pollution. Then their outrage was "dressed up as policy." Both the vice president and the chairman were irritated by the incoherence of those arguing in favor of subsidies for expensive new fuels while refusing permission to open the Arctic National Wildlife Refuge, and endlessly harping on about America's increasing dependency on imports. The administration's hope for some resolution of those differences in the Energy Policy Act in July 2005 proved forlorn. Although a proposed windfall tax on oil companies was defeated, and Congress approved a clause to overcome Florida's refusal to lease its eastern regions for drilling by redrawing the borders of the Gulf of Mexico and assigning an area under Florida's jurisdiction to Alabama and Louisiana, it forbade the construction of new refineries. This refusal was in part influenced by popular suspicion about Cheney's relationship with the oil industry. Soon after the 2004 election, senators had demanded that executives from the oil majors come to Congress to explain their contribution to an energy task force chaired by Cheney in 2001 and 2002. Some executives appeared to be concealing their collaboration with Cheney to promote policies favorable to Big Oil. The politicians' suspicions were, in Raymond's opinion, self-defeating. America's capacity to refine oil had fallen by 10 percent since the last new refinery was built in 1976, but demand for oil products had risen by 20 percent. Although existing refineries had been modernized and expanded, the reduced investment threatened shortages and inevitable price rises.

A series of hurricanes in the Gulf of Mexico stoked more suspicions. In 2004 Hurricane Ivan snapped underwater pipelines, and then on August 26, 2005, after two other minor hurricanes, Hurricane Katrina's 220-mph winds hit rigs and eight refineries. Four weeks later, before the devastation was repaired, Hurricane Rita damaged deepwater platforms. A quarter of US oil production and nearly a third of its refining capacity were hit. Fuel prices began rising. In the 27 years since the last energy crisis, an entire generation of Americans had grown up unaware that cheap fuel was not a God-given right. Responding to their clamor, Washington's politicians accused Big Oil of profiteering, or "price-gouging." Republican senator Pete Domenici's summary was crude: "Most Americans think that someone rigs these prices." "What [Americans] don't want," said the president, joining the attack, "and will not accept is manipulation of the market. And neither will I." There were suspicions about a decline in OPEC's production, and outrage that at $50 a barrel the price of oil was too high. Saudi Arabia was suspected of manipulating the markets over the previous four years. Recent research had shown that after the collapse of prices in 1998, Middle East oil revenues had fallen in one year by $42 billion, or 29.7 percent, but in 2002, by carefully controlling supplies, the region's revenue had doubled. Raymond was convinced that speculators added $20 a barrel to the price of oil. Accusations of price manipulation, the oil companies knew, were a convenient cover for the more complex causes of the crisis. Although taxation, unrestricted consumption and the refusal to allow offshore drilling in American waters contributed to the perception of a shortage, the 15 percent increase in China's demand in 2004 gave substance to the EIA's projection of global demand increasing by 54 percent between 2001 and 2025. "It's the end of cheap oil," announced David O'Reilly, Chevron's chief executive, "and the beginning of a bidding war between East and West for Middle East oil."

News stories about high prices, greater demand and the threat of a shortage did not persuade Raymond to increase Exxon's investment in exploration. Like the chairmen of his rival oil majors, he feared that oil prices would fall, and investment would undermine the corporation's share price. The politicians were surprised. The oil companies had received huge grants to explore in the Gulf of Mexico, but the billions of dollars had merely been banked as increased profits. In their attempts to understand the industry's conservatism, the lawmakers seized on an official study produced in 2004 showing that the market concentration following 2,600 mergers in the US petroleum industry since the 1990s had led to "higher wholesale gasoline prices" by about 2 cents a gallon, and over 7 cents a gallon in California. The approval of the Exxon–Mobil merger, Senator Chuck Schumer of New York complained, had been "a great mistake." The oil majors had hoodwinked Congress during hearings in the late 1990s about the benefits of mergers. At the time, Raymond and other witnesses had dismissed concern about lack of competition. Six years later the politicians regretted being persuaded by Raymond that Exxon without Mobil was not big enough. The mergers were blamed for increasing prices, underinvestment in refineries, record profits and allegations of collusion among retailers. In 2005 Exxon paid $23.2 billion in dividends and share buybacks, but spent only $17.7 billion on exploration and production. Raymond had boasted that his corporation had earned a 31 percent return on capital in 2005, beating BP, with 20 percent, into second place. With those profits, many in Congress suspected the industry was using the loss of a million barrels a day of oil after Hurricane Katrina as an excuse for extra profiteering. Schumer blamed President Bush for being "addicted to oil companies."

On November 9, 2005, Lee Raymond replied to those accusations at a Senate hearing. With a weary air of "While politicians come and go, ExxonMobil continues in the same course," he explained some realities: ridiculing the idea of oil companies rigging the market, he pointed out that ExxonMobil produced only 3 percent of the world's oil; the market had worked perfectly after the summer's price rises; oil prices had fallen as supplies were restored, especially after Saudi Arabia increased production to compensate for losses from the hurricanes; ExxonMobil invested the same amount in good times and bad, because each project took so long to mature — the proof was Sakhalin's development, which would take 10 years, but would produce oil for 40 years. Polite and self-assured, Raymond gave the impression of regarding the congressmen as puppets.

His manner failed to allay the suspicions held by some about the extraction of oil from federal land. About 25 percent of all oil and gas produced in America came from publicly owned terrain, earning the government $8.65 billion in royalties in 2005. Just prior to the hearings, reports surfaced about a corrupt relationship between government officials and some oil companies to underpay royalties for oil extracted from federal land. This alleged "outright cheating" was calculated to have lost the government about $10 billion over five years. The oil companies have disputed the allegations, and the case is still to be heard in the Supreme Court, but the public's simmering distrust of oil was increased.

To counter the suspicion, John Hofmeister, the president of Shell in America, appealed to the chief executives of Chevron and other oil companies to regain public sympathy by launching an "outreach" program. Hofmeister, formerly Shell's director of human resources in The Hague, had been chosen by Jeroen van der Veer to restore the company's image, rated as probably the worst among the majors in America. To please Dick Cheney, Condoleezza Rice and several senators, he had advised van der Veer to obey the American embargo and abandon the development of an LNG plant in Iran. "The Americans should be kept in their box," Malcolm Brinded, the head of exploration and production, had protested. "You'll get nasty hearings," warned Hofmeister, and Shell retreated. With his new authority, Hofmeister sought to spread Shell's influence. The forum for the industry's discussions, the American Petroleum Institute (API), was dominated by ExxonMobil. "Whatever we do or say will be a waste of money," Lee Raymond told Hofmeister, and the outreach initiative was stillborn.

Talk about global warming, peak oil and alternative fuels irritated Lee Raymond. Nature, not man, he said, was responsible for climate change. In his view, the costs and regulations of the Kyoto Protocol were "unworkable, unfair and ineffective." Calling for more research in 2001, he blamed natural sources in part for the greenhouse effect. The benefits of Kyoto's regulations, he said, "have yet to be proven. Their total impact is undefined, especially because nations are not prepared to act in concert." John Browne's polished speeches to the United Nations and elsewhere about the environment appalled Raymond, who welcomed one congressman's snipe: "We're addicted to oil, and they're addicted to profits." Profit was Raymond's bedrock. "I don't make moonshine," said Rex Tillerson, articulating the reasons behind Raymond's funding of the Competitive Enterprise Institute, a group dedicated to denying that fossil fuels endangered the environment. Nevertheless, Raymond was prepared to chair a major study organized by the National Petroleum Council about the world's dependence on oil. Exxon research showed that between 1980 and 2030 an additional two billion people would inhabit the planet, and the demand for energy would increase by 50 to 60 percent. Providing that energy would cost $17 trillion, and 40 percent would be supplied by oil. Among the obstacles to supplies were some oil-producing countries, none more so than Venezuela, which possessed the largest oil reserves outside the Middle East.

Born in a mud hut and an avowed admirer of Fidel Castro, Venezuela's socialist president Hugo Chávez was elected in December 1998 with 56 percent of the vote, and was committed to redistributing Venezuela's oil wealth among the working classes. His targets were the "rancid oligarchs" and "squealing pigs" who managed PDVSA, Venezuela's national oil company. The executives of the badly managed company were accused of purloining the oil profits to fund their expensive lifestyles. Repeatedly, Chávez highlighted their misuse of PDVSA's luxury villas and private jets at the very time that Venezuela's oil prices had fallen to $8 a barrel and inflation had topped 800 percent, plunging the country into temporary destitution. For their part, the oil executives accused Chávez of "stealing" the presidency. Success in the battle for power depended on the fluctuating oil prices. Chávez's popularity among the impoverished workers rose as prices hit $30 a barrel in March 2000, but dropped at the end of 2001 when they fell to $15. In an attempt to reverse the trend, Chávez courted popularity by promising to seize control of PDVSA. His first step was a presidential decree raising taxes on foreign oil companies from 16.7 to 30 percent, and ordering the transfer to PDVSA of a 51 percent stake in every venture operated by a foreign company. PDVSA's managers, including those appointed by Chávez, opposed these proposals. Foreign investment, they feared, would be deterred, and only the oil majors were capable of doubling Venezuela's production to five million barrels a day. Chávez wanted the opposite. If prices increased, he calculated, production could be cut.

To resolve the impasse, he fired the managers. PDVSA's engineers staged a mass walkout. Oil and food supplies dwindled, and the country became paralyzed. When protestors were shot, Chávez was accused of killing innocent workers. On April 12 the beleaguered president was under arrest, and believed to have resigned. His overthrow was publicly welcomed by President Bush: 60 percent of Venezuela's oil was sold to the USA, and the administration was outraged by Chávez's flirtation with America's bitterest enemies, especially Castro. "Chávez deserved what was coming to him," commented Otto Reich, Bush's Latin America adviser. Two days later Chávez was liberated, and returned to Caracas convinced that Bush had plotted his demise. Unrest erupted again in October 2002. Senior PDVSA executives demanded Chávez's resignation and abandoned their work, reducing oil production from 2.5 million barrels a day to less than 100,000, just as oil prices were rising. Strikes and the use of inexperienced workers were damaging the oilfields. In the war of attrition, Chávez fired 13,000 PDVSA strikers and routed the middle-class protestors. By March 2003 he had regained control.

Amid the turmoil, PDVSA's fate remained undecided. The corporation needed expertise and finance, and the only sources of these were the foreign oil majors. Although he lacked any evidence, Chávez was convinced that the oil companies, especially ExxonMobil, had collaborated with President Bush in the failed coup against him. But with oil prices at about $30 a barrel, he could not bring about the confiscation of the oil companies' Venezuelan assets. Instead, he appeared to support an appeal by Ali Rodriguez Araque, the new director of PDVSA, to lure American banks and oil companies to return to Venezuela. In Houston on May 5, Rodriguez Araque described Venezuela to Exxon-Mobil, Shell, Chevron and ConocoPhillips as open to foreign investors prepared to contribute toward a $43 billion program to increase production to 5.1 million barrels a day. The chairmen of the four oil companies were wary. One day, they noticed, Chávez ridiculed President Bush and threatened to reduce oil supplies to America, and the next he greeted Chevron's executives in Caracas to discuss granting a lease to produce natural gas and oil in the "Deltana," a pristine area. His apparent indecision — "blowing hot and cold" was one executive's description — reflected insecurity and ignorance. Rather than rebuild PDVSA to resemble Brazil's Petrobras, Chávez wanted it to serve his political agenda, and to punish the oil majors.

The rise of oil prices to $42 a barrel during 2004 encouraged that plan. PDVSA and the country were awash with cash. PDVSA was ordered to employ extra staff, causing inefficiency and even higher prices. Chevron and other foreign companies were told to await news of whether their offers to invest had been accepted. Watching oil prices rise to $53, Chávez became convinced that the free market in oil was being permanently strangled. Unlike textiles and steel, the fate of oil, Chávez believed, could be controlled by the national oil companies causing shortages to increase prices and their revenue. Although PDVSA was unable to guide horizontal drills with seismic sensors through 8,000 feet of rock or sandstone in the search for crude oil, he preferred to humiliate the foreign oil companies rather than use their expertise to enrich his country. Victory in a referendum in August 2004 stiffened his resolve to trounce the oil majors. Before turning the final screw, he had to decide whether to use the 268 billion barrels of bitumen oil in the Orinoco Belt to transform Venezuela.

Ten years earlier, Chevron had pioneered the conversion of the semi-solid sludge in the Orinoco into oil. On the basis of an agreement to pay 1 percent in royalties, the corporation had invested $1 billion in the complex project, and in 2004 began shipping 180,000 barrels a day to the US. At that moment Chávez breached the contract by increasing taxes on the Orinoco oil to 16.6 percent. "It's absurd they were so low," said Venezuela's energy minister, as the government's annual income notionally rose from $46 million to $766 million. Chevron's chairman protested. By reducing the corporation's income, he said, its investment to secure more oil in the Orinoco would decline. Chávez had no sympathy for this argument. The executives of the oil majors, he felt, had consistently refused to treat him with respect. He believed that rising oil prices endowed him with sovereign power.

In January 2005, after ordering a review of Venezuela's contracts with oil companies, Chávez flew to Beijing. After the four-day trip he announced that sales of Venezuela's oil would be switched from the US to China. "There have been 100 years of domination by the United States," he explained. "Now we are free and place this oil at the disposal of the great Chinese fatherland." China was given access to 15 oilfields, and in return offered credits to build housing for poor Venezuelans. Shipping oil to China, a 45-day trip that added $5 a barrel to the cost, and China's need to build new refineries for Venezuela's heavy oil, would previously have been dismissed as fanciful, but increasing prices had transformed the finances.

Posing as "Castro with oil," Chávez set out to compensate for Fidel's failure to incite a revolution across Latin America. Offering cheap oil and technical help to Uruguay, Argentina, Brazil and Ecuador, he gathered support among the socialist and nationalist fraternities across the continent to challenge the capitalists. In May 2005 he announced on television that foreign oil companies in Venezuela owed $2 billion in unpaid taxes. Those who failed to pay by January 2006 would be compelled to leave. Simultaneously, he ordered the foreign oil companies to hand over ownership of their operations to the Venezuelan government. Chevron, Shell, ExxonMobil and Conoco risked losing investments worth $8 billion. The roller coaster gathered momentum. In April 2006, as Arjun Murti's year-old prediction that oil would rise to $100 a barrel was gaining credibility, Chávez seized 32 foreign-owned oilfields, declared his intention to take control of operations in the Orinoco, and forecast that royalties on foreign operators would be hiked from 33.3 percent to 50 percent.

By any reckoning, the oil majors were fighting for survival. If prices hit $100 a barrel, Chávez would be liberated from Big Oil. O'Reilly believed that Chevron could not afford to argue with the president, or to retaliate. In any case, the rising oil prices had made that less necessary. Chevron's engineers had struck oil in Big Foot and Knotty Head in the Gulf of Mexico. A triumphant partnership with Shell at Perdito, a cutting-edge project in a previously uncharted area using new light steel down to 16,000 feet beneath the sea's surface, encouraged Chevron's executives to revel in their superiority over BP: "Our seismic processing is a strong indicator of our success." In 2006, O'Reilly increased Chevron's capital expenditure for the following year by 20 percent, from $16 billion to $19.6 billion. After four years of agonizing, a Chevron director sighed, "We're choking on our success." In those circumstances, brinkmanship against Chávez was unnecessary. Rolling over would pay dividends in the future. Chevron, O'Reilly agreed, would remain in Venezuela on Chávez's terms. Shell also capitulated.

By contrast, Rex Tillerson, chairman of ExxonMobil since January 1, 2006, followed the company's standard procedure. Contracts were sacrosanct, and too much was at stake elsewhere in the world to allow Chávez victory. Those who attempted to avoid their legal obligations were always challenged by ExxonMobil. There was, Tillerson believed, no choice. Chávez needed to be punished, not only for acting illegally but for his foolish contempt for the oil cycle—"What goes up, always comes down." Tillerson's announcement provoked Chávez to confiscate ExxonMobil's assets in Venezuela. In San Ramon, O'Reilly's executives gloated over ExxonMobil's fate. Forced out of Venezuela, nearly kicked out of Indonesia and squeezed in Newfoundland and Alaska, their rival was, they agreed, losing the plot.

During the oil industry's 1995 "Hard Truths" debate about the future choices it faced, one participant noticed that "Raymond's public persona as a bigoted tyrant had become invisible. He was normal." Raymond had not anticipated that global warming would be introduced as a major topic. His antagonism to the subject was well known. In 1997 he had urged developing countries to avoid environmental controls and ignore the arguments about climate change, which hindered the development of the oil industry. Renewables, he knew, would only provide 1 percent of the world's energy over the next 20 years. But during the early debates, Raymond was persuaded to change his mind and recommend the limitation of carbon emissions.

The "Hard Truths" report was published in 2007, two years later than originally planned and a year after Raymond's retirement. No one leaves the oil industry poor, but his $398 million package guaranteed unusual comfort. If he had been unloved by his employees, Raymond was appreciated by shareholders. Rex Tillerson, his 54-year-old successor, was accused of "living on his inheritance from Raymond, bringing no extra value to Exxon; and he's just as nasty as Raymond if he doesn't get his way." But Tillerson brought a mood change to the company, and a willingness to listen to new ideas following the simultaneous departure of several senior Raymond loyalists.

John Hofmeister seized on the transition to revive his quest to win public understanding for the oil industry. The discussion among the top executives, he decided, should not take place on Exxon's turf. The host would be Jim Mulva, in ConocoPhillips's offices in Houston. Tillerson was persuaded to fly from Dallas, and David O'Reilly from San Ramon, California. Tillerson was noticeably less adversarial than Raymond. Greenhouse emissions, he admitted, were, despite "significant uncertainty," one factor affecting climate change; and while there was no change to Exxon's principal purpose to increase oil and natural gas supplies, and not to chase alternatives, he did reluctantly agree to Hofmeister's proposal to spend $100 million on a "goodwill" program. "But it's only going to be about oil and gas," he insisted. "We're not going to advertise windmills and sunshine."

Chapter Twenty

The Backlash

————◄◌►————

GRUBBY BATTLES with the dictators of oil-producing states, and the image of the oil industry, had become less important to John Browne. The detail and even the substance of exploration and production had become of less than passing interest. Only the prospect of a historic event, the ultimate deal, excited him. While maneuvering to delay his retirement three years beyond 2008, he contemplated becoming chairman of the world's biggest oil corporation by merging BP with Shell. His entourage listened in respectful silence, but his discussions with Peter Sutherland, BP's chairman, provoked arguments. Sutherland and his fellow directors had tired of Browne's destructive self-importance preventing a proper debate in the boardroom about his successor. Sclerosis, in their opinion, was contaminating Browne's achievements. Instead of acknowledging that even great men must accept the inevitability of a final curtain, Browne appeared addicted to the spotlight.

For some critics, Browne's visit to New York in February 2003 to present BP's annual results had been a turning point. Securing the sympathy of Wall Street's oil analysts was important for all chief executives, but Browne treated these events with special significance, and was particularly anxious to glamorize his presentations. His podium was constructed to enhance his stature, and music and flashing lights combined to produce a Hollywood-style show. Browne's antagonists compared the performance to a circus, reflecting sizzle rather than substance. A question of trust and credibility had arisen.

No one doubted Browne's achievement. In New York he was able to report that his decision in July 1999 to reduce costs by $4 billion, sell assets worth $10 billion and invest $26 billion by 2002 had been a spectacular success. Between 1995 and 2002, BP's value had risen from £20 billion to £90 billion. Only a fifth of the company's capital was based in Britain, and that proportion was declining. Within three years, BP had converted itself into an American company. Nearly 50 percent of the company's capital and 44 percent of the profits were American. BP's ascendancy had been masterminded by upstream cadres, the handpicked and glamorous set who explored for and produced oil. In the Gulf of Mexico, BP was the largest acreage-holder, owning a third of all the reserves, and over the next decade Browne had committed to drilling between four and seven wells every year in the Gulf, at a cost of $15 billion. His 5.5 percent growth target had depended on their success and on radically cutting costs. "Pull the trap," Peter Davies, BP's chief economist, had advised, implying that large numbers of employees should be dismissed, especially as oil prices could fall to $10 a barrel. Unlike in ExxonMobil and Shell, those involved in BP's downstream—refineries, chemicals, marketing and gas stations—were regarded as second-class citizens by Browne and his entourage. Peter Backhouse, a downstream expert, and other potential successors to Browne had lost favor by opposing the relentless pursuit of cost-cutting, especially in BP's refineries.

Expensive, unglamorous and complicated, refineries had failed to earn guaranteed profits for over 10 years. Beleaguered by constantly revised environmental regulations, political opprobrium and the price swings between crude oil and the distillate products, refining was the Cinderella of the oil business. Recruiting engineers was difficult, and the cost of maintaining the metals and valves for a process involving temperatures reaching 540°C was hurtful. "A Shakespearean tragedy" was one description of refining during the era of surplus oil and low prices in the 1990s. Although America's "addiction to oil" depended upon the ugly, smelly processing plants, President Clinton encouraged Carol Browner, his environmental secretary, to increase controls through the Clean Air Bill and prosecutions for infringing laws to produce cleaner fuels. An additional chastisement was the regulations being drafted by the Environmental Protection Agency to reduce sulphur in diesel, a seemingly innocuous technical adjustment that had unforeseen consequences. Browning's laws hit the refineries just as the historic protection for selling gasoline and diesel was disintegrating. Supermarkets were undercutting the oil majors with cheaper fuel obtained from independent suppliers. BP and the other majors regarded the erosion of their profits as a hammer blow.

John Browne was particularly antagonistic toward refineries. Dirty and encumbered by old working practices, they offended "Beyond Petroleum." BP's engineers, he believed, had become an unnecessary burden. Financially, refineries were also toxic. Beyond America and Europe, the production from new refineries exceeded demand, depressing prices. "Worldwide refining capacity is continuing to grow faster than demand," Browne believed in 1999. "Our aim is to reduce refining coverage by about one third." He hoped to save about $1 billion annually. The selling off of BP's refineries was encouraged by Robert Pitofsky's flawed one-size-fits-all rules about competition, which stipulated their disposal as a condition of allowing the mergers with Amoco and Arco. The oil industry, the FTC chairman failed to understand, was unique. Splintering ownership of refineries among independent operators removed the industry's collective intelligence to anticipate bottlenecks.

Browne's attitude was not welcomed by Doug Ford, the ex-Amoco executive responsible for refineries. After the merger, Ford had arrived in London, and a series of intensive conversations had persuaded Browne that he should be appointed to BP's main board. Ford had refrained from criticizing Browne's underestimation of BP's new acquisition. Amoco's five refineries, he believed, ranked among the world's best, and the company's technology center in Naperville, Illinois, was outstanding. Ford was relieved when, after an interview with Browne, Amoco's head of refining Al Kozinski replaced BP's existing expert. But shortly after, Browne ordered the sale of the profitable Alliance refinery in Plaquemines, Louisiana, and the closure of Naperville. At the time, these moves seemed astute. After September 2001, refining profits slumped, recording the greatest losses in 30 years. Neither Ford nor Kozinski protested. They did not foresee that profits would recover in 2003. "What do you know about refining?" Rodney Chase challenged Kozinski, in his customary manner. "You don't know how to run the business properly."

Discomfited, Kozinski did not rebut the challenge, and began a tour of BP's refineries. His visit to Grangemouth in Scotland was revelatory. "Everything's wrong," he said. The maintenance was inadequate, not least because the dykes around the storage tanks were full of oil from the previous year. "BP doesn't have people who understood how to run refineries," he realized. He was unsurprised by this, as his predecessor had been a trader, not an engineer, a symptom of Browne's cost-cutting by buying in or outsourcing expertise. "We know how to find and trade crude," Chase told Kozinski during their next meeting, "but we're not that good at running refineries." Browne and Chase had little intention of improving their skills in that vital area. In 2000, after three incidents within two weeks, including a serious fire and the fracture of a high-pressure steam pipe, Grangemouth was temporarily closed. An official investigation revealed breaches of safety regulations linked to reduced maintenance costs.

Ford and Kozinski drew their conclusions about the cultural differences between Amoco and BP. They were unaccustomed to Browne's confrontational style in formal meetings. His challenges to cut costs, improve performance and increase profits were accompanied by the unspoken threat "You'd better deliver what you've promised." This was alien to those nurtured on Amoco's bloated regime, characterized by Nick Starritt, a former Mobil executive employed by BP, as "process, bullshit and bureaucracy." Browne's brazenness bewildered the

Americans. "Refineries," announced Chase in one meeting, "are an expensive way to play the crack spread." He meant that BP used the intelligence it gleaned from refining to speculate on Nymex and the OTC market. "This is a clash of big egos," concluded Kozinski, confused whether targets were aspirations or orders. He and Doug Ford assumed the latter, and clicked their heels in obedience rather than challenging Browne.

One critical item for discussion was Amoco's refinery at Texas City, one of five refineries owned by BP in America (two Amoco, two Arco and a BP). Rebuilt in 1934 on the site of a 19th-century Rockefeller refinery, Texas City had been repeatedly expanded to become Amoco's flagship, producing six refined products. In 20 years, America's most complex and high-cost refinery had only twice earned a profit. Amoco had repeatedly cut budgets since 1992 to increase shareholder returns, and BP wanted more savings. Despite the industry's excess capacity, the temptation to close the refinery had been resisted because it would have been more expensive to restore the site's purity to meet environmental regulations than to reduce activity and limit maintenance. Accordingly, rusting pipes pockmarked the unmodernized plant. Browne did not intend to change course. At their first budget meeting in 1999 he ordered Ford to cut the costs of the American refineries. In his discussion with Kozinski, Ford mentioned Texas City, the most complex and inefficient of the refineries, as a target for savings. Kozinski would say that he was not given sufficient money to improve Texas City over the long term, but that in Amoco's traditional manner, he never compromised safety. In Browne's opinion, he was simply demanding better competence without taking risks. In 2001 BP spent about $400 million on repairs and new technologies in the American refineries, but limited the costs by not installing the most modern equipment. Browne did not appear to place priority on those details.

Nick Starritt, BP's director of human resources since 1998, had for some time warned Browne that BP's culture was unsuited to overriding the traditions of Amoco and Arco and integrating the acquisitions within a single corporation. The new circumstances, Starritt repeated in memos, personal conversations and group discussions, required special measures to cope with the dramatic expansion and cost-cutting. BP, he warned, remained a collection of companies without the glue of a common culture and adequate "diversity of management experience." Starritt's ideal model was ExxonMobil, but Browne preferred devolution to centralization. By the end of 2002, although Starritt had been acclaimed by professionals as "Human Resources Director of the Year," his relationship with Browne was unwittingly fractured. Anyone expressing dissent was vulnerable. Browne executed his control through Rodney Chase. "It's time for a change," Chase told Starritt, who was bewildered by his fall. His sin was misunderstanding Browne's management culture, an error shared by Ford and Kozinski. Both were required to scrutinize Browne's cost-cutting challenge, and if necessary to reject the targets. Acceptance meant delivery, or else. Ford was unable to adapt to that style. "Rodney wants him out," observed a board member. "Ford's got on the wrong side of the beast. Rodney's fingerprints won't be on the knife, but it will be stuck between Ford's shoulders."

In 2002, Ford retired. John Manzoni was appointed his successor on the board of directors, although he had never managed or operated a refinery. Manzoni, a tireless accountant and marketing expert, obedient to Browne's mantra and methods, was enthralled by the expensive advice showered upon BP by consultants. "I've never seen a budget which can't be cut," he had told Kozinski toward the end of one of his regular 18-hour days. Bedazzled by figures, Manzoni rarely saw the big picture, but as a conceptual thinker he could translate Browne's thoughts into a coherent argument. At presentations in BP's New York headquarters he praised "Beyond Petroleum," and in the conflict between oil and Washington, he criticized the API as the "hopeless bad guy" and the oil industry's "own worst enemy." US audiences were impressed by his candor compared to American oil executives. He had been told that either he or Tony Hayward, a geologist in charge of exploration, would emerge as Browne's heir.

Unwilling to work with Manzoni and Chase, Al Kozinski had taken early retirement. He was replaced as the vice president of refining by

Mike Hoffman, an American engineer in tune with Manzoni's requirements. Since he intended to run the refineries from London rather than regularly visit the sites, Manzoni expected a constant stream of updated data reporting every refinery's performance. Browne's priority, he knew, was to cut the losses. In 2004, Browne decided to sell Grangemouth, BP's star refinery. He expected an independent to be the purchaser. In America, between 1998 and 2003 Valero had increased its daily output from 170,000 barrels to two million by purchasing refineries from the oil majors. Browne wished he could unload all of BP's refineries. His hostility toward them had internal repercussions. "It sends a bad signal to the remaining refineries," admitted one of Browne's aides. "They find it hard to get more finance." Among the casualties was the unloved outpost of Texas City. By never visiting the 1,200-acre site, Browne remained unaware of the state of BP's inheritance, which Exxon or Shell would have instantly remedied.

Texas City was not pretty. Except to the engineers who perfect the distillation of crude oil into fuels and chemicals, refineries are rarely attractive. But even they regarded Texas City as "decrepit." Yet, despite the rust and the lack of paint, 460,000 barrels of crude were processed there every day into 11 million gallons of gasoline. On a rare visit by Manzoni in July 2004, Don Parus, the latest manager, presented a slide show showing the consequence of Amoco's underinvestment in the refinery. His plea for more money was denied. Instead, in 2005 he was ordered to cut costs by 25 percent. Manzoni would say that that figure was a "challenge," and not a "directive." Parus never met the "challenge," because he departed, the fifth manager of Texas City to leave within six years. Before he went, he told the staff, "The Texas City site has undoubtedly experienced the best profitability ever in its history last year." Part of that success was due to BP persuading local regulators in Texas not to enforce stricter environmental controls. An exemption from upgrading monitoring equipment in all five of its American refineries contributed to BP saving about $100 million. The history of the company's thrift would become contentious after the departure of Ford and Kozinski.

In August 2002, Mike Hoffman had written identifying financial cuts as responsible for Texas City's eroding infrastructure and "poor state." His report passed up the chain toward Manzoni. Hoffman's complaint was not new. Maintenance spending, on some disputed calculations, had fallen during the 1990s under Amoco, but BP's imposition of a 25 percent cut on fixed costs soon after the merger had compounded the deterioration. And then BP added more cuts. Hoffman had not protested about the reduction of the refinery's central training staff from 30 in 1997 to eight in 2004, or about those eight instructors also being assigned to other duties. Nor did he complain about the widespread use of casual labor: to save money, about 800 contractors regularly worked on the site alongside the 1,800 permanent staff. He would deny that Parus had specifically asked for extra money, but he accepted Manzoni's congratulations for contributing to a reduction of the maintenance budget.

A warning to BP about the consequences of the cuts at Texas City was commissioned and delivered in January 2005. After surveying the employees, Telos, a firm of security consultants, reported that BP was judged by them to be most interested in money and least interested in people. "Production and budget compliance gets recognised before anything else at Texas City," Telos reported. "We have never seen a site where the notion 'I could die today' was so real... This had a profound impact on us all." The existing culture, Telos found, was of "things not getting fixed," and BP was accused of suppressing notification of accidents. The report was discounted by BP as lacking credibility. Although Browne's overall directive was to cut costs, to improve the safeguards for all the refineries, expenditure had been increased between 2001 and 2004 from $40 million to $115 million. Nevertheless, within the overall budget of a $90 billion corporation, the additional money was less relevant than the prevailing attitude toward the albatross.

John Browne was unaware of Texas City's disintegration. Greg Coleman, the vice president of BP's health, safety and environmental programs, was later even prepared to allege publicly that Browne had "no passion, no curiosity, no interest" in safety issues. Browne's indifference toward the unprofitable Texas City, as he would later admit, meant silence about safety instead of a "clamor for funding." He relied on his management model: Mike Hoffman was responsible for meeting the contract or the targets agreed upon with Manzoni. The flaw was Manzoni's failure to supervise Hoffman's performance, or to notice his inability to pinpoint the precise responsibilities of Texas City's section managers. Hoffman recognized his own impotence. Weakened by the departure of a succession of managers, he hired consultants to compose a request to Manzoni for extra funding. The submission revealed that engineers were employed for 12-hour shifts over 30 continuous days, and also commuted long distances. On the site, alarms and gauges were broken, and some critical equipment was known to be malfunctioning. There was, Hoffman's report would mention, widespread noncompliance with standard procedures. Secluded in London, relying on jargon-splattered reports, Manzoni could not connect the facts to arrive at the obvious conclusion that judgment and discipline at Texas City had withered. "They were being led by the blind," a BP executive would later admit. With hindsight, what followed was inevitable.

On March 23, 2005, a group of contractors completed the maintenance of an isomerization unit, used to boost octane levels in gasoline. To save money, obsolete "blowdown stacks" had once again been installed. Other refineries had phased out that device, but Amaco's and BP's engineers had rejected incorporating new and more expensive safety equipment during three previous renovations, in 1995, 1997 and 2002. To reignite the unit, the engineers were expected to follow strict guidelines, including evacuating the area around the unit. On that day, all the rules were broken. Unaware of the imminent refiring, a group of contractors remained inside a nearby trailer while unsupervised engineers ignored every safety procedure. Unseen, a drum overfilled with fuel and overheated. An experienced engineer would have spotted the pressure increasing and stopped the process. Instead, the contractors allowed 7,600 gallons of flammable liquid to vaporize. Within two minutes the gas spread across the refinery, until it was ignited by an electrical spark. The ground shuddered as a huge clap of thunder was followed by a surging fireball. In the rush to escape, employees scrambled over barbed wire fences. Left behind among the 15 fatalities were James and Linda Rowe, childhood sweethearts. Their daughter Eva had difficulty identifying their charred corpses. Over 170 workers were injured in America's worst industrial accident for a generation.

John Browne arrived at Texas City the following day. Waiting for him at the Four Seasons hotel were Manzoni and Hoffman. Others at the meeting heard Manzoni complain that he had been summoned from his vacation. "This has cost a precious day of my leave," he would repeat in an e-mail. Browne had thought long and hard about his own conduct. A spot check had already blamed human error for the explosion. "Surprising and deeply disturbing" mistakes by employees was the company's conclusion within weeks. Browne recalled the consequences for Exxon's public image of the company's stubborn manner after *Exxon Valdez*. To avoid similar damage to BP, he decided on unusual candor. After moving among the bereaved and the survivors, engaging in four-minute conversations with large men with calloused hands and sunburned faces, he emerged to address the media. His shareholders, and more importantly his erstwhile admirers and critics, he knew, would pass instant judgment on his performance. Contingency lawyers, ambulance chasers, trade union officials and Big Oil's enemies were waiting to pounce.

"The dead are contractors, not employees. What can you do for them?" he was asked. His reply was disarming. "In BP we are responsible for what happens within our facilities, and we will make amends. We cannot repair the past, but because of BP's resources we can make the future a bit more secure. We will conduct a full investigation." Browne's candor, he hoped, would defuse some antagonism. In the darkest hour, his friends noted, he was at his best. After all, explosions and fatalities had occurred at other refineries. In 1989, Phillips had paid $4 million to settle charges after an explosion in Texas killed 23 people; in 1990, Arco had paid $3.5 million to settle violations of laws after an explosion killed 17 in Channelview, Texas; Shell had had two fatal accidents—in May 1988, seven had been killed and scores injured after gas escaped from a corroded pipeline, and the company had paid $160 million to settle 17,146 claims, while in 1990 a fire at the Shell refinery at Deer Park, Texas, had caused injuries and widespread damage. Single fatalities also happened regularly.

But times had changed. The inevitable fine and even the compensation would be untroublesome to a corporation whose annual profit in 2004 had been $15.73 billion, while the first quarter net profits in 2005 were $5.49 billion, 29 percent up, largely as a result of the oil price rise to $57 a barrel. The mood in Texas was unforgiving toward Lord Browne, whose mantra had been "more for less." His announcement that BP had set aside $700 million for compensation to victims, including $20 million for each fatality, and had agreed to pay a $21.4 million fine with limited admissions of liability, was received with suspicion. A bitter blame game erupted between BP and Amoco's retired engineers and directors. The Americans regarded BP's executives as paper-pushing accountants, obsessed by analysts but ignorant of maintaining refineries. BP's directors referred to the "cultural misunderstanding" between the American and British staff, and claimed that Amoco's staff were negligent. "The British manager was a bully," complained one of Amoco's senior directors. "I left when they appointed him my boss."

Outside Texas, the damage was greater. "Beyond Petroleum" was transformed into a spectacular bonehead play, and its originator was recast as a star pupil caught cheating. Not only was BP's image marred, but the financial repercussions were serious. To save money the company had decided not to insure against losses of the kind it would incur as a result of the Texas City explosion. The minimum cost in lost profits and compensation would be $4 billion. Avoidable catastrophes such as this one reinforced the antagonism toward a hated but essential industry. "BP is too arrogant in the way it wanted to profit," commented an oil trader in New York, a victim of the "Cushing Cushion." Beyond his contrite public appearances, Browne felt the pressure in London from Peter Sutherland and his fellow directors. There was talk of internal inquiries, criticism of his integration of Amoco and suggestions of "slashing his remuneration." His good fortune was BP's care to avoid publicly compounding its legal liabilities. But eventually, he knew, there would be a reckoning. First, there were formal rituals he could not avoid.

The official investigation was entrusted to the Chemical Safety Board (CSB), led by Carolyn Merritt, a chemical engineer. Browne's hopes that the small agency with its low-paid staff could be persuaded to show understanding were dashed. Merritt focused on "systemic lapses" by BP's management and, in a preliminary report, warned that BP's managers, scarred by "a cultural issue," still posed "an imminent hazard" to safety. Browne feared the worst. Merritt was heading toward damning conclusions. Her preliminary conclusions blamed BP for budget cuts that knowingly left "unsafe and antiquated designs...in place, and unacceptable deficiencies in preventive maintenance were tolerated." Even though supervisors knew that instruments were unreliable or broken, she discovered, repairs were deferred or abandoned to save money. Merritt's criticism temporarily united the warring BP and ex-Amoco directors to accuse her of ignorance and prejudice.

Browne needed to limit the damage. The CSB's staff, he complained, was unqualified to understand the complexities of Texas City, and he believed Merritt was grandstanding. While willing to accept that human error was to blame, he wanted to demolish the idea that a policy of reducing costs and increasing profits could have caused the accident. Merritt, he rightly feared, was ignoring BP's case that people, not budgets, were responsible. He preferred not to understand her criticism that under his direction, BP's money had not been spent on curing "problems [that] were many years in the making," and that there had been a failure to invest in equipment and the processes of ensuring safety. The CSB's final report, he decided, was certain to be flawed. "Our enemies are stoking up the criticism," he complained to aides. "They want to create antagonism toward us in America." Convinced of a conspiracy, Browne had no alternative but to mitigate the CSB's condemnation by commissioning BP's own report. James Baker, the former US secretary of state and a Texan oilman, agreed to conduct an inquiry financed by BP. Browne believed that Baker would be receptive to his reason and intellect. His ploy of an independent investigation did not receive universal approval.

Brent Coon, a Texan lawyer famed for his hostility toward industrial corporations, had introduced himself into the saga, representing hundreds of claimants on a contingency basis. "I hate big corporations," Coon announced. "They cheat wherever they can and put people's lives at risk. BP has blood all over their hands. I'd like to depose Lord Browne over a roasting pit. You're not supposed to die if you work in a petro-refinery plant." Even before its publication, Coon criticized Baker's report as "a whitewash written by Browne's friend." He was not entirely wrong. While criticizing BP's lack of "safety leadership" as a "corporate blind spot," Baker acquitted BP of "purposefully withholding resources" and shortchanging safety. In a direct contradiction of the CSB report, which quoted BP's challenge to refinery managers to cut budgets by an additional 25 percent, Baker declared that there was no evidence of BP cutting costs. Citing John Manzoni's own admissions, Coon scoffed that Baker's report was specious. "Fixed-cost reductions," Manzoni had told the CSB's investigators, contributed to the "situation" at Texas City. Some BP insiders understood that Manzoni was alluding to a "lack of trust, motivation and sense of purpose" within the corporation.

Two months later, the CSB's report unequivocally criticized BP: "Warning signs of a possible disaster were present for several years, but company officials did not intervene effectively to prevent it." The CSB found "a broken safety culture at BP," and that equipment maintenance was based on "run to failure." Manzoni later acknowledged that he had not properly carried out his responsibilities. Despite visiting Texas City, he said, he "did not know which questions to ask, did not ask the right questions and was not told about the plant conditions."

Having failed to achieve his purpose, Browne's interest in Baker's report was cursory, and he never read the CSB report. "I wasn't aware of that," he repeated when he was eventually asked by Brent Coon about the maintenance failures. His detachment was reinforced by an explosion at Total's storage facility at Buncefield in Hertfordshire on December 11, 2005, in which 43 people were injured in Europe's biggest explosion since the Second World War. Total admitted negligence, but unsuccessfully disputed the £1 billion claim for damages by denying that it could have foreseen the extent of the harm. In contrast, BP paid most claims promptly, contributed $32 million to public institutions in Texas, paid a $50 million fine, and eventually settled with Eva Rowe and other claimants after Coon's aggressive campaign. Mike Broadribb, BP's international director of process safety, conceded, "Texas City was a preventable incident, a totally preventable process failure, a management failure and a culture failure...If we learned one thing from this incident it's the need for humility." Browne refused to go that far. The evidence of BP placing profits above safety, he knew, had aroused antagonism in America. Internationally, the company was also damaged. Foreign governments would take a jaundiced view of BP when it pitched for licenses. There was, Browne knew, limited time to repair the damage. Peter Sutherland's skepticism had become undisguised. Concerned about his fate and frantic to restore his reputation, Browne introduced a myriad of processes into BP's operations to impose safety checks. Called "The Green Book," his plan reversed a lifetime's agenda to delegate and to decentralize BP's management. This somersault, costing about $3 billion according to internal estimates, evoked screams of protest about excessive bureaucratization. Less than a year after the explosion at Texas City, it did not prevent a new calamity.

In the Western Operating Area of Alaska on March 2, 2006, 5,000 barrels or 267,000 gallons of oil seeped out of GC-21, a 34-inch pipeline carrying crude into a gathering center for shipment through the Trans-Alaska Pipeline System (TAPS) to Valdez. The leak was the largest ever on the North Slope. Browne knew that BP was vulnerable. The company had been fined $716,000 in January 2005 for an explosion in a well bore caused by inadequate procedures. The latest incident would be exploited by former oil broker Chuck Hamel, who since his financial ruin at their hands had dedicated his life to scrutinizing the oil corporations' activities.

Hamel had become the recipient of regular complaints from Alyeska's employees, especially after a congressional committee heard workers' complaints about the pipeline's management during hearings in 1992 and 1993. Later in 2005, Hamel heard about two more blowouts and an oil spillage, and encouraged the Environmental Protection Agency to launch a criminal investigation into BP's conduct. Alaska, once BP's Elysian fields, had become a millstone. Gloom permeated its operations there as production fell to just 35 percent of the peak in 1988, and the cost of maintenance increased proportionately. BP was the pipeline's sole operator, but the anticipated financial benefits of merging BP's operations with Arco's had evaporated after the compulsory sale of Arco's share to Phillips, and Exxon's refusal to renegotiate the financial contracts.

About 2,000 miles of pipelines crisscrossed the oilfields on the North Slope of Alaska, terminating at Prudhoe Bay, a desolate area with a permanent population, according to the official census, of five people. From there the oil was piped 16 miles to the 800-mile Trans-Alaska Pipeline, which carried it south across the Alaskan tundra to terminals at Valdez. Those pipelines, built on stilts to allow animal movement and to protect the permafrost, had been hailed as a success when oil started to flow in 1977. Construction, however, had been beset with difficulties. To save money, the pipeline was not built from stainless steel, but from cheaper carbon steel, which was more liable to corrosion. Sand and water had accumulated at the bottom of bends in the pipeline, and when this had combined with sludge deposited by the crude oil, the trapped liquid had turned into acid. The normal treatment to prevent the acid corroding the metal was the introduction of a chemical inhibitor into the pipes, but that was partially abandoned once technicians realized that the sludge was preventing the inhibitor from oozing across the metal. The first sign of corrosion was spotted in the early 1990s. The usual solution would have been to introduce a so-called "pig" inside the pipes: a traveling machine that scraped the sludge and the microbial bacteria that caused corrosion off the metal tube. But, conscious of Browne's edict to cut costs, the engineers abandoned that option, relying instead on exterior checks using ultrasonic equipment. Even that precaution was limited. The contractors were working under the supervision of Richard Woollam, BP's corrosion manager. Known for refusing to suffer fools gladly, Woollam regularly deferred maintenance in order to reduce costs, and threatened contractors that he would replace them with a cheaper rival unless their prices were reduced.

Browne had understood the pipeline's politics and finances ever since he had worked as a low-paid field engineer in Alaska during the 1970s. BP's relationship with Alaska's politicians had always been fraught, especially because of unpaid royalties and taxes. In 1995 the company paid Alaska $2 billion to settle tax disputes as part of a sweep that earned the state an additional $3.7 billion. Behind the widespread public mistrust of the oil industry was fear. Experts estimated that the North Slope could be exhausted by 2012, and would be shut down by 2018. The state's only hope of salvation was to open up the Arctic National Wildlife Refuge to oil exploration.

Some oil companies, encouraged by Alaskan politicians, were convinced that 10.3 billion barrels of recoverable oil lay beneath the wilderness of the ANWR. There were also an estimated 35 to 100 trillion cubic feet of natural gas across north Alaska. Recovering both would require huge investment, not least to run a natural gas pipeline 1,800 miles from the Arctic through Canada to the US. The latest estimate of the cost over 10 years was $20 billion, the biggest private venture in oil's history in America. Browne did not believe in the existence of huge oil reserves, but was required to pose as hopeful that the historic antagonism about developing the ANWR would be reversed after blackouts rippled across California in January 2001. President George W. Bush, entering office against a background of soaring electricity prices and the potential bankruptcies of utility companies, was keen, like any Texan oil operator, to satisfy the industry's high expectations. His appointment of Vice President Dick Cheney, the former CEO and chairman of Halliburton, to forge a new energy policy encouraged the oil industry's hopes that the ANWR would finally be opened up, not least to reduce America's dependence on imports.

Their fears that President Clinton would decree in his final days that Alaska's coastal plain was a national monument, thus preventing development, had not been realized. Clinton was advised that the decree would be pointless, although the temporary distress caused by discussions on the subject increased the oil industry's anger toward the Democrats. Permanently antagonistic toward Big Oil, the party was accused of focusing on conservation, renewables and lies about the industry. Bush's vocal support for opening up the ANWR and his zeal for deregulation aroused hope, and then disappointment. Despite the need for reliable supplies, the majority in Congress appeared to regard oil as a "dirty business." Bush and Cheney realized that even a Republican Congress would not be able to overcome the support for environmental regulations or counter the argument that opening the ANWR was not a substitute for a proper energy policy. The administration's energy bill was delayed by opponents of its plan to allow drilling in the ANWR. The only bright spot was BP's operations at Seal Island, a man-made gravel atoll off the Beaufort Sea. Drilling for the first time through the Arctic ice into water began in December 2000. The crude would flow six miles through the first subsea pipeline in the Arctic to the Trans-Alaska Pipeline. Contrary to expectations, the drilling failed to improve BP's share price or profile. The next hope was to win approval and a special tax agreement to transport the natural gas to the USA. Without that breakthrough, the continuing decline of Alaska's oil production would sap BP's commitment.

The task of Steve Marshall, BP's president of exploration in Alaska, was complicated by defeatism. The decaying infrastructure and the 10 percent reduction in maintenance spending was accompanied by a deteriorating relationship with the workforce, especially after Richard Woollam canceled their monthly lobster dinner to save money. Contracting out work, Marshall admitted, had contributed to "poor" relations, deterring the contractors from advocating "pigging" the pipelines. Relying on the assurance that no leaks had been found in the 16 miles of connecting pipes, Marshall was unaware that the steel was gradually being reconverted into iron ore. He did know, however, that

as a precaution the pressure in the pipes had been steadily reduced from
800 psi in 1992 to 80 psi in 2006, but he was oblivious to the consequence that as the pressure was reduced, more water entered the system,
speeding corrosion. Marshall's options were limited, but in response to
Chuck Hamel's warnings he did commission Coffman Engineers to
investigate the warnings of corrosion.

Like all whistleblowers, Hamel was hated by BP's executives. In
late 2000 he had received calls from trade union officials about the
company's poor safety record. In particular, Robert Brian, employed
in Alaska by BP for 10 years, complained about oil spills, suspected
corrosion in the pipelines and BP's "orchestration of inspection procedures" to rectify any faults in advance of official checks, a charge of
which BP said there was no evidence. After serving for six years as the
union representative on BP's health, safety and environment committee, Brian claimed that a BP executive was placing him under pressure
to remain silent. Several hundred pressure valves, he wrote, had not
been checked, some for nearly three years.

Both Woollam and Hamel knew that BP's attitude toward safety
was influenced by the state government's lethargy. To please the oil
companies, Alaska's enforcement agencies, reflecting the attitude of the
local politicians, were ineffective. The Department of Environmental
Conservation's budget had been cut by 55 percent since 1991, and its
request for $500,000 to monitor pipeline corrosion had been rejected
after the oil companies retorted that they were monitoring it themselves. The state employed just five inspectors to check 3,500 wells
across a vast area. Obligingly, these inspectors notified BP about the
time of each check, allowing the company to repair any faults before
they arrived. By comparison, California, which produced 40 percent as
much oil as Alaska, employed 40 safety officers. Despite their meager
resources, the Alaskan inspectors did reject 10 percent of the safety
shutoff valves during the first quarter of 2001, a considerable increase
over previous years.

"Chatter" about Robert Brian's complaints was heard by Chris
Knauer, the chief investigator of Democratic Congressman John

Dingell's House Energy Committee. To BP's good fortune, the Republican leadership, focusing on Enron's collapse and the arguments in favor of nuclear energy, was uninterested in a trade unionist's complaint about a multinational led by Lord Browne, an international hero. A disappointed Hamel was unaware that Richard Woollam had finally received a damning report from Coffman Engineers. The pipes, Coffman's team discovered, had not been inspected internally for 14 years, and were probably suffering some corrosion that could be removed using smart pigs. The existing anticorrosion program, they wrote, had been inadequate. BP was accused of providing insufficient information and of being unhelpful. Vehemently, Woollam rejected the report. He accused the consultants of being "biased and unduly negative," and of producing an error-strewn report that ignored BP's successful record of containing corrosion. "We will never expose our workforce to unnecessary risk," wrote BP to its consultants, requesting that the final report be rewritten. Steve Marshall, BP's manager in Alaska, denied in 2006 that he had seen the Coffman report at the time.

BP ignored the advice to deploy "pigs" inside the transit pipelines. The cost was excessive, production would have been interrupted and the part-use of pigs in 1998 had been, BP believed, sufficient. The decision was supported by senior staff. "This place is getting safer and safer," said George Blankenship, BP's manager at Prudhoe Bay. The pipeline, agreed George Nelson, Alyeska's manager, was "an example of industry's excellence" and a demonstration of "how well the oil industry can co-exist in a fragile environment." The only visible environmental damage had been caused 20 years earlier by exploration trucks driven across the tundra and by abandoned building sites. After those were removed, said Bob Malone, BP's former president of the pipeline service, the corporation would be seen as a white knight for cleaning up the wilderness. Marshall did however urge his staff to focus on safety, "as if our lives and our future in Alaska depend on it. Because they do." His exhortations were not taken seriously. "Window dressing" was the reaction of the new trade union leader, not least because Richard Woollam's intimidation of the staff had not been curbed.

Steve Marshall's freedom of action was constrained by John Manzoni. Some blamed Manzoni's excessive zeal for his niggling demands to reduce costs, which extended even to BP's refusal to continue paying for films, ice cream and the annual workforce barbecue. Manzoni's self-conscious certainty might have had its origins in his family's distinction. Alessandro Manzoni's 18th-century novel *The Betrothed* is regarded as the greatest novel ever written in the Italian language, while Verdi dedicated his *Requiem* to a Manzoni. From that lofty perch, Manzoni imposed the same "fixed-cost reductions" on Alaska as at Texas City. In the squeeze between profits and performance, he was responding to the culture to increase profits.

On August 16, 2002, A-22, a BP oil well on the North Slope, exploded, burning and maiming a worker, Don Shugak. The company denied failing to inspect the well, but nevertheless closed 137 of its other wells for tests, losing 50,000 barrels a day, a significant financial loss to the Alaskan economy. BP announced that a mechanical integrity test had previously been conducted on the damaged well, and that it had shown no cause for concern. Robert Brian accused the company of lying, complained that it was seeking to silence him, and then resigned. "BP," Chuck Hamel wrote to the company, "appears to be engaged in a cover up of its operations violating regulations... and safe industry standards." In November 2002, Robert Brian testified to a congressional committee about the pressure to be silent placed upon him before he left BP. "There is no doubt that cost-cutting and profits have taken precedence over safety and the environment," he said, asking rhetorically whether, rather than "Beyond Petroleum," BP actually stood for "Beyond the Pale." A few weeks later, just before Christmas, an explosion in Alaska killed Rodney Rost, a BP employee. BP would be fined $1 million by Alaska's regulators for failing to prevent these accidents. Steve Marshall had reason for concern. An internal report about the A-22 accident revealed that production at the well had, as Brian alleged, restarted "without taking adequate safeguards," making BP responsible for the explosion.

Kip Sprague, working in Richard Woollam's anticorrosion team, blamed a lack of money. Despite the criticism of BP for poor engineering practices, the improvements to methods of inspecting the pipelines for corrosion advised by the Coffman report had barely been implemented. The company still relied on ultrasonic equipment and visual checks, which missed existing corrosion points. On April 10, 2005, three weeks after the Texas City explosion and three years after the first accidents in Alaska, Sprague protested to Marshall about his refusal to approve an enhanced anticorrosion program. "After 20 years of minimalist resources and maintenance," Sprague e-mailed, delivering an adequate program "overnight" was impossible. "Reliable funding and resources is a yo-yo, accurate scheduling activities is a joke…we are sitting on a huge backlog of over 1,000 locations." Twenty years of parsimony, both Sprague and Marshall knew, could not be rectified overnight. Expertise and understanding, and not just extra money, were needed.

John Browne was unaware of the safety problems in Alaska. His political priority was seemingly to lobby in Washington for access to the ANWR, although his support was not genuine. George Bush and the Texan oil lobby, he knew, were eager to exploit any oil reserves that would minimize America's reliance on Saudi Arabia. Although environmentalists raised the specter of ruining a 19-million-acre wilderness, Browne expected that rising oil prices, and his own star status as North America's biggest oil and gas producer, would spur the administration to win sufficient support for legislation in Congress. The mood was only soured by the state of Alaska's decision to launch an antitrust suit against Exxon and BP for conspiring to withhold its natural gas from US markets. Depressed by the state's declining fortunes and frustrated by the delay in building the $20 billion pipeline, Alaska's politicians accused the two oil companies of "corporate greed" in their efforts to reinforce their power over North Slope supplies.

Alaska's legalistic gripes were inconsequential compared to the simmering unease in BP's headquarters about technical incompetence. Memories of Texas City had not faded, although John Manzoni decided to leave the corporation, when, in July 2005, in the wake of Hurricane

Dennis, Thunder Horse, BP's $1 billion status symbol in the Gulf of Mexico, was discovered to be listing 20 degrees. Production would be delayed for at least a year, and repairs would cost at least £250 million. The alarm within BP after Texas City had been mild; after Thunder Horse the stakes were raised. BP's technical incompetence, rather than Hurricane Dennis, had caused the listing. Lacking in-house experts to scrutinize the platform's design and construction, BP had relied on freelance contractors. The internal inquiry revealed the consequences of Browne's apparent indifference toward maintaining engineering excellence within BP. He had deflected criticism by presenting Peter Sutherland with an expert's report describing Thunder Horse as the victim of cutting-edge engineering. Sutherland was unaware that Exxon's, Shell's and Schlumberger's specialists ridiculed that conclusion, but the chairman knew that, at the age of 57, Browne could not change course, and he was disturbed by the irritation aroused among the staff by Browne's introduction of the "Green Book" to improve safety. Then, on March 2, 2006, while in London he heard about the oil leak in Alaska. Across the state there was outrage, not only about the spill but, more important, about the gift to environmentalists opposed to opening up the ANWR.

In Washington, the Department of Transportation ordered a team to start intensive testing of the pipelines, and Chuck Hamel lost no time in contacting the staff of Congress's two energy committees. The pipes, he asserted, had not been cleaned or examined with a "smart pig" since 1992, and were filled with sludge. Congressman Dingell, an avid conservationist but no longer the chairman of the House Energy Committee, did not wait for an official report or replies from BP. After the committee's investigators had spoken to BP's staff, he wrote to Bob Malone: "I have concluded that BP's cost-cutting was severe enough that some BP field managers were considering measures as draconian as reducing corrosion inhibitor to save money." He asked BP for all the records of pigging from 2000 to 2005, and to provide all e-mails referring to any reduction of corrosion inhibitor.

A review of the e-mails unearthed Kip Sprague's opinion that the testing was "a joke." Reluctantly, that e-mail was included in the bundle sent to the committee, but other more incriminating ones were excluded. Before their delivery, Dingell had received from Norman Mineta, the secretary of the Department of Transportation, an interim report about the pipelines' maintenance. Some pipes, it appeared, had not been cleaned with a smart pig since 1990. The tests, according to BP in that year, had been "terminated due to high volumes of debris present in the pipeline." That policy, Mineta told Dingell, "does not represent sound management practices for internal corrosion control." By early August the Department of Transportation's inspectors had found 16 "significant anomalies" of "unexpectedly severe corrosion" at 12 locations, and even a small leak in the eastern sector of the Slope. The logs of BP's staff, reported the inspectors, exposed a haphazard regime of checks at intermittent intervals, missing corrosion by "a few feet." In London, Browne agreed to increase the maintenance budget in Alaska by 15 percent, to $74 million. The following day, August 7, another pipeline in Alaska split. Within 24 hours the US government ordered the closure of the whole Alaskan oilfield. Eight percent of America's entire daily oil supply, 400,000 barrels, was cut off. BP immediately agreed to replace 16 miles of pipeline suffering "severe corrosion," at a cost of $150 million, and to increase maintenance spending in 2007 to $195 million. The company's carefully nurtured reputation was being shredded.

In BP's London headquarters at No. 1 St James's Square there was anger. Not with Browne or the managers in Alaska, but about the "exaggerated hyperbole over a 30-square-foot area of contamination," the regulatory agencies' motives, American politics, and BP's mistake in overreacting. "We're not dealing with an apocalypse," shouted one director. "All this talk about cost-cutting is an absurd exaggeration." The board, naturally enough, relied on what Browne told them about the nature of the incident, and about safety and cost-cutting in Alaska, and did not undertake their own investigation.

Five days later, under pressure from the oil industry, the US government agreed to reopen half of the pipeline. Browne tried to dismiss the three successive catastrophes—Texas City, Thunder Horse and Alaska—as "a series of unrelated events." Compounding BP's predicament, at the end of June the CFTC announced that the company was to be prosecuted for manipulating the prices of propane gas in February 2004. "It would have been better if July had been canceled," Browne quipped when unveiling BP's interim results in August. His hopes of limiting the damage by admitting responsibility, offering generous compensation and joking were being undermined by Chris Knauer, an investigator for the House Energy Committee. During a trip across Alaska, Knauer met Maureen Johnson, responsible for BP's maintenance on the North Slope. Her hesitant replies to his questions persuaded him to report to Joe Barton, the chairman of the House Energy Committee since 2004, "It doesn't smell right, why did it happen?" BP's predicament was unappealing. Over the previous years, Browne and his lobbyists had plied their charm and the company's narrative across the capital, endlessly presenting the chief executive to congressmen, senators and the most senior politicians in the White House as the European equivalent of the legendary GE chairman Jack Welch. Methodically, Browne had worked through the celebrity list, especially Senator Jeff Bingaman, the chairman of the Senate Energy Committee, John Dingell, Joe Barton and Dick Cheney. All were friends of Big Oil, and all felt betrayed by Browne.

Congressman Joe Barton's distrust of Browne and BP's staff was growing. Recalling FTC chairman Robert Pitofsky's suspicions that BP wanted to dominate the West Coast oil supply by total ownership of Alaskan production, Barton wondered whether there was a link between the shutdown in Alaska and new allegations of BP's manipulation of propane prices. He suspected Big Oil of plotting a conspiracy, a suggestion Browne ridiculed. Chuck Hamel bolstered this theory when he delivered a copy of Coffman's five-year-old report. With hindsight, the consequences of BP's cost-cutting appeared irrefutable to Barton, who was persuaded that the company's "chronic neglect in Alaska [had] directly contributed to the shutdown." On August 11 he wrote to Browne accusing BP of dishonesty. The company's past assurances that the leak in March was an anomaly, he wrote, and that the corrosion control program was working, were untrue. The evidence of corrosion, he continued, "contradicts everything the Congress has been told. The fact that BP's consistent assurances were not well grounded is troubling and requires further examination."

Steve Marshall was summoned to testify to Barton's committee on September 7, 2006. In the high-ceilinged, wood-paneled room, Barton looked down upon Browne's representative. Accused of systematic neglect, Marshall's defense was, "I am not a pipeline expert." As he looked up at the politicians on the dais, he must have grasped the seriousness of BP's plight. Five years after BP had received the Coffman report and refused to use "smart pigs" despite the danger of corrosion, events suggested his culpability. "I am even more concerned," Barton told Marshall, "about BP's corporate culture of seeming indifference to safety and environmental issues. And this comes from a company that prides itself in their ads on protecting the environment. Shame, shame, shame."

Eleven days later, Marshall's reputation was again attacked. Chris Knauer and members of the Oversight and Investigations Subcommittee working for Barton obtained a report written by Vinson & Elkins, a firm of lawyers, describing the atmosphere for those employed by Richard Woollam as "fraught with workplace intimidation and harassment from senior management." Congressional staff had found an e-mail dated February 5, 2003, and headed "Authorisation for Expenditure" in which the proposal for a permanent pig had been rejected as too expensive. BP, the congressmen quipped, no longer stood for "Beyond Petroleum," but for "Big Problems," "Broken Pipelines" and "Bloated Profits." Recalled to testify again, Marshall was told by Democratic Congressman Bart Stupak, "We are now learning that there were a number of troubling personnel problems in BP's corrosion management program...over the past several years...These problems apparently caused a 'chilling' atmosphere in workers' ability to report health and safety issues," deterring the report of potential problems. Why, Marshall was asked, had he tolerated the intimidation of the Corrosion

Inspection and Chemicals Group (CIC)? Congressmen were renowned for using crude hyperbole on occasions such as this, but that was little consolation for Marshall. He had no defense. After arriving in Alaska in 2001, he admitted, "We had a poor relationship with our workforce. I worked hard to change that. We have done a lot, but we still have a long way to go." His attempts to improve matters in 2003 had failed. Woollam had been ordered to undergo "sensitivity training," but had resisted the instruction to cease intimidating employees, and in 2004 he had been reassigned to Houston. Woollam refused to testify to the committee by pleading the Fifth Amendment, as protection from self-incrimination. "It's a corporation rotting from within," commented a member of Barton's staff. In London, Peter Sutherland and his fellow directors were convinced that BP had become the victim of grotesque political opportunism. The company was suffering its *Exxon Valdez* moment. Nevertheless, Sutherland again concluded, John Browne, once an asset, had become a liability.

"Beyond Petroleum" was haunting the company. The unmistakable green-and-yellow sunburst logo promoting BP as the world's second largest solar producer, with plans to invest $8 billion in wind generation in southern California and Scotland, was vocally lambasted by Chuck Hamel as "greenwash." Oilmen, Hamel told the politicians and others flocking to his home by the Potomac in Washington, "play Russian roulette like you wouldn't believe." Browne complained that BP was a victim of Hamel, but Robert Malone, a Texan, was realistic: "I don't believe in bad luck. For many the shine has come off BP over the last year as we have stumbled operationally." To halt BP's relentless slide, Browne appointed Malone as BP America's chairman, and once again sought out a celebrity to limit the damage. Judge Stanley Sporkin, a distinguished retired lawyer, was hired to review BP's systemic disregard of employees' complaints. Barton was unimpressed: "You apparently have done a 180-degree turn and are doing now what you should have done years ago ... Your former corrosion manager took the Fifth Amendment ... But I find it hard to imagine that no executive further up the chain of command was aware of what was going on." Even Browne's agreement to spend $550 million over two years to improve integrity management program in Alaska could not forestall the backlash. BP's lawyers had discovered incriminating internal documents and e-mails about cost-cutting and the nonuse of inhibitors in Alaska. These could have remained undisclosed if, in March 2007, Congressman John Dingell had not planned a hearing to allow BP an opportunity to describe its success in clearing the oil spill and describe progress on the repairs. Dingell believed that BP would be denying the allegations of drastic cost-cutting, but the box of new documents delivered to the committee on the lawyers' insistence made that impossible. In public, BP's spokesman murmured, "These documents are not really important," but the evidence reignited hostility. "Everything in these documents," said Joe Barton, "suggests that BP is trying to do the right thing in public while fighting like a tiger in private." BP had become a punching bag. Its rivals believed that the corporation was heading toward a temporary graveyard, and blamed BP for increasing the general mistrust of the oil industry.

John Browne was fretting, although not about BP's problems in America, but his own fate. He was due to retire in February 2008 at the age of 60. If Exxon could extend Lee Raymond's contract for two years, he thought, BP could do at least the same for him, and extend his by three years. To avoid the issue, he had delayed the nomination of a successor. Ignoring Peter Sutherland's impatience, he hoped to remain in place, an attitude that was supported by several important advertising executives and bankers enriched by contracts from BP, who agreed that the corporation needed Browne's talents to steer a route out of danger. With Browne's encouragement, the *Financial Times* had some years earlier unprecedentedly published eulogistic descriptions of his achievements on three consecutive days. Their glowing effect still lingered. Sutherland's public expostulation about "this madness" had not deterred Browne from approving Merrill Lynch circulating a note to clients on Friday, July 21, 2006, describing Browne's retirement as a "medium-term risk" for BP. Investors were advised to write to the company suggesting that Sutherland step down to allow Browne to inherit the chairmanship.

Browne's long campaign to remain was addressed by Sutherland on the same day. The whole board had agreed at its recent meeting in Venice that Browne should leave in summer 2008, the hundredth anniversary of BP's first oil strike, after his sixtieth birthday in February. Browne was summoned to Sutherland's office. The directors, he told Browne, had agreed that he should announce his retirement. The appropriate moment would be the following Tuesday, at the end of his announcement of BP's latest results. Visibly angry, Browne refused. The next day, Anji Hunter, Tony Blair's trusted former assistant whom Browne had hired to assist with BP's public relations, was marrying a well-known television presenter at St James's church in Piccadilly. Among the 360 guests, including much of the Labour Party establishment, would be sympathetic journalists. For years Browne had relied on these admirers to sustain his heroic profile, and latterly to campaign for his indispensability to BP. To sabotage Sutherland's timetable for his departure, Browne decided to launch a public campaign to embarrass the directors. At the wedding Browne met journalist Andrew Neil, now editor-in-chief of a small publication, *Sunday Business*, for the first time. Coincidentally, the following day the paper featured a report describing a popular demand that BP's board retain Browne beyond his retiring age. "That's put the canary among the pigeons," sighed a BP executive early on Sunday morning as he awaited Sutherland's irate telephone call.

Sutherland summoned Browne to his office on Monday morning. "You'll say unambiguously tomorrow that you're going," he bristled. "I'll think about it," replied Browne. When the Irishman hit back with a sharp ultimatum Browne, visibly shaken, agreed to withdraw. He only asked that his departure should be delayed until the end of 2008. Sutherland agreed to the seven-month extension. In return, Browne would cooperate to find his successor. Since John Manzoni had left BP under a cloud, Tony Hayward was the obvious candidate. But Sutherland remained dissatisfied. While Lamar McKay, a former Amoco manager, was negotiating across Washington to resolve BP's major crisis — or, as Browne's critics snapped, "to clear up the mess" — Browne's continued presence would be damaging. Ten years earlier, David Simon, the former chairman, had said to an executive, "There are two things you don't want in BP. First, to work for John Browne; second, to have John Browne work for you." Sutherland had been an admirer of Browne's relentlessness. Ten years later, he was no longer prepared to play along.

Chapter Twenty-one

The Confession

————◄○►————

E VEN VLADIMIR PUTIN, usually cool and unflappable, was shaken by the news. Revenge was inevitable. Jeroen van der Veer, the chairman of Shell, had confessed on July 21, 2005, that the cost of developing Sakhalin 2, the giant oil and natural gas operation in the Pacific, had spun out of control. The reports to the Kremlin suggested extraordinary duplicity.

Two weeks earlier, on July 7, Alexei Miller, Gazprom's chief executive, had met van der Veer at the Shell Centre in London to sign a new agreement for Sakhalin. Ever since the original PSA was signed in 1994, the Kremlin had felt cheated. At that time, oil was $15 a barrel, and few imagined that prices would go much higher. Facing bankruptcy, Russia had swallowed the extortionate conditions demanded, giving generous tax advantages and, more remarkably, Marathon and then Shell permission to own 55 percent of the project without any Russian partner. Eleven years later, to moderate the imbalance, Miller had enticed Shell to improve Russia's dividend. After months of negotiations, Miller had been persuaded that the revised PSA agreement for Sakhalin 2 was fair. In exchange for transferring 25 percent plus one share to Gazprom, Miller agreed that Shell could retain 30 percent of the operation, and receive a license to develop a 50 percent interest in Zapolyarnoye-Neocomian, a giant gas field in western Siberia. The crunch for Miller and Dmitri Medvedev, the head of Gazprom, was Shell's $10 billion costs to develop Sakhalin. Once that money was spent and Sakhalin was operational, Gazprom would receive 25 percent of the profits. The natural gas would also enhance Gazprom's power to play West against East through the pipelines, and gain a profitable foothold in Asia.

Unknown to Miller, one aspect of the new contract, buried in the small print, removed Gazprom's power to veto board decisions. Just before the deal was signed in London, Shell and its Japanese partners had changed the rules. To veto any board decision, a shareholder needed more than 25 percent of the shares. Van der Veer had not told Miller about that change before they signed the contract. Gazprom, he knew, was a corrupt and inefficient Goliath. Worth about $360 billion and controlling about 20 percent of the world's natural gas, potentially 1,200 trillion cubic feet, Gazprom had been strengthened by rising energy prices, but Miller's ambitions were restricted by the company's inability to meet the demand for gas. Plans to develop new sources in the Yamal Peninsula, Shtokman on the Arctic shelf, eastern Siberia and Kovytka relied on foreign expertise and expensive pipelines, which were beyond Gazprom's finances. Without Shell, Miller knew, Sakhalin 2 could not be developed. "The deal is fair," van der Veer had said, feeling no need to guide Miller to the passage in the small print that had apparently been missed by Gazprom's lawyers in Moscow. He would later say that he was unaware of that clause.

Alexei Miller arrived late at Shell's headquarters on July 7. London was paralyzed by that day's suicide attacks on the city's public transport system. After exchanging brief comments with van der Veer about the carnage across the capital, the two men signed the agreement, held a brief press conference, shook hands and launched themselves back into the chaos to cross the River Thames.

Ever since Phil Watts had squeezed the estimate to build Sakhalin under $10 billion in order to win the directors' approval, Shell's accountants had grappled with endless studies in their attempts to control the spiraling costs. Shell's mistake was to have adopted Marathon's original plan, dividing operations between the oil in the north of the island and the LNG plant in the south, connected by a 500-mile pipeline. No Shell director had understood the complexities of building and joining two offshore platforms, a plant to liquefy natural gas into LNG and a condensate refinery, with a pipeline crossing 1,086 streams and rivers inhabited by shoals of salmon. Shell's predicament was caused by self-deception. To secure the directors' support and to satisfy Phil Watts's hunger for reserves—Sakhalin would add at least one billion barrels of oil and 17.3 trillion cubic feet of gas to Shell's reserves—the costs had been artificially squeezed. Watts's justification was that Sakhalin would be a "sweetener deal" to secure more contracts in Russia. Walter van de Vijver had been the first to spot the discrepancies, and after his departure rigor mortis set in. Malcolm Brinded, the new head of exploration and production, had spoken about restoring control over the costs, but to little effect; Shell's accountants sought to avoid censure for their original underestimate; and Paul Skinner, the downstream director, suggested that the company should abandon Sakhalin and instead spend the money on refineries and developing the Canadian tar sands.

Shell's internal disputes were aggravated by the constant rotation of the executives responsible for the development. Frequently, after a manager at Sakhalin or in The Hague had mastered its complexities, he would be transferred to another post. The only certainty was the dramatic increase in the costs. The financial review undertaken in Moscow by Shell's manager for Russia John Barry revealed the principal reasons. Steel prices were soaring, the ruble was appreciating in value, and labor costs had exceeded the estimates because Shell's engineers often refused assignments to the frozen wasteland. Hired contractors, filling the gaps among the 18,000 workforce, showed limited loyalty. Completion was postponed from 2006 to 2008. Van der Veer expected the latest cost review to be completed in September 2005. Alexei Miller had been warned to expect a $3 billion increase, but the accountant collating all the studies called van der Veer on July 20 with unexpected news. "We're going to be $10 billion over," he said. "The final cost will be $20 billion." The discovery of the $10 billion discrepancy, van der Veer knew, exposed fundamental weaknesses within Shell, but there was a more urgent crisis to resolve. "My goodness," he said. "If this leaks, the Russians will be very angry."

Legally, there was no immediate reason to disclose to Gazprom that the costs had doubled, but on the eve of the merger between Shell and Royal Dutch, the corporation was obliged to give any new information to the stock market. The Kremlin needed to be warned in advance. Van der Veer placed a call to Alexei Miller. Shell's indispensability, he hoped, would persuade Gazprom's chief executive to accept that his ignorance of this new development in London two weeks earlier had been genuine. Miller was on holiday, so a desperate van der Veer telephoned Ivan Solotov, the translator for the Russian department of energy. "Ivan, I've got a real problem," he said. "Sakhalin's costs have increased by 100 percent. It's not $10 billion but $20 billion. Believe me, I knew nothing about it." Solotov drew in his breath. "I believe you, Jeroen, but no one else will." He passed the message on to Miller and Dmitri Medvedev. Both were furious. Shell, they were convinced, had deliberately cheated. Gazprom would need to earn another $10 billion by selling oil and natural gas before it received a share of the profits from Sakhalin.

Jeroen van der Veer's offer to fly to Moscow to deliver a full explanation was consistently rejected. He would have to wait until Putin's state visit to Holland on November 1. Surrounded by other Dutch businessmen, he met the president in the mayor of Amsterdam's official home. Spotting van der Veer, Putin unleashed his anger about Shell's untrustworthiness. Van der Veer's attempts to reply were brushed aside. "I have to meet him separately and alone," he begged a Dutch politician watching his humiliation. Before leaving the mayor's house, Putin agreed. In a side room, without interpreters, the two men argued in German for 20 minutes. Van der Veer assumed that Putin understood the background. Shell's technology, project management and marketing skills were far beyond Russian competence. The Russians working on the project had so far failed to master the skills to produce 3D seismic, horizontal drilling or LNG or to design rigs able to withstand earthquakes and tsunamis. Gazprom's rawness was equaled by that of the officials at the ministry of natural resources in Moscow. Without Shell, Sakhalin would not be developed, and based on an oil price of $30 a barrel, Russia could expect to earn $80 billion over the project's lifetime. "You will," said van der Veer, restraining his emotion, "still get the same, good-quality house, but our initial budget was too low. We won't do a lousy job, we'll do the same good job, but it will just cost more." Putin appeared amenable, but wanted to revise the deal. Negotiations, they agreed, would start in Moscow. Insensitive to the Russians' instinctive suspicion of the West, van der Veer was convinced of Putin's gratitude to Shell. The root of his mistake was self-inflicted.

In 1999, Shell had moved its Russian headquarters from Moscow to Sakhalin, isolating the company at the very moment that Putin was reasserting the Kremlin's authority across the country. Reporting to The Hague via an Italian executive based in the regional office in Dubai, John Barry, Shell's manager for Russia, was ordered to promote Shell by sponsoring the Bolshoi Ballet, building schools and community centers, and offering foreign trips to influential Russians. Public relations was not a substitute for cultivating relations with government officials and potential enemies, but his excuse was truthful: "In Nigeria I could always get to see the president. Here, I can never see Putin and not even Sechin. It's impossible to get into that small clique." Detached from Putin's revolution, no one in The Hague understood the real background to Alexei Miller's request to renegotiate the PSA agreement, or that Mikhail Khodorkovsky had persuaded Putin that PSAs were ruinous. In 2000 Putin had received a memorandum from the Russian finance and tax ministry outlining the disadvantages of the PSA agreement with Shell. "The contract should be annulled," he was advised. Putin disagreed. Russia, he insisted, would always adhere to contracts. He added, "We'll put Gazprom in and increase our income through Gazprom."

The original orders from Phil Watts to Malcolm Brinded and Rein Tambeuzer, a Shell executive sent to Moscow, about their negotiations with Gazprom suggested that there was no hurry, even though Watts accepted that Shell would eventually hand over a stake to Gazprom. Assuming that to delay would be shrewd, van der Veer continued drawing out the discussions. After his meeting with Putin in Amsterdam in November 2005, he did not change his tactics, and the negotiations with Gazprom in Moscow were inconclusive. By May 2006 the lack of progress, first by John Barry and then by Chris Finlayson, his successor, was irritating Miller and Putin. The Dutch, Putin was told, were haggling. He decided that Shell should be embarrassed. Officials in the state's agencies did not require specific instructions to understand the Kremlin's requirements. No one, especially foreign corporations, could be in Russia for more than a few days without breaking the law. Survival depended on negotiating protection from prosecution with government officials. Until then, Shell had benefited from that understanding, but Putin's edict removed the shield. "We'll hit them first, then they'll listen," was the order. Shell's cozy world was about to implode.

Ever since the tsarist era, the Kremlin had shown little concern for Russia's environment. Siberia's plains were polluted by oil spills, and the Aral Sea between Uzbekistan and Kazakhstan, formerly the world's fourth-largest inland body of salt water, had been transformed by drainage for irrigation into a virtual desert. Like his predecessors, Putin had been silent on this subject until van der Veer's bombshell and the excruciatingly drawn-out negotiations. Government lawyers, responding to Putin's request for a lever against Shell, spotted that under the original PSA agreement signed with Governor Farkhutdinov, Shell was exempt from Russian taxes and the regulatory courts, but, by omission, was not immune to Russia's environmental laws. No one in Shell or the Russian government had even thought about the issue. In the customary Russian manner, the Kremlin's direction to low-level bureaucrats at the country's environmental protection agency, part of the natural resources ministry, surfaced mysteriously, and was perversely negative: the agency was to cease turning a blind eye to Shell's activities in Sakhalin. The official deputed to enforce the law was Oleg

Mitvol, the agency's deputy director, an intelligent 40-year-old electronics engineer, previously employed in space research, who had also studied law and accountancy.

In June 2006, Mitvol received an unexpected telephone call from Dmitri Kisitsin, representing Environmental Watch over Sakhalin, a Russian organization previously unknown to Mitvol. Kisitsin expressed his concern about the island's fate. Until that moment, Mitvol's office had been unaware of the activities of Shell or ExxonMobil 10,000 miles away, but soon afterward Mitvol received calls from Greenpeace and a Russian representative of the World Wildlife Fund. In unison, they complained that Shell, in constructing the 500-mile pipeline, had felled thousands of trees, causing soil erosion, and had used the wrong concrete. Worse, to save money the contractors had not raised the pipeline over the 1,086 rivers, which was preventing the salmon spawning. The environmental groups also claimed that another pipeline to the offshore platforms was jeopardizing 126 rare gray whales breeding around the island. The death of a single breeding female whale, said the WWF spokesman, would be "catastrophic." In normal circumstances, Mitvol's negative response to these protests would not have surprised the environmentalists. For more than a decade their pleas to Russian officials to save the whales had been ignored, but now that changed, and Mitvol, echoing the Kremlin, expressed distress for the mammals' plight. Very publicly, Shell began to receive a stream of criticisms of its conduct.

Ian Craig, Shell's director in Sakhalin, had not anticipated the environmental problems. Nor had Malcolm Brinded. As part of Shell's expensive public relations campaign in Russia, the corporation had published the pledge, "The biggest contribution to sustainable development will come from finding environmentally and socially responsible ways to meet the world's future energy needs." Quoting that pledge, Mitvol enjoyed his first contact with Craig. Shell, he said authoritatively, had failed to abide by Russian laws. By this time environmental groups were demonstrating in Moscow, and similar protests were mounted in New York and Zurich. Unusually, an environmental group,

SEW, had appealed to a Russian court to order Shell to cease damaging the environment. In early July nearly 50 groups were "spontaneously" protesting against the impact of Shell's operations on the whales and fishes, and even objecting to their harmful effects on the indigenous population. The emotional onslaught sparked doubts among officials at the European Bank of Reconstruction and Development (EBRD) about its provision of a $300 million loan to finance the development. Including the bank in the lending consortium had been a Dutch idea to give political stability to the project. Instead, the EBRD's officials in London began an "environmental and anthropological impact assessment" to evaluate how Sakhalin's development would affect the island and its inhabitants. Shell had merely wanted to borrow money from the bank, but instead found itself encumbered by social engineers promoting "international standards of good governance" and concerning themselves with "the environment and democratic norms." Although Russian politicians had consistently ignored such matters, the Kremlin was now using them to harass Shell. Van der Veer and his subordinates were struggling in uncharted waters.

Oleg Mitvol was unaccustomed to the spotlight, but in his own words, he felt the hand of destiny urging him to fight for a noble cause. "We want international investment," he puffed at the outset, "but we don't want to be made into a banana republic." Fearful that the Russian government might revoke the permit to develop Sakhalin, and that Mitvol was more than a Kremlin puppet, Shell's directors had organized a steady stream of callers to his office. Among the first were Russians employed by the Sakhalin Energy Corporation, financed by Shell. "Shell is a friend of Russia," was a phrase Mitvol heard frequently. Two British diplomats had called to question Mitvol's relationship with the Kremlin. "I'm completely independent," he replied. Shell's directors arrived last. "You don't realize," Mitvol told them, "Russia has changed. This new generation of Russian bureaucrats is different." They did not believe him.

The squeeze on Shell enhanced Vladimir Putin's authority during his welcome of foreign political leaders to the Group of 8 (G8) summit in Saint Petersburg in July 2006. Rising oil and gas prices had inflated Russia's importance. Dick Cheney was annoyed about his host's new-found swagger. Putin's mood, he observed, had changed during 2005. Russia was no longer just a volume player, but also a price player. Oil had transformed the country's obsolescent agricultural and industrial economy. Now, with just 3 percent of the world's population, it controlled 34 percent of the world's gas and 13 percent of its oil. Cheney confirmed Putin's belief that his strategy, outlined 10 years earlier in Saint Petersburg, had been successful. Russia, said Cheney accusingly, was using energy as "a tool of intimidation or blackmail," and backsliding on democracy.

Scenario planners in Washington, considering the options for oil supplies and energy security, questioned a meeting between Mahmoud Ahmadinejad, the Iranian president, and the leaders of China, Russia and eight other central and south Asian nations to discuss "more effective ways of cooperating" in energy. On March 7, 2002, Richard Armitage, the deputy secretary of state, had warned, "We will not stand idly by and watch [Iran] pressure their neighbors." That threat had proved to be empty. Now, four years later, Putin intended to exploit the Iranian crisis caused by America's embargo. To prepare for his visit to Tehran for a Caspian summit in 2007, Lukoil and Gazprom were encouraged to offer Iran help against America. At the same time, he was pondering how to undermine the BTC pipeline running from Baku to Ceyhan. Caspian oil, Putin believed, should run through Russia. The new bellicosity emerging in Moscow, Cheney feared, projected through the "Shanghai Cooperation Organization" and the forthcoming Caspian summit, was uncomfortably reminiscent of Hugo Chávez's threats to endanger Western oil supplies.

Determined that Russia should no longer be compared to Nigeria but to Kuwait, Putin had won applause by declaring during a meeting of Russia's Security Council on December 22, 2005, that the Motherland was "back on top," playing a key role in the world. Impervious to protests, he pledged that Russia would act solely in its own interests. Rising oil prices had reinforced the *siloviki*'s self-confidence. The Saint

Petersburg crowd of former KGB and military officers including Igor Sechin, Dmitri Medvedev, the deputy prime minister and chairman of Gazprom, and Viktor Ivanov, formerly of the KGB and now chairman of Aeroflot, the country's biggest airline, enjoyed reminding the oligarchs after Mikhail Khodorkovsky's conviction in May 2005 that they had to obey Kremlin Inc. Reasserting control over Russia's energy was an essential ingredient of Putin's plan. Sechin had staged a phony auction of Yukos assets during 2005 and organized their transfer to Rosneft, the state-owned company he chaired, embracing 21 percent of Russia's oil production. Negotiations were under way for Roman Abramovich to sell his stake in Sibneft to Gazprom on January 1, 2006, for an estimated $13 billion. Questions were later asked about whether Abramovich kept all of the money. Billions, some alleged, were earmarked for Putin and his close associates. Energy supplies, prices and profits had transformed Putin's Saint Petersburg thesis into a crusade. The coercion emerged in Ukraine.

During November 2005, Gazprom announced price rises for natural gas delivered to Ukraine from $50 to $160 and then $230 per unit. The Ukrainian government refused to pay the increases, but continued to take gas from the pipeline crossing its territory from Russia to western Europe. On January 1, 2006, Gazprom turned off the taps. Within hours, in the midst of winter, the pressure fell by 30 percent. European countries were shocked, and their leaders used a G8 meeting about energy security, coincidentally held days later in Moscow, to persuade Putin to agree a compromise and supply Ukraine at $95 per unit. Supplies were restored, but Putin's demands remained unequivocal. During a meeting with EU officials in the Black Sea resort of Sochi in May to negotiate a European Energy Charter, he demanded that in return for Western investment in Russia, Gazprom should be allowed to invest in western Europe. The EU's negative reaction, described as "near hysterical" by the Kremlin, confirmed Putin's suspicion of "totally unfair" discrimination, and he rejected the proposed charter. In retaliation for Europe's "unprincipled competition," expressed in its merely "considering" Gazprom's request to invest, Putin directed the sale of

Russia's energy to Asia. Russia, he announced, was turning to the East. Sakhalin's oil and natural gas would be part of the new prize for China. In Saint Petersburg, Cheney and the other leading G8 politicians were powerless spectators of Putin's revenge.

In July 2005, Hu Jintao, the Chinese president, had visited Kazakhstan. For 200 years the Kazakhs had sought alliances with Russia as a defense against the Chinese. Despite that history, a 620-mile pipeline had been completed from Kazakhstan to Alashankou in northwest China, and the Chinese now wanted to buy a stake in Kazakhstan's oil reserves. The offer was rejected as a step too far. Hu Jintao's next stop was Moscow. Despite disappointment over Putin's cancellation of the 2003 agreement with Khodorkovsky, China had lent Rosneft $6 billion in February 2005 to buy Yukos's Yuganskneftegas production unit, in return for which a million barrels of oil a day would be delivered to China through a new pipeline to be completed by 2008. Seeking more supplies, Hu Jintao arrived in Moscow just after Putin had returned from a trip to Japan accompanied by 100 Russian businessmen. In Tokyo, the Russian president had reaffirmed his commitment to build a 2,600-mile pipeline to the Sea of Japan costing $11.5 billion. "A new Russian gas region has started to emerge," wrote Viktor Khristenko, Russia's industry and energy minister. To balance that pledge, Putin told the visiting Chinese president that 2006 would be "the Year of Russia in China." During his visit to China in March 2006 with 1,000 Russian business leaders, Putin witnessed the chairmen of Gazprom and CNPC sign a cooperation agreement to pipe natural gas from Kovytka. The ceremony was a charade. Putin lacked the necessary $14 billion to develop the Kovytka field (62 percent of which had been licensed to BP), or funds to finance the pipeline's construction, but no one dared to call his bluff.

To further tantalize the West, Putin dangled the possibility of the oil majors developing the giant Shtokman field in Russia's Barents Sea. Only Western oil companies possessed the expertise to exploit the field, 342 miles from the shore and containing an estimated 141 trillion cubic feet of natural gas. Ever since its discovery in 1988, Russia had dithered about granting a license. Five Western companies believed they had won the Kremlin's approval in 1990, but two years later the agreement was canceled. In 1995, Norsk Hydro of Norway, Conoco, Total and others were encouraged to consider an LNG terminal in the Barents Sea, but that was also abandoned. In 2000, Gazprom signed agreements with Exxon, Conoco, Chevron and Shell, and in 2005 invited their bids to build an LNG plant, but the following year Putin excluded all the American companies.

By October 2006, in the midst of the argument with Shell, Putin had decided that Russia would develop Shtokman itself. Since the country lacked the expertise to do this on its own, nothing could happen, yet Putin flaunted his intention to supply 10 percent of the US's LNG consumption. He enjoyed the confusion this created, but later agreed to allow Total and ConocoPhillips to develop the field, but without any stake in the gas itself, a deal that other majors rejected for sound financial reasons and, like previous proposals, would probably crumble. The continuing uncertainty annoyed Dick Cheney and President Bush. At the G8 summit in Saint Petersburg they found Putin joyfully dismissing questions about energy security as "misguided colonial-era arrogance." Unrepentant about cutting off the gas in Ukraine and frustrating Chevron's plan to export more oil from Kazakhstan by opposing the expansion of the pipeline through Russia or the construction of another one under the Caspian to Azerbaijan to link to the BTC, Putin presented the summit as his personal triumph. "Oil delirium is influencing everything," Cheney snapped, angry that Putin spurned any agreement that would benefit the West. Putin's Asian alliance with China and Iran was, said an American visitor, "OPEC with nuclear bombs."

During the summit and over the following weeks, Jeroen van der Veer and Shell's lobbyists urged Putin and Igor Sechin not to jeopardize the West's biggest investment in Russia. Rebuffed, van der Veer asked ExxonMobil's Rex Tillerson to join Shell in a common front. Pleased not to be in the firing line, Tillerson refused. "I don't know what's worse," he commented. "Having a cost blowout like that, or not knowing it's going to happen." Shell's directors were scathing about ExxonMobil being "content to let others hang out to dry." The company, Phil Watts had discovered, had been similarly unhelpful during Shell's arguments with the SEC about its booked reserves. While Shell was being pilloried by the Kremlin, ExxonMobil had not suffered for its increase of costs at Sakhalin 1 from $12.8 billion to $17 billion. The difference, van der Veer knew, was Shell's political weakness. For Putin to challenge ExxonMobil would be akin to confronting the American government. Shell lacked similar political patronage. The outcome was the suspension by the Russian general prosecutor's office on September 16, 2006, of Shell's environmental license to develop Sakhalin 2. Corrupt Russian officials at the ministry for natural resources, said the prosecutor, had been bribed by Shell in 2003 to approve the agreement. No evidence for the claim was ever produced. A court would decide whether to revoke the license because of Shell's "unqualified project decisions."

Nine days later, on September 27, Oleg Mitvol was sitting aboard a 160-seater government-owned Ilyushin jet with journalists and representatives of the environmental groups flying toward Sakhalin. The journey bore the hallmarks of government sponsorship. "I needed a plane to deliver some papers to Sakhalin," Mitvol explained disingenuously, "so I thought, if it's empty why not take some journalists as well?" Standing on Sakhalin's muddy hillsides surrounded by his guests, Mitvol identified upturned trees and landslides as examples of Shell's disregard for the environment. Pointing at two dead fish in a stream, he pronounced the corporation guilty of pollution. "Those roads," he said during a helicopter ride, "were built without permission; the pipeline is destroying the salmon fisheries, and look at the dredge Shell is dumping into the bay." Back on land he quipped, "If someone had done this in America, he'd be in jail. Here, he's sitting in a Mercedes." He concluded, "This construction cannot go on. We must stop the project and start over again. We want criminal cases for every damaged river. I think I'll open a criminal case for every tree they cut down. If criminal cases are opened for everything, the company will come to its senses and stop the barbarian activity. Shell could be fined $50 billion."

Mitvol returned to Moscow triumphant. Although Shell made serious allegations against him, he confused the company's directors by presenting himself alternately as an independent one-man band and as a mere cog in the government's machine. Only Vladimir Putin, van der Veer knew, was empowered to halt him. Van der Veer mobilized supporters including Tony Blair, Japan's prime minister Shinzho Abe and Gerhard Schroeder, the former German chancellor, who was employed by Gazprom, to personally contact the Russian leader. Pressure from Rex Tillerson, Shell suspected, deterred the US State Department and the White House from offering similar help.

Unusually, ExxonMobil's failure to become involved proved self-destructive. Soon after Mitvol's return from Sakhalin, ExxonMobil had been told by the ministry of natural resources that the first shipment of crude oil, due to leave Sakhalin 1 in October, would be halted for health and safety checks. Robustly, the company declared that its tanker the *Viktor Titov* would sail as planned. Just as the Shell directors had anticipated respect for their achievements, ExxonMobil's managers were proud to have completed a seven-mile horizontal well in 61 days, 15 days ahead of schedule, linking Sakhalin 1's oilfield to a terminal. A nearby Italo-Russian energy project to build an onshore pipeline, ExxonMobil's executives noted, was disastrously delayed. Without ExxonMobil or one of the other oil majors, the Texan engineers smiled, the Russians seemed unable to complete difficult projects in their own country. Their self-congratulation was to bear a price. ExxonMobil's bluff was called by Sechin, and the *Viktor Titov* remained moored while negotiations began to extract a better deal from the company. Simultaneously, the ministry questioned whether Total's PSA agreement in Kharyaga in the Arctic remained acceptable. The Kremlin had calculated that $18 billion had been invested in Sakhalin 1, Sakhalin 2 and Kharyaga, but that Russia had received only $500 million in revenue. "Laughable," was the translation of the Kremlin's reaction. In Putin's coterie, the commercial sanity of repaying the investment before taking any profits was derided. The fates of gray whales and spawning salmon were useful tools. "International pressure on Shell's behalf," Mitvol said, "will not have an effect. We have to care for Russia's environment and reputation."

Fearing that Shell's investment, worth about $5.5 million at that stage, would be lost, Chris Finlayson visited Alexander Medvedev, Gazprom's deputy chairman. "I must tell you," said Finlayson, "if you kick us out, the scrap metal will be worthless." Medvedev remained impassive. "The oil," continued Finlayson, "will only flow in 2008." Medvedev's expressionless face betrayed no hint of Shell's fate. The possibility of replacing Shell with Schlumberger had been discussed, but the contractor's services were limited to production. Schlumberger lacked experience in interpreting and finding solutions in exploration, and only drilled on the spot identified by the customer. It could not operate alone, offshore. Gazprom had no choice. Only Shell possessed the expertise to exploit Sakhalin. Restricted by those facts, serious negotiations started once Medvedev was persuaded that Shell "would not give in." Initially, Medvedev appeared content to accept Shell's offer of a 25 percent stake but at a lower price. However, during the following two-hour negotiating bouts, intermediaries told van der Veer that the Russians wanted 50 percent. He panicked. Fearing disruption and huge costs if Shell's staff was expelled, he did not bargain for a compromise, but utterly capitulated. There was no point, he told his fellow directors, in making small moves that would be rejected. "We have to take a big step to get a fundamental solution and make Gazprom and the Russian government happy," he explained. Without testing Gazprom's resolve any further, van der Veer flew to Moscow to settle the deal with Miller. At the end of his surrender speech, he realized that Miller did not understand. Only gradually, with more help from the Dutchman, did Miller recognize that he had won. "Alexei cannot believe his luck," observed one of his assistants. "Miller made a lot of progress that morning," agreed van der Veer. The remaining details were how Gazprom would pay the historic costs, and whether in cash or oil.

Inching toward a financial settlement, Shell rectified its environmental blunders. Two subsea pipelines were relocated away from the gray whales' breeding grounds, and the pipelines across the rivers were rebuilt. By December 8, the deal between van der Veer and Medvedev was sealed in Moscow. Shell would sell 50 percent plus one share of the Sakhalin development to Gazprom for $7.45 billion, and Gazprom would not be liable for any extra costs in the future. The oil price was calculated at $4 a barrel, lower than for any other Russian project. Shell's own share in Sakhalin was reduced to 27.5 percent. Once the deal was agreed upon, Shell's environmental problems suddenly disappeared. Out of step with the new politics, humiliated by his demotion and roundly scorned, Oleg Mitvol complained that his new superior, recruited from Saint Petersburg, was "corrupt." Mitvol was demoted. "I've been replaced by Ludmilla, a former cleaner in Saint Petersburg's best venereal disease clinic," he complained. To parade his triumph, Putin insisted that van der Veer return to Moscow and appear on television with him and the chairmen of Gazprom, Mitsui and Mitsubishi to confirm that Sakhalin 2 was under Russian control. Van der Veer agreed. Appearing with Putin on television from the Kremlin would guarantee the contract. "We need a strong signal about the environment," he said to Putin in German. "Leave it to me," replied the president. "I'm pleased," Putin told Russian viewers on December 21, "that our environmental services and the investors have agreed on the way of resolving the ecological problems." Van der Veer would consider the humiliation worthwhile. Sakhalin was safe, even if Shell had debooked half of its reserves and lost billions of dollars of profits.

Defeating Shell inspired Miller. To raise oil prices, in November 2006 OPEC began to cut production by 1.5 million barrels a day. The small decrease in shipments from the Gulf was noted by the tanker-trackers, the agencies that employ spotters at the oil terminals to physically count tankers and their loads in and out. The creeping increase of oil and natural gas prices, and the growing apprehension in both Asia and the West of energy shortages, animated Miller's ambition to dominate Europe's gas supply. With Putin's encouragement, he set about establishing Gazprom's presence in Germany, Italy, Hungary and Britain, and then looked to Africa. There was a piquant headiness to challenging Shell in Nigeria, its troubled heartland. The country's oilfields were in turmoil, and Russia, Miller believed, could appeal to Nigerians as fellow victims of Western exploitation. Disregarding the Soviet Union's misunderstanding of Africa's tribal conflicts during the Cold War, Miller felt Gazprom could offer the Nigerian government an alternative to Shell.

New estimates had increased Nigeria's oil reserves to 140 billion barrels, half of Saudi Arabia's, and Nigerian oil was cheap to produce. Notionally, Shell's reserves could have been boosted by the discovery of 1.6 billion barrels in the Niger Delta, but constant strife was hampering production. Shell's conundrum, Jeroen van der Veer recognized, was its failure to increase its overall oil reserves. The industry was spending five times as much annually as it had less than 10 years earlier—from $80 billion in 1999 to over $400 billion—and it was expected to spend $300 trillion between 2005 and 2050 to produce energy; but Shell was stymied. Local governments, like Nigeria's, with sufficient reserves to more than double production to six million barrels a day, were standing in the way of its ambitions.

Production in the delta had intermittently restarted in 2003, but gasoline pumps across the country had run dry, projects had been abandoned and Shell was losing its dominant position to ExxonMobil, which had chosen safer offshore locations. The arrest in London in September 2005 of Die-preye Alamieyeseigha, the governor of the oil-rich Bayelsa state, on charges of corruption and money laundering, was hailed as a new opportunity to clamp down on Nigeria's rampant sleaze. The chance was lost. Allowed bail on a bond of £1.5 million, Alamieyeseigha escaped from Britain in October 2005 dressed as a woman. On his return to Nigeria, his tribesmen, encouraged by antigovernment politicians, launched attacks against Shell's installations. Repeatedly during 2006, militants in the delta used explosives to destroy pumping stations, sabotaged pipelines and, despite foreign engineers traveling in armored cars, abducted over 150 workers. Retaliation for the kidnappings and the killing of soldiers was messy. Shell paid ransoms to rescue the workers, and suspended operations, while the government dropped bombs on the gangs stealing oil from the pipelines and allowed rampaging troops to burn homes and shops. In those conditions, neither Shell nor the other oil majors could invest to stop spillages. Playing a cat-and-mouse game with the government, Shell halted work on a $6 billion project to end the flaring of about two billion cubic feet of natural gas from 1,000 wells every day, and on the construction of a power station using that gas. In theory, Shell was also committed to building an LNG plant, but Nigeria's finances were anarchic. In 2006 the oil companies handed about 90 percent of their oil revenue, or $25 billion over three years, to the government, yet the politicians refused to pay Nigeria's share of the oilfields' development. "The situation is grave," admitted Basil Omiyi, Shell's first Nigerian chairman, appointed in 2004. More than Shell's entire profits from Nigeria had been reinvested in the country, yet production had fallen. Shell's local chairman lamented the Nigerian leaders' cynicism. Similarly, the civil servants, enfeebled by military dictatorships, had lost the tradition inherited from the British of loyalty to the state. On reflection, Shell's executives realized, their error after 1958 was to assume any responsibility for funding community projects. Participation had become a poisoned chalice. Hospitals had been built by the corporation, but there were no doctors, and Shell's school buildings stood empty because the state had provided no teachers. Omiyi could not solve the contradiction of compelling the government to use its petrodollars wisely, but not interfering in the country's affairs. The anarchy of Nigeria could only be solved, all the oil companies realized, by erecting a physical barrier between the oil and the militants.

Shell's fate depended on Bonga, an offshore oil rig, isolated from the delta gangs, costing $1 billion, producing 150 million cubic feet of natural gas and 225,000 barrels of oil a day, 10 percent of Nigeria's production. Shell hoped that its revenues would be used by the Nigerian government to repay $3.8 billion owed to Shell and their joint venture. On each $100 barrel of oil, the government took $90, leaving Shell with $3; the remainder was used to finance production. An agreement became unlikely after the arrival of Alexei Miller in Nigeria in early 2008. This challenge to Shell on its own terrain aroused unusual curiosity. Offering "strong technical expertise and financial resources," Miller was on a mission of revenge against Europe's decision to restrict Gazprom's investments. Since Europe received 25 percent of its natural gas supplies from Russia, and anticipated demand doubling by 2030, Miller's dream was to control western Europe's energy supplies. He predicted "a great surge" in prices to a "radically new level," with Gazprom rather than OPEC becoming the world's biggest energy provider.

His first call had been on Colonel Gaddafi. Western sanctions had been removed from Libya as part of a wide-ranging settlement negotiated with John Browne's advice by Tony Blair. Fifteen corporations, of which 11 were American, obtained exploration licenses in the first round of bidding in January 2005. To reassure himself of Libya's reliability, Rex Tillerson met Gaddafi before finally signing for Exxon-Mobil's rights. In 2006, Gazprom also won a license. By 2008, Miller, Tillerson and Jeroen van der Veer were aware that progress was expensively slow in Libya. Importing equipment and staff into the country was a protracted process, causing costs to soar, and Gaddafi reopened negotiations about the profits. Unlike the Western corporations, Miller was uninterested in such complications. He told Gaddafi that Gazprom would buy all of Libya's oil and gas. He then flew to Lagos and offered to buy all of Nigeria's natural gas.

By 2016, Nigeria was expected to be able to supply 20 billion cubic meters of natural gas a year. Delivery would be either by Shell's expensive LNG plant or through a 2,670-mile pipeline across the Sahara to the Mediterranean. Miller secured Nigeria's agreement to cooperate with Gazprom to build the pipeline for $21 billion. In the wings, rival delegations arrived from China and India. Even the Nigerians, accepting a £2.5 billion loan from China, recognized their new creditors to be "quite aggressive," secretly seeking to oust the major Western oil companies and control no less than six billion barrels of oil or 15 percent of Nigeria's production. ExxonMobil, Shell and Chevron would be under financial pressure to match China's generous financial offers, not only in Nigeria but also subsequently in Ghana. Omiyi, however, was not distressed. The Nigerians would, he believed, be uneasy about the danger of Chinese insensitivity, especially in the delta, and as for the Russians, they could also learn the hard way. If Miller chose to ignore Nigeria's history and believed in irreversible high energy prices, Omiyi would be pleased to witness him learning the lesson of "cycles." The initial contract, signed in June 2009, would pledge Gazprom to invest $2.5 billion in Nigaz, a joint venture to develop Nigeria's gas reserves. In The Hague, there was wry amusement when just after Nigeria had been relieved of repaying $35 billion of debts by Western governments, the government had started borrowing again. Unwilling either to become involved in maintaining law and order in Nigeria or to offer any suggestion of withdrawing from the country, Shell reluctantly agreed to lend at least $3 billion to the government to finance essential exploration. Fifty years' experience in the country amounted to tolerating costly confusion.

The year 2007 was Shell's hundredth anniversary. Although profits had risen by 8 percent, and the corporation planned to spend $33 billion in exploration and production, the volume of oil produced had fallen by 5 percent, and would drop again in 2008, the sixth annual fall in succession. The big projects in Sakhalin, the Athabasca tar sands, Qatar natural gas, Kazakhstan, Australia, Norway, Nigeria and the Na Kika field in the Gulf of Mexico with BP promised to improve Shell's replacement rate to 124 percent, but its current production was barely keeping pace with depletion. Individually, each enterprise, relying on high oil prices to justify the investment, was typical of the industry, but combined, they presented an offputting picture of the corporation. Compared to ExxonMobil and even BP, Shell did not present a self-confident image. Ever since the failed merger in the late 1990s and the deflated merger after the reserves scandal, the corporation's culture of compromise rather than decisiveness had suggested a bloated, uncertain management. Even Shell's self-confidence about the flagship "green" agenda had withered.

In 2002, Phil Watts had congratulated Shell on achieving a target set in 1990 to reduce its greenhouse emissions by 10 percent. Without that initiative, said Watts, emissions would have increased by

30 percent that year. To match BP's financial commitment to an environmental program at Stanford University, Watts contributed $3.5 million toward a Shell Center for Sustainability at Rice University in Houston. That had been followed by announcements of investments to develop solar power, grow marine algae in Hawaii to produce vegetable oil as a biofuel, inject carbon dioxide emitted from a Norwegian power plant into offshore oil wells, and build wind farms, especially a giant project deploying 341 turbines in the Thames Estuary to supply 1,000 megawatts of electricity per hour, or 25 percent of London's needs. Shell's investments in renewable energy were costing it about $1 billion over five years. All those projects, condemned by Rex Tillerson as "moonshine," were offered as proof of Shell's commitment to the environment. Tillerson's skepticism was based on Shell's development in 2000, with Chevron as a minority partner, of the extraction of oil from tar sands in northern Alberta, Canada, at an initial cost of $3.5 billion. Tar sands were the world's most expensive and environmentally damaging source of oil, and depending on oil prices, Shell expected to dispatch at least 100,000 barrels a day from Alberta to the United States. The company had simultaneously promoted a $750 million program for developing solar energy. Shell needed another dose of rebranding its green credentials in the aftermath of the Sakhalin battle, but the expensive competition among the oil majors over "greenness," van der Veer knew, had become skewed. Europe's target to use 20 percent of renewables for its energy by 2020 was undermined by the refusal of China and the US to price carbon on similar measurements as the European Union. In any event, van der Veer realized, producing renewable energy actually damaged the environment. The use of biofuels involved deforestation, damage to the countryside and increased food prices. The problem was how to stage an elegant exit from the competition.

Since 2000, Chevron, like other oil and gas companies, had established subsidiaries to develop renewable energy sources. The company had spent $1.5 billion on hydrogen, biofuels, advanced batteries, and wind and solar technologies. In a huge advertising campaign to project

Chevron as a supporter of human rights, a fighter against illnesses and a friend of the environment, the corporation promised to provide oil "more intelligently, more efficiently, more respectfully," and never to stop looking for alternatives. Millions of dollars were spent on a film commercial recorded in 22 locations in 12 countries and showing images including an amputee athlete, mountaineers and a child taking its first steps, with the slogan, "Imagine that, an oil company as part of the solution." Not to be outdone, BP, the inventor of "green oil," had invested about $1.5 billion every year on solar cells, wind power, hydrogen and biomass ventures. "This is not an obscure policy issue, it's about the future of the world," said Vivienne Cox, BP's head of alternative energy, in 2005, committing BP to spending $8 billion over 10 years without any prospect of profits. Even Exxon-Mobil, despite Rex Tillerson's antagonism, cofounded the Global Climate and Energy Project at Stanford University and spent $712 million in 2005 on "green" issues. Spending only a small fraction of its profits on "greenness" reflected Tillerson's conviction that renewables would contribute no more than 2 percent of total energy supplies by 2030, and even then only with state subsidies. Unlike ExxonMobil, which Tillerson believed existed solely to produce oil and natural gas, Shell was navigating through irreconcilable extremes. There was an incongruity, Shell executives realized, about oil companies campaigning to wean their customers away from oil. Commercial relationships were being undermined, and in a lopsided world, promoting heavily subsidized, uncommercial substitutes for oil could prompt a decline in consumption, and thus undermine investment to find more oil. The prospect of financial losses persuaded van der Veer to contemplate a partial retreat from environmentalism to focus on hydrocarbons.

For some months Shell's analysts had anticipated that the world's demand for oil would grow from about 85 to 100 million barrels a day. Shell's dilemma was its inability to profit from increasing demand by supplying conventional fuels. Production from its own reserves had not increased, and the replacement of those reserves had fallen from 158 percent in 2006 to 17 percent in 2007. Van der Veer speculated about "smart technology" providing new reserves from the Canadian tar sands, but the contradiction of a "green" oil corporation relying on that source was obvious. During the summer of 2007, van der Veer questioned whether renewables could provide even 20 percent of global energy by 2050. The public, he believed, had been misled that wind and waves could provide an alternative to fossil fuels. Posturing for publicity purposes was harmless, but relying on the profitability of renewables was foolish. His thoughts coincided with Al Gore's global-warming documentary *An Inconvenient Truth* winning an Oscar, the publication in Britain of the Stern Report on the inevitability of climate change, and President Bush abandoning his former skepticism to declare that the US should be "actively involved if not taking the lead" in limiting emissions. Oil traders across the world were thrilled by the convergence of these arguments. The more Gore speculated about a world without oil or becoming "less dependent on unstable and threatening oil-producing nations," the more the fear of oil supplies peaking grew. Although Gore liked to quote Sheikh Yamani that "The Stone Age came to an end not for a lack of stones, and the Oil Age will end, but not for a lack of oil," the difficulties of an oil company simultaneously anticipating an oil shortage and global warming increased Shell's confusion. Van der Veer decided to limit Shell's commitment to protecting the environment. First, in December 2007, the costly plan to inject carbon dioxide released from power stations into North Sea oil wells was abandoned. One week later, to the distress of its subcontractors, Shell sold its solar business. Next van der Veer abandoned Shell's 33 percent stake in the London Array, the giant North Sea wind farm, the cost of which had increased from £1 billion in 2003 to £2.5 billion. The British government's subsidies were insufficient, and the cost would barely justify the electricity generated. Only America, van der Veer realized, provided sufficient subsidies. By March 2009 he would abandon all new investment in wind and solar renewables. Shell had lost too much from what he called "technology baths." Oil companies would not lead the renewables market. Alternative energy was not a substantial business for oilmen, casting doubt for some on the future of the industry.

The sharp differences between the oil majors were rapidly disappearing. BP, van der Veer was pleased to see, was similarly modifying Browne's commitment to the environment. Its plan to inject carbon dioxide into oil wells near Aberdeen had been abandoned, the land banks for wind farms in the US, China and India were being reassessed, and BP was buying a 50 percent stake of Husky Energy to develop a $5.5 billion project in Canadian tar sands over eight years. BP's renewables headquarters in County Hall was closed, and the green commitment was effectively abandoned. Like Shell, BP was undeterred by Greenpeace's accusations that it was "dishonest and irresponsible," "a climate villain" committing "climate crime." Accusations of "sheer greed" were ignored. In March 2008, as oil prices were heading toward $140 a barrel and Shell's net reserves remained stubbornly at 11.9 billion barrels, van der Veer pledged to expand Shell's daily production of 155,000 barrels of oil from tar sands by 500 percent. In the end, as Lee Raymond had argued, oil companies existed for a single purpose, and their fight for survival was hard enough without pandering to unprofitable sidelines.

Chapter Twenty-two

The Oligarch's Squeeze

———◄○►———

ACCUSTOMED TO ARGUMENTS, Mikhail Fridman and the oligarchs assumed that relations with BP's executives would improve after their bruising negotiations during 2003 to finalize the shareholders' agreement. Instead, the wounds barely healed.

During the first months, the Russian partners appreciated the imposition of corporate processes to merge a multinational organization with a disparate Russian company. By spring 2004, the reorganization of TNK's oilfields, refineries and sales organization by BP's experts was increasing turnover and profits. Deciding it was time to cash in, Fridman flew to London to meet John Browne. The anniversary of their agreement would trigger the first of three annual payments in BP shares. The first payment was worth about $1 billion. Fridman explained to Browne that the oligarchs wanted all three payments immediately. Markets were booming, and they wanted cash to invest. At no cost to BP, Fridman had arranged that the Deutsche Bank would advance $4.2 billion against all the shares. To avoid a personal confrontation with a noisy Russian with more money than sense, Browne made what appeared to be a positive response: "Okay, I understand your point, Michael," he said. On his return to Moscow to finalize the deal with the bank, Fridman was told by Robert Dudley: "John said you both agreed not to change the agreement." Fridman was outraged. In his opinion, Browne's behavior displayed the arrogance of a large corporation. Despite the 50/50 agreement, their partnership was not equal. Soon afterward, Dudley repeatedly excused himself from his weekly lunches with Fridman.

John Browne only partially understood the oligarchs. His talent with politicians and businessmen was to listen and empathize, sensing the mood and adjusting his approach accordingly to cut a deal, but his flair was eclipsed by Fridman. Browne regarded his Russian partners as tough opportunists with shady pasts and unpredictable futures. He assumed that, like all oligarchs, they measured success by humiliating those who failed and by indulging themselves in the kind of baubles he also personally valued — yachts, private planes and unlimited luxury. But Browne did not grasp Fridman's unemotional attitude toward the oil industry and its engineers. Twinkling, "I've never lost a battle yet," Fridman regarded oil as just another merchandise — like the computers or furs he traded in the dying days of the Soviet Union — by which to earn money. His pleasure was the intellectual challenge of constructing a business empire like John D. Rockefeller's or J.P. Morgan's. Anyone obstructing his accumulation of wealth, regardless of his skills, was disposable.

Robert Dudley, Fridman decided by late 2005, fell into that category. BP's blameless ambassador did not seem to embrace his new environment. His deference to notions of scrupulous governance offended the oligarchs' mind-set; it extended even to rewriting the minutes of board meetings to slant decisions in BP's favor, which seemed no more than a childish game to men disdainful of bureaucracy. In return, Dudley complained that Fridman never expressed his personal thanks for BP's impressive progress in Siberia. But the unbridgeable chasm was their attitude toward money. While the American was delighted to earn about $2 million a year, the oligarchs focused on increasing their personal fortunes by billions of dollars.

Browne told Dudley to pay little heed to his Russian partners. The shareholders' agreement specifically entrusted BP and Dudley with complete management authority over TNK-BP. Browne and Dudley believed that rather than controlling the company, the Russians had, in the image of one BP executive, "discovered a 600-pound gorilla sitting in their front room. They had not bargained for what they're getting." In an uncertain truce, Viktor Vekselberg had reported a bruising encounter with Browne about money. "Don't forget BP is very big," Browne had shouted. "Don't try and push me around!" He appeared to have forgotten just how the oligarchs had accumulated their wealth. "Shouting," Vekselberg replied, "only works if you have real power." Browne quoted the shareholders' agreement to verify BP's impregnable status. His own obstinacy was also rooted in money. The combination with TNK had transformed BP into the world's second-largest private oil producer, against the trend of static or declining production suffered by Shell, ExxonMobil and Chevron. For 13 consecutive years, BP had replaced 100 percent of its production. With new agreements in Azerbaijan, Egypt, Angola, Libya and Oman, BP's reserves guaranteed that it would maintain its size for over 40 years. That achievement, Browne hoped, would save his reputation despite his failure to deliver 5.5 percent annual growth. He ignored the weakness of BP's position in owning an equal 50 percent share with partners with different interests. Complicating the relationship was Kovytka in Irkutsk, which was estimated to contain two trillion cubic meters of natural gas, America's consumption for three years. Vekselberg was insistent that the concession should be exploited, and Browne agreed. The only hurdle was building a 3,000-mile pipeline to China, the nearest market, which would cost at least $14 billion.

Tony Hayward, BP's head of exploration and production, supervised the negotiations with the Chinese, and awaited permission from the Russian government to build the pipeline. To his surprise, Gazprom ridiculed the deal. Hayward, complained Alexei Miller, "had opened up all his cards at the outset and sold the gas to the Chinese cheaper than necessary." Without a satisfactory contract, Gazprom would not allow a pipeline to be built. Hayward assumed he could persuade Miller to change his mind, but he misunderstood the situation. Miller was not a mere state employee pursuing Russia's national interests, but a state oligarch assumed to have amassed a personal fortune. Behind his back, executives employed by Gazprom were using inside information to tender for capital projects on behalf of private businesses that included profitable padding in the contracts. Within Gazprom's family, exposure for corruption was deemed to be "unacceptable." Having successfully sabotaged German Gref's plan to break up Gazprom, Miller was intent on securing Kovytka as another jewel in Gazprom's expanding empire. Kovytka's situation would be familiar to admirers of the novel *Catch-22*. The Russian government complained that since TNK-BP was failing to produce sufficient gas as prescribed by its license, the concession would be revoked. Entering into an *Alice in Wonderland* world of mirrors, TNK-BP appealed that decision in a Siberian court, but lost after the judge decided he lacked jurisdiction. A bewildered Hayward accepted this as "just one of those bumps in the road" and agreed to sell Kovytka to Gazprom for between $700 million and $900 million, only to discover that Vekselberg intended to invest on TNK-BP's account over $200 million in a pilot project in Irkutsk, while simultaneously insisting that Gazprom would have to pay a much higher price for TNK-BP's stake. "Vekselberg's as cunning as a fox," Miller complained. The stalemate contributed toward the worsening relations between BP and the oligarchs.

Assuming that by the end of the four-year agreement there would be an argument, Browne wanted to grab the profits as oil prices rose. Dudley's management was making the plan work. TNK-BP was a petroleum engineer's dream. Within two years, its production had risen by 30 percent and profits had increased twentyfold. In 2004–05, BP and its Russian partners earned $9 billion in dividends. Unlike Shell, ExxonMobil and the other oil companies in Russia, BP was making billions of dollars without the risk of being crucified by the Kremlin. The only hitch was Fridman and his partners. To Browne, it seemed they only wanted more money, and were looking for excuses to express their dissatisfaction. Their first complaint was about TNK-BP's staffing. Over 100 BP employees and their families had been sent to Moscow at TNK's expense for short tours. The temporary staff did not impress the oligarchs. "They're not committed to TNK-BP," German Khan complained, despite the remarkable results achieved in the oilfields. The oligarch who shadowed Dudley as the unappointed chief executive griped, "We're not expanding. We're losing entrepreneurship and opportunities." Browne and Hayward preferred not to listen. Despite the tension, Browne's meeting with Putin on April 22, 2005, was reassuring. On the eve of visiting Germany to boost Gazprom's 40 percent share of the country's gas supplies, the president supported BP's presence in Russia. If BP was secure, wondered Browne, what was the oligarchs' fate? Ever since Yukos and Sibneft had been seized and Putin's intention to reassert control over Russia's natural resources had become explicit, Browne had monitored speculation about the fate of the oligarchs' shares in TNK-BP. Briefed by his advisers, he assumed a cozy understanding between Gazprom and the Kremlin that the partners would be presented with a fait accompli to sell their stake. Fridman, he guessed, was in the midst of secret negotiations.

By 2006, there were new strains in relations between BP and the oligarchs. The latest obstacle was BP's interpretation of the shareholders' agreement, treating TNK-BP as a subsidiary, even in the corporation's accounts, rather than an independent, jointly owned company. While Browne assumed that the partners' receipts of $18 billion in dividends would placate Fridman, he misunderstood the oligarch's agenda. In Fridman's mind, oil was Russia's only resource that could be the foundation of a global empire, and he was in a prime position to become his country's Rockefeller. BP, he hoped, could help him realize that dream, but instead his ambitions were being obstructed by Browne. The only solution was to weaken BP's control. Fridman began composing a narrative to portray the Russian partners as BP's victims, and equating the oligarchs' interests with those of the Russian state.

Robert Dudley, the oligarchs complained, was resisting their plan for TNK-BP to invest outside Russia. Browne banned TNK-BP from opening gas stations in Germany to compete with Aral, the biggest distributor in the country, because it was owned by BP. He did reluctantly agree to TNK-BP opening offices in Venezuela, Turkmenistan and Kazakhstan, but blocked Fridman's proposals to drill for oil in Cuba and Kurdistan. "That's against American government regulations," said Dudley. "But we're not an American company," countered Fridman. "We are Russian. You're treating TNK-BP like a subsidiary." Dudley closed his ears, as did Browne. BP wanted to reenter the Iraqi oilfields, and any involvement in Kurdistan would alienate the government in Baghdad. One year later, Fridman would fume that Hunt Oil, an American company, had snatched the Kurdistan contract, but in June 2009 BP would secure the first Western contract to develop Iraq's oilfields. "We're stagnating," Fridman complained after Dudley explained his reinvestment plan to earn profits over the next 25 years. "We're not interested in investing for 45 years' time. We want a good return—now." He was a player, and felt that Dudley did not understand business and risk. Fridman wanted a buying spree like the one Browne had executed in the 1990s, with TNK-BP buying the equivalent of Amoco, Arco and even Sidanco to become a global corporation. Dudley's only concession was to spend $900 million on gas stations across Moscow.

The partners discussed their irritation with government officials. Igor Sechin was aggrieved that TNK-BP's chief executive was an American who spoke only pidgin Russian. Dudley was never invited to official functions, and could not address the 60,000 Russian employees. "He doesn't respect the reality of Russia," Sechin felt. "He can't work with the government or in the regions." The issue was governance—if TNK-BP's chief executive was Russian, the company would be part of Russia. Fridman agreed. Securing the Kremlin's support against BP was easier in the aftermath of the journalist Alexander Litvinenko's murder by polonium poisoning in London in November 2006, and sympathy for Britain eroded after a former KGB officer was publicly accused by the British government of the murder.

Mikhail Fridman's relationship with the Kremlin was misunderstood by Browne and Hayward. Although he met the oligarchs at board meetings, Hayward relied on two Americans for precise information — Dudley and James Dupree. Dupree, a Texan oil engineer, had arrived in Moscow from BP's operations in Angola, giving the impression that BP regarded the two civilizations as not dissimilar. The Russian oil engineers found him arrogant and patronizing, and complained they were ignored. They interpreted his attitude as "I'm from Texas and we know best." "He's a low-level guy and not intelligent," was their self-interestedly distorted opinion. For his part, Hayward was unsympathetic to the oligarchs. He trusted Dupree despite his inability to speak Russian and his disdain for Russian culture. Dupree, with Dudley's support, convinced Hayward that by December 2007, at the end of the four-year lock-in period between BP and Alfa-Access-Renova, the oligarchs would be inclined to sell their 50 percent stake. "They want cash," Hayward was told. "I was always trying to figure out their strategy from so many inconsistent actions," Dudley told his boss. "But they just want money." The oligarchs were complaining because, unless someone paid a slam-dunk amount for their stake, they wanted more profits from their investment. "Something like 50 percent of $60 billion," was the sum Hayward mentioned following a conversation with Vekselberg in Davos. BP valued TNK-BP at between $38 billion and $45 billion, so Hayward thought there was room for negotiation. But Vekselberg, renowned for spending $75 million on the Forbes family collection of Fabergé eggs, had unintentionally confused Hayward. In the oligarchs' minds, any emotional attachment to a business was self-destructive. If someone made a stupendous offer, there was no alternative but to sell; but so far they had no intention of quitting the oil business.

The misjudgment of their Russian partners by Dudley, Hayward and Browne had fateful consequences. In their own scenario, BP's future in Russia would be better served with Gazprom. Not only were BP executives accustomed to dealing with governments rather than businessmen, but a relationship with Gazprom would offer other opportunities. They were encouraged by Alexei Miller and Alexander Medvedev. TNK-BP's assets and Gazprom's business, agreed the state oligarchs, were "well matched." But the triangular relationship among

Fridman, Gazprom and the Kremlin did not, as Browne imagined, lead to an obvious conclusion.

Fridman feared Miller's ambitions to own the lucrative oilfields, and he suspected that Browne and Hayward were encouraging Miller's appetite. Miller, Fridman knew, had asked Browne and Hayward whether they would oppose the purchase of the oligarchs' stake. "Talk to the partners," Hayward told Miller. Miller agreed to do so, hinting that the invincible Gazprom could create problems if its ambitions were stymied. Hayward placed himself, as he later told his directors, "in receiving mode." So long as BP's interests were protected, he decided, the Russians should resolve their own dispute. Vekselberg's friends at Gazprom informed him of the conversation. "What happened at Gazprom?" Vekselberg asked Hayward and Dick Olver, John Browne's deputy, at a board meeting in Moscow. "We had a broad conversation, nothing specific," replied Hayward, convinced that he was required neither to be wholly candid nor to offer Vekselberg reassurance. The Russians disliked Hayward's obfuscation. Shortly after, Fridman bumped into Miller at the private jet terminal at Vienna airport. Both had flown from Moscow to watch a football match. Fridman challenged Miller about his conversation with Hayward. "How could you know what happened at the meeting?" asked Miller. "I was by myself with Hayward, and there were no interpreters." "We'll only sell our stake for an attractive offer," Fridman volunteered.

Tony Hayward hoped to win the Kremlin's support against his Russian partners. One of BP's advisers fancifully reported that although "the road map to seize Khodorkovsky's assets was organized by the Alfa Bank" (Fridman's major asset), the oligarchs' relationship with the Kremlin had deteriorated after Khodorkovsky's arrest. Other advisers, including ex-ambassadors, academics and Mark Heathcote, a former senior MI6 officer, suggested that the oligarchs' access to the Kremlin had become restricted. They could no longer buy more assets, Browne was told, but they needed to pay officials more bribes to keep their empires. Evidence of the oligarchs' declining influence, said BP's advisers, was Roman Abramovich's agreement to sell Sibneft to Gazprom for

$14 billion. That money, it was speculated, would be deposited in an off-shore bank to be divided between Abramovich and Alexei Miller. Only John Gerson, another former MI6 officer, argued the opposite, that BP was "in for big trouble." All those reports, except Gerson's, encouraged Hayward to calculate that the Kremlin would direct Fridman and the partners to be similarly paid off. If anything, BP's executives believed, their relationship with the Kremlin was protecting the oligarchs.

Hayward, however, was misinformed by unsubstantiated fantasies. Lamar McKay, BP's only eyewitness to Fridman's personal relationship with Putin, had been withdrawn from Russia to resolve the corporation's problems in America. In 2004 and 2005, McKay had accompanied Fridman to meetings with Putin and Sergei Priodka, the president's assistant on foreign affairs. He had observed how Fridman reported to the president every detail about TNK-BP's production, reserves and corporate behavior, and reported that Putin clearly approved of the company's direction. Fridman also did not believe that Abramovich was compelled to sell Sibneft. The sale, Fridman believed, was only made to finance his lifestyle, which included giant yachts, countless houses and the ownership of Chelsea Football Club.

Handicapped by that misunderstanding, in early 2007 BP began briefing the media that AAR wanted to sell its stake, and that the Kremlin supported Gazprom's bid. Inevitably, Fridman became irritated by the apparent conspiracy between Miller and Hayward. He urged Peter Aven to discover the truth from the Kremlin. After his visit to the Citadel, Aven reported that so long as the partners remained politically loyal and neutral, neither Igor Sechin nor Putin intended to expropriate their ownership. Indeed, Sechin had advised Putin against compelling the oligarchs to sell. Fridman began stirring. Newspaper correspondents in Moscow reported mischievous rumors that Sechin, as chairman of Rosneft, was competing with Miller to buy TNK-BP. Hayward was cast as an unsubtle executive adrift between the competing state oligarchs, destabilized by the Kremlin and failing to finesse his partners to execute an elegant solution.

Having cast the company as duplicitous for secretly negotiating with

Gazprom, Fridman sought to exploit BP's new weakness. Since 2003, Tony Blair had been BP's ally. The corporation could rely on the prime minister to telephone Putin and caution him about the consequences of any argument between BP and Russia. On the eve of Blair's resignation, in March 2007, Hayward joined Blair's successor Gordon Brown in Moscow. He hoped Brown would persuade Putin to halt the spiral into conflict. The argument was convincing. BP's investment had been beneficial for Russia. With advanced technology and an incentive scheme for workers, production had been boosted by 13 percent every year, taking over 30 percent of the oil from the ground compared to 25 percent in the Soviet era. Booked reserves had increased to 10.5 billion barrels. TNK-BP had paid over $50 billion in taxes, levied at 90 percent on oil companies' revenue, not profits. Putin had recovered the losses from the assets privatized during the 1990s by tax rather than renationalization, and Hayward assumed he would not want to harm Russia's financial interests or international reputation. Reassured by Putin's agreement to meet and listen to his comments, and then by the president's smiles and pleasantries, Hayward would have been shocked by the officials' scoffs after his departure: "Why should Putin care about Brown?"

Hayward had also assumed that he would secure Sechin's support by agreeing to bid in the auction for Yukos's assets in March 2007. As the chairman of Rosneft, Sechin had asked Hayward to bid as part of a charade. Under Russian law, a sale by auction was only valid if there were at least two bidders, so to maintain good relations BP agreed to act as the patsy. BP opened the bidding at $7.5 billion, and immediately dropped out after being outbid by a mere $97 million. At the same meeting with Sechin, Hayward also agreed to buy a 1 percent stake in Rosneft for $1 billion when the company was floated in London in June. Rosneft, Hayward knew, was Sechin's pride. With 170,000 employees and $9 billion capital expenditure every year, it had, since its employment of former Morgan Stanley banker Peter O'Brien as chief financial officer, sought international acceptance. A billion dollars seemed to Hayward a fair price to gain access to the Kremlin's oil oligarch.

Convinced of the strength of his position, Hayward encouraged

Alexei Miller once again to bid for Alfa's 50 percent stake in TNK-BP. Spotting Fridman at a meeting in the Kremlin, Gazprom's chief made an offer. Peter Aven called Putin to ask for his reaction. Putin was again unambiguous. Despite his reservations, he liked the 50/50 shareholding as a good example of cooperation, and he wanted BP to stay in Russia. "Tell Miller that Alfa's share is not for sale," he said. In Fridman's opinion, Hayward's tactics had failed. Events in London would reinforce his skepticism.

Ever since Christmas, John Browne had been struggling to prevent a London newspaper publishing an embarrassing account by Jeff Chevalier, his former boyfriend, about their relationship. Browne's homosexuality had never been concealed from BP's directors. Many of his male friends had attended BP functions, and Chevalier was a frequent guest at parties and weddings, and even had traveled with Browne to Peter Sutherland's home in Spain. The end of their relationship, Sutherland and his fellow directors had heard during 2007, was messy. Browne was threatened with embarrassment by Chevalier's account of their relationship, which he was offering to British newspapers. "I can't talk about it," explained Browne. "There's a massive injunction which even covers myself not to speak." Some within BP speculated that an agent employed by the oligarchs, BP's estranged Russian partners, had led Chevalier to Associated Newspapers in a plot to destroy Browne, but there was no evidence for that conspiracy theory. Others were convinced that Peter Sutherland and Roddy Kennedy, BP's influential head of media relations and an éminence grise, found Browne's sexual behavior offensive to their Irish Catholicism. Neither man appeared willing to offer support after Browne's vulnerability was amplified in February 2007, when Wall Street and the City of London had been spooked by his admission that BP would miss its five-year production targets. The "Sun King's" shine had evaporated. Despite Browne having clawed BP back from the dark days, its shares had risen just 4.5 percent, compared to 15 percent by Shell and 36 percent by Exxon. In July 2006, BP's and ExxonMobil's shares were both trading at about $76. By February 2008, ExxonMobil's would be $90, while BP's were $66.

Browne's estrangement from his fellow directors had deepened. In America, Lamar McKay was negotiating settlements for the criminal damage in Texas City and Alaska, but could barely mitigate BP's embarrassment by pleading guilty and paying about $70 million in fines. Those incidents had permanently tarnished Browne's professional reputation. His impatience, bravado and unsettling certainty about his decisions could no longer be tolerated. Resolutely, the directors decided he should leave in July 2008, five months earlier than agreed, but cushioned by $32 million in benefits. That decision changed on May 1, 2007. Over breakfast with Sutherland, Browne revealed that his application to prevent a newspaper publishing Chevalier's account of their relationship had spectacularly imploded: a judge had criticized Browne for submitting an untruthful statement to the court. Browne hoped to remain at BP, but later that morning was advised by two trusted media specialists that his position was untenable. After 41 years' service, he abruptly resigned, losing about £15 million of his expected benefits.

The premature transition from Browne to Tony Hayward was, Fridman lamented, a lost opportunity to make a deal, but also a chance to squeeze some advantage from BP. Hayward had become critical of Browne, even though Thunder Horse and Prudhoe Bay had become Hayward's ultimate responsibility. "We have a management style," Hayward had said on BP's website in December 2006, "that has made a virtue out of doing more for less. 100 per cent of the task completed with 90 per cent of the resources." Browne's penny-pinching, Hayward admitted after becoming chief executive, compounded by "dreadful" bureaucracy and inefficiency, had impaired BP's performance. "I've had to eat a bit of humble pie," he confessed. "I've had to recognize my failings." BP's profits in 2007 would fall 22 percent to $17.2 billion, compared to Shell's $28 billion and ExxonMobil's $40 billion. The cure, said Hayward, would be a "fundamental shift in the way BP works." Four layers of bureaucracy would be removed, the "Green Book" dumped, the adversarial management style abandoned, and, as a sign of the new austerity, the Armani-style beige suits worn by BP security men would be replaced by traditional dark gray. But there would be no change in BP's relationship with the oligarchs. Hayward's wishful thinking coincided with their decision to take advantage of the upheaval in London and intensify the pressure.

In June 2007, German Khan announced that Kris Sliger, an American responsible for business development who blocked TNK's developments in Cuba and Kurdistan, should be dismissed. The oligarch's next missile was aimed at Robert Dudley, who Khan insisted must be replaced by a Russian. "If 50 percent of the shareholders have no confidence in the CEO," said one of the partners, "he should go." Khan's philosophy, Dudley knew, was always "Win, win, win on every issue, big or small." For Russians, fighting was natural. In October 2007, Khan delivered a personal message to Dudley: "The government is going to squeeze you to leave the country." "I've ignored it," Dudley reported to London. Hayward accepted Dupree's advice to resist the Russians.

In Moscow, the oligarchs had arrived at a profound conclusion. Their treatment as inferiors by BP's executives reflected colonial arrogance, an unshakable confidence in the superiority of American and British culture over Russian. "Of course we're different," said one of the oligarchs, "but we're smart enough, and their culture doesn't work here." "I've read many more books and know more about art and music than Hayward," said another oligarch, "so I don't know why he thinks he's superior." Their grievances echoed identical discontent with the oil majors voiced over previous decades in Venezuela, Mexico and elsewhere. Unintentional arrogance soured relationships between oil majors and oil producers. In the bond between BP and TNK, the repercussions were unpredictable.

Tony Hayward did not anticipate any unusual difficulties at TNK-BP's regular board meeting at the Eden Roc hotel in Antibes in November 2007 at which Fridman and the other Russian partners were present. Before the meeting began, Fridman approached Hayward. "We have a big problem," he said. "President Putin says we cannot have a non-Russian running a Russian oil company." Skeptically, Hayward replied, "Thank you, Michael, but I would have to hear that from the president personally." "This is a problem that will not go away," warned

Fridman. "Hayward has refused to fire Dudley and does not want to cut a deal," Fridman reported before the board met. In that potentially incendiary atmosphere, Dudley presented a plan to expand TNK-BP in Russia, funded by a 75 percent cut in the annual dividend. The Russians were furious. Oil prices and profits were rising. They wanted to expand across the world, not to spend money in Russia. "This is war," one partner announced as they flew back in a private jet from Nice. "But first we must talk to our friends." Nothing could be done without the "emotional support" of the Kremlin.

Peter Aven approached Sechin, the point man. Through takeovers during the previous four years, Sechin had increased Rosneft's oil production from 200,000 barrels a day to over two million. He could interpret the president's opinion. "Sechin," it was afterward reported, "has approved harassment going ahead without breaking relations." Nothing however would happen until after the presidential elections on March 2, 2008. Following the smooth transfer of presidential office to Dmitri Medvedev, Sechin's authority was confirmed. Appointed a deputy prime minister, he moved with Putin from the Kremlin to the White House, the prime minister's offices. Sechin was now acknowledged as all-powerful, and extended his rule over Gazprom. To confirm his new authority, he visited Gazprom's 34-floor headquarters in Cheryomushki, south of Moscow, and was escorted by Alexei Miller to view one of Russia's holiest secrets: the giant switchboard controlling the whole of the country's gas supplies, from the oilfields into the pipelines to the towns and cities. Sechin did not stay for tea.

The mood at TNK-BP's office in Moscow noticeably changed. From their conversations with Sechin and Igor Shuvalov, the partners understood how they were to put pressure on BP. The fulcrum was Dudley's contract with TNK-BP, which Khan had allowed to expire. Without a contract there would be no work visa, and without a work visa Dudley's residence visa would not be renewed. In London, Hayward admitted that "Understanding the soup in Moscow is hard." He was not helped by Browne's decision to downgrade BP's office in Moscow, appointing a non-Russian-speaking representative isolated from

Moscow's power brokers. Fridman's message about survival was mis-understood. "Fridman doesn't know what he wants," a senior BP direc-tor sniffed. Isolated from Russia and relying on Hayward and Dudley, BP's board of directors assumed that Fridman and his fellow oligarchs were wholly unreliable. "Negotiating with these people is impossible," said one director, unaware that Fridman's demands were crystal-clear: TNK-BP was to be Russified, and the oligarchs wanted more money from the business. Hayward had not anticipated how much pressure the partners would put on BP. The next move was unexpected.

The enforcement of Russia's laws and regulations, as BP's execu-tives knew, was random. Everything was prohibited, and every accused could effortlessly be found guilty if prosecuted by the country's reg-ulators, police and government departments. In that jungle, the tax authorities, the ministry of the interior, the FSB or security police (the successor to the KGB), and the agencies responsible for granting visas to alien nationals could, depending on the circumstances, take the initiative against foreigners working in Russia. Since 2003, Khan had shielded BP's staff and had managed the bureaucrats. By establishing "an accommodation" with the government's officials to "overlook" any transgressions and obtain protection from pressure, Khan had protected BP's staff from valid accusations of lawbreaking. Contrary to Russian law, many BP staff were receiving their salaries offshore, and inevitably all of them had failed to comply with the complicated employment regulations. Khan, quite legally, now "withdrew" the shield, and the government officials ceased to turn a blind eye to the transgressions. Overnight it became open season for badly paid officials who regarded businessmen as "low life" to snap hungrily. Aware that the rigorous enforcement of punitive laws would now be praised, the tax collectors and senior officers of the FSB considered themselves at liberty to harass TNK-BP's expatriate employees. Without any explicit order from the Kremlin or even the oligarchs, senior FSB officers proceeded to do their job: to enforce Russia's laws. Both the oligarchs and the Kremlin would later deny having any role in what happened next.

On March 19, 2008, none of TNK-BP's Russian employees reported for work. Soon after Dudley arrived, about 50 officers from the FSB and the interior ministry barged in, demanding to see TNK-BP's employment contracts. Safes were drilled open and beds installed for the security officers to rest in. In the raw battle for power, the pressure point had been identified. Khan failed to seek the renewal of visas. Simultaneously, Vekselberg refused repeated requests to sign the renewal of Dudley's contract. Both could stand back and allow the law to take its course. In April, 146 work permits were suspended. BP's operation would inevitably be paralyzed by the expulsion of Dudley and the 146 expatriate specialists and their families. Dudley was on a timetable for departure.

On May 20, there was another security raid on TNK-BP's offices. "Let them suffer," said an oligarch. Other state organizations, smelling action approved by the Kremlin, took the opportunity to join the campaign against a corporation plainly guilty of "mistreating" Russia's oil production. Incandescent, Dudley asked Fridman, "Are you responsible?" "No," replied Fridman. "What's the real agenda?" asked Dudley. "We want TNK to be independent of BP," said Fridman. "Is this your work?" Dudley asked Khan. Khan denied it. The Russian culture, Dudley believed, was to deny responsibility. Fridman and the other partners later admitted to trying to remove Dudley because, on their account, he was biased toward BP and was causing the company to underperform, but they denied accusations of harassment. Dudley, in turn, denied allegations of underperformance, adding that on his watch TNK-BP's performance had been "stellar." He was unaware that Sechin, receiving regular reports, had agreed not to interfere. "Fridman just wants to cause chaos," a BP director told Hayward. Hayward's conviction about the existence of a coordinated plan amused Fridman. "When I wake up and I think I know what will happen tomorrow in Russia," he smiled, "I lie down and wait for the feeling to pass."

BP's troubles were reported across the globe. Most observers forecast that the company would lose its entire investment. No less than 20 percent of its reserves, 25 percent of its production and 8 percent of its profits would disappear, along with its annual gross earnings in Russia of $5.7 billion. The chance of returning to normality, BP's spokesmen admitted, was slim. The company's plight seemed consistent with so much of the world's reserves becoming closed to the oil majors or becoming subject to the state-to-state deals perfected by China with African and Latin American governments, which distrusted the majors.

"There's always an element of risk," Hayward was told by an adviser. "There's something going on, but I don't understand all the moving parts." Hayward had still not given up on Alexei Miller. On April 26 he met the Gazprom chairman at Stamford Bridge to watch Abramovich's Chelsea play Manchester United. Hayward hoped that Miller would finally agree to buy the partners' 50 percent stake. Instead, he was told that Gazprom would need another month to consider the options. In frustration, Hayward decided to stage a media blitz. He hoped to embarrass the Kremlin into putting the oligarchs under pressure. On May 26, Dudley described in a newspaper interview what he regarded as the "sabotage" perpetrated by German Khan. BP, he said emotionally, had reached a "watershed." "The road ahead is risky and will likely test the strength of Russian institutions." The outburst surprised the oligarchs. Dudley, they agreed, had been "misjudged." "His screaming is not good for us," said a Russian. In London, Hayward was upset by the interview and the repercussions.

Things only became heavier for TNK-BP. Two Russian brothers, educated at Oxford and employed by TNK-BP, were arrested and charged with industrial espionage for stealing documents from Gazprom. Soon after, Tetlis, an obscure broker with a nonexistent address, won a temporary injunction in a Siberian court against TNK-BP for illegally employing specialist contractors for inflated salaries. "Fridman seems able to find friendly judges in the remotest places," complained a BP director. Astutely, the partners had gotten around a term in their original agreement with BP prohibiting the use of Russian courts to resolve disputes. The British courts, they knew, would be unsympathetic. "Of course we're using the Russian legal system," Sechin was told. "It's the only system we have." The Kremlin was content to ignore the treatment being meted out to BP because the Alfa partners appeared transparent and, unlike Khodorkovsky, never sought to squeeze any advantage from the president. "They've got a back road into the Kremlin," one of Hayward's advisers lamented. "We can't see what's happening. All we can do is go through the front door."

To discover the truth and embarrass the oligarchs by securing Putin's support, Hayward arranged a meeting with Sechin. The opportunity arose during Rosneft's annual meeting. "I'll hear both sides of the argument," Sechin told Hayward when they met in a Moscow palace. Like a Godfather, Sechin listened to Hayward's complaint about the oligarchs. Hayward appealed to him in English and, if he did not understand an expression, in Spanish, "We just want a level playing field to limit the risks." Recalling how Yukos had been dismembered and Khodorkovsky imprisoned on contrived charges, Hayward urged, "Please don't allow any fabrications to help those guys." To illustrate his complaint, he added, "What they're playing isn't rugby but Australian rules football." Sechin looked puzzled until Hayward explained the style of the "dirtiest contact game in the world," then laughed and opened the trap by asking in English, "Whose interests do you represent?" "I'm BP's chief executive," replied Hayward. "TNK-BP is a subsidiary and my responsibility." Sechin smiled. Hayward had exposed exactly the conflict of interest that the Alfa partners criticized. The judgment was merciful. "I must tell you, this is a corporate conflict and the government does not want to be involved. I urge you to find a solution." Hayward left without the comfort of Sechin's support against Alfa, but content that Putin wanted BP to stay for the long term. "They could do a lot worse to us," he reported in London. Sechin, he believed, would control the dogs.

Irritated by Miller's procrastination, and unaware that Putin had excluded any purchase by Gazprom, Hayward felt he could at least rely on Sechin's assurances when he arrived in Larnaca, Cyprus, on May 29 for a board meeting. In Moscow, the deadline for the departure of BP's employees was approaching. At the outset of the meeting, Hayward was challenged. "You should withdraw Bob," said Fridman. "He's not independent. The interview he gave was unacceptable." "We're not prepared to give you Bob Dudley's head," Hayward replied. The Russians regarded Hayward with a mixture of fondness and disdain. Compared to the preening John Browne, Hayward was balanced, but unlike Browne he was tactically weak. In Hayward's place, Browne would have been realistic, struck a deal and carried his board; but Hayward, they believed, did not listen, and feared returning to London to announce a fair surrender. The partners stormed out of the meeting. A new frenzy of speculation had been triggered about BP's fate.

Six days later, Hayward arrived in Moscow with an armory of threats. At Rosneft's annual general meeting, he spoke from the floor to warn Russia to obey the rule of law. From Moscow, he flew to the World Economic Forum in Saint Petersburg. BP's employees in the capital were packing to leave, while Dudley, questioned for five hours by officials about his taxes, accepted his powerlessness. Beset by "intolerable" working conditions, an "orchestrated campaign of harassment" and telephone tapping, he knew fighting against Fridman, Khan and Vekselberg was pointless. "They understand the wiring diagrams inside the Kremlin," he told Hayward. "They have thrived because of their intimate relations with individuals in the Kremlin." Unpersuaded, Hayward sought out Sechin in Saint Petersburg for reassurance that BP's investment would not be confiscated. Sechin enigmatically advised him not to worry. Hayward appeared relieved, but did not fully understand. Sechin had indicated during lengthy conversations with the oligarchs that the Kremlin would not interfere with the campaign to remove the foreigners. Western newspapers quoted government officials hinting that Mikhail Fridman and his group "are only starting the offensive. In the near future, it will be worse."

Flying on to Kuala Lumpur, Hayward became outraged by news of intense pressure on BP's staff. Sechin's assurances, he realized, could not be taken at face value. Those encountering Hayward in Malaysia noted, "He's spitting blood." In London, Peter Sutherland heard from BP's advisers that the Kremlin was taking note of Exxon-Mobil's lawsuits against Hugo Chávez for nationalizing the oil majors' assets. Threatening to cast the Russian government into the wilderness as an international pariah, Sutherland thought, would halt Putin in his tracks. Priding himself on "never moving away from a fight," Sutherland was attracted to the idea of shaming Putin by spreading the message, "Russia is not yet ready for prime time—Russia does not have the *Good Housekeeping* Seal of approval for investment." Hayward preferred not to engage in a public brawl, but after a few days' reflection Sutherland decided to "open a new front." Unlike Shell's capitulation over Sakhalin, he would not roll over. "I won't stomach it," he told his fellow directors. On June 12 at a conference in Stockholm, Sutherland publicly accused Fridman of resorting to "corporate raiding activities that were prevalent in Russia in the 1990s," which was "very bad for Russia"; and he ridiculed the supposed attempt by Dmitri Medvedev, the new president, to reform Russia's culture of "legal nihilism."

Sutherland was pleased by the worldwide publicity. The oligarchs, he believed, were thugs. Their traditions were different, but now they knew his style, and his determination to win. "Damn his grandiose view of the world," commented one of the oligarchs traveling outside Russia. "He thinks he's a master of the universe." The partners in Moscow felt the heat. Telephone calls from Sechin reflected Putin's unease. Sutherland had organized telephone calls from Gordon Brown, the head of the European Union and Washington. Most important, Angela Merkel, the German chancellor, had agreed to contact Putin personally. In that testing moment, Fridman and Aven could count on Sechin's ideology. Convinced that the West was engaged in a conspiracy against Russia, the purist bureaucrat believed he was the bulwark to protect Russia's interests. Russia, Sechin decided, would not be publicly embarrassed by threats, especially from Sutherland, whose sole interest was BP's profits. Nevertheless, a balance would be struck. Although Sechin preferred state to private ownership, he liked Hayward, and accepted the partners' assurances that once the foreigners had left, normality would be restored. Sutherland had raised the stakes, but had failed to tip the balance.

In one of many telephone calls between Fridman and Hayward,

the Russian was blunt: "You're playing a dangerous game of fighting and not listening. We are not angels, we are businessmen. We play a hard game." To prove that the battle would be fought with no hostages, Fridman called a press conference and commissioned his staff to write an article for the *Financial Times*. Denouncing TNK-BP's performance as "dismal," he described BP as "arrogant" and "wild barbarians," and ridiculed Sutherland for acting "in the best traditions of Goebbels' propaganda" and "insulting the leadership of Russia." BP, he claimed, thought oligarchs were only interested in "having fun on yachts, private Boeings and football clubs." But the Alfa group was only focused on earning money. There would, he concluded, be no compromise about Dudley's departure, and he added that the Kremlin was not involved in the dispute. Oligarchs expressing themselves in fluent journalese in the capitalists' journal surprised Sutherland and Hayward. Neither could grasp the relationship between the oligarchs and the Kremlin, but they hoped that Gordon Brown's meeting with Dmitri Medvedev at the G8 summit in Japan on July 7 would resolve their plight. To Hayward's dismay the president confirmed his detachment to Brown: "Their argument is commercial. It has nothing to do with the Kremlin. The Russian state is not involved." Inside Putin's White House, however, there was apprehension. "The dispute is destroying the image of Russia," commented Kirill Androsov, responsible in Putin's office for Russia's economic development. The free-market technocrat with a master's degree from the University of Chicago's business school disdained the corruption of low-level officials to use against BP but, despite President Medvedev's public condemnation of those officials, the process was unstoppable.

Two weeks later, BP's staff left Moscow "in a hurry." Newspapers reported, "BP's control of TNK-BP is slipping." Hayward was invited to discussions by Alex Knaster, an associate of the oligarchs, at his London offices in Park Lane. "We can't carry on with Dudley," said Knaster. "Nor Dupree." Hayward accepted the advice to withdraw Dupree and recall Lamar McKay, the trusted American who had originally forged the deal. In Moscow on July 24 Dudley received an unambiguous message. The next day, he was told, he would be arrested. Instantly, he drove straight to the airport, and flew to Paris. At BP, he was universally hailed as a hero. Over the following weeks he would work from hotels in France, Belgium and Holland. In London, Hayward was flattened. Until the last moment, he had believed that Dudley would not be forced to flee. In that moment of "greatest" uncertainty, his aide admitted, "The Russians didn't blink." BP's fate in Russia appeared to be terminal, and the corporation's very survival as an independent oil major was in doubt. "It's a hairball," smiled an oil trader in New York, likening BP to a cat swallowing its own fur.

Two days later, Lamar McKay phoned Fridman to suggest a meeting. Fridman chose to meet in Prague, his favorite city, which reminded him of Lvov, his birthplace. "We're on the way to buttoning this up," he was told by an associate in London. Hayward arrived on July 30 in a good mood. BP's board had agreed to end the war. Fighting in Russia had proved to be too dangerous. The unemotional meeting was notable only for Hayward's denial that TNK-BP was ever run as a subsidiary. "Tony, we must remove the conflicts of interests," smiled Fridman, satisfied by his victory. In a five-page "memorandum of understanding" the two men agreed to appoint an independent Russian chief executive, to be found by BP, and to float 20 percent of the company after 2010. BP's staff would not return to Moscow.

Boarding his jet for the return flight to Moscow, Fridman felt flush. Oil prices were rising, and with more control over TNK-BP, his hopes of building an empire were no longer a dream. He now owned a $20 billion business that included a bank, a grocery chain and a telecom giant (he was simultaneously squeezing his Norwegian partner to secure control of their joint cell phone company), and after winning many battles he had prevailed over an international oil giant. The next step would be to appoint a new chief executive. He would go through the motions to seek an independent expert, but his ultimate objective—realized in 2009—would be the appointment of himself.

Unaware of the oligarch's plan, Hayward returned to London feeling relaxed. The new agreement gave BP stronger legal protection right down the chain over all the subsidiary companies, and it could report record quarterly profits, up 56 percent to $8.6 billion, including 14 percent in dividends from TNK-BP. The outcome could have been so much worse. BP had retained its share of a profitable business. Its stake in Russia was safe, just as oil prices were booming.

Chapter Twenty-three

The Frustrated Regulator

————◆◇◆————

THE COLD CONCENTRATION of oil traders across the globe evaporated in the spring of 2008. Frenzied speculation was changing the oil price during the course of a single day by dollars rather than the usual cents. On March 18, it rose by $3 to $109 a barrel as hedge funds and index speculators returned to the markets. The following day, it slipped back to $103. By May 1 it had soared through $120, having doubled in just nine months. Breathlessly charting the increases, radio and television bulletins reported the fate of the world's lifeblood with the same emotion as eyewitnesses described an army marching toward Armageddon. During a brief pause, Fadel Gheit, a veteran oil sage based in an anonymous Wall Street tower overlooking the Hudson River, recalled the previous surge six months earlier. On November 21, 2007, prices on Nymex had nudged over $99, meriting special attention in the *New York Times*. In reality, taking inflation and the fall of the dollar into account since 1980, when oil was at $40, the price should have been $120. Oil was still cheap, but sentiment suggested the opposite.

Forecasting oil prices in an industry plagued by secrecy, inaccuracy and unexpected disasters, Gheit knew, was a fool's game, yet regular uncertainty had once again been replaced by volatility. Convinced that the supply and demand of oil had not changed during those months, Gheit blamed speculators feeding scare stories to the markets and the media. Technical commodity experts using algorithmic models to eliminate risk had been joined by casino patrons. Index speculators had increased their investments by $55 billion during the first 52 days of the year. Over the previous five years, their investment had increased from $13 billion to $260 billion. With average profits during those years of 118 percent, more investors were betting on crude prices rising in the future. Those bets were directly influencing the daily "spot" price of physical crude.

Oil prices were no longer merely dependent on how much consumers would pay to suppliers, but in an eerie hiatus, traders and speculators were spreading uncertainty to fan the flames. Experts, responding to summonses from the television studios, confirmed the imminence of doom. One week after prices vaulted through $120, the doomsters' fears were endorsed by Arjun Murti, the Goldman Sachs analyst famed for boldly forecasting in 2005 that oil would reach $105 a barrel "in a few years." With oil at a record $122, Murti anticipated another "super spike." Oil, he said, would hit $200 "within the next six to 24 months." His opinion was endorsed by Chakib Khelil, the president of OPEC, and by T. Boone Pickens, who predicted $150 by the end of the year. Speculators, guessing that Pickens had wagered $1 billion on oil's scarcity, assumed he was "talking his book," but nevertheless began betting that the price would hit $200. Two weeks later, on May 21, the price had shot through $130. On that day, the directors of Nymex noticed, speculators held 849,472 contracts for 849 million barrels of oil, over 10 times the world's daily production, and 18 times bigger than the "futures stockpile" in 2003. The sheer volume of money was forcing prices upward. This perfect storm, some suspected, was a conspiracy among speculators.

Verifiable facts embellished by rumors were stoking panic about catastrophic shortages: demand had surged, OPEC was refusing to pump more oil, supplies of non-OPEC oil were diminishing, biofuels were not prospering, the demand for electricity in the Third World was soaring, some nuclear power plants in Japan had faltered, LNG and coal had reached record prices, and the dollar was weak. Even a drought in Chile was blamed — the lack of water meant that a fall of electricity generated through hydroelectric dams required the country to use oil as a substitute. There was a general sense of despair — despite $300 billion spent every year, the supply of oil had barely increased in the previous five years. Pleas to OPEC to increase production were rebuffed. Mesmerized by their self-destructive decision in Jakarta in November 1997 to increase production by 10 percent, which prompted oil to fall to $9 a barrel, the OPEC producers preferred to keep prices high. Authoritative observers corroborated the foreboding about a new era. Despite Schlumberger using horizontal drills to tap western Siberia's vast reserves, Russia's oil production had fallen in April by half a percent, the first drop in 10 years. Leonid Fedun, the vice president of Lukoil, confirmed that Russia's production had "probably peaked." Saudi Arabia's refusal to pump additional oil caused the banker Matthew Simmons to surmise that Ghawar was exhausted; an opinion dismissed by the Kingdom's habitually secretive rulers, but partially confirmed by Sadad al-Huseini, the former head of production at Saudi Aramco. Saudi Arabia, al-Huseini revealed, had inflated its "proven" reserves by 300 billion barrels, and the old reservoirs in the Gulf — 70 percent were over 30 years old — were 41 percent depleted. OPEC, already supplying 44 percent of the world's consumption, could not, he said, compensate for declining supplies from elsewhere. Matt Simmons predicted that oil production had peaked in 2005 at 85 million barrels a day, and was on a permanent decline to 60 million barrels a day by 2015.

Some facts were irrefutable. Production in Venezuela, Mexico and the North Sea was falling just as demand in China, India and especially Saudi Arabia and the Gulf states was rising. Guy Caruso, head of EIA, suggested that stocks of crude had fallen, that the world could not immediately produce more than 85 million barrels a day, and that prices could be high "for years to come." BP's statistics confirmed the shortage. In 2007, production of conventional oil had fallen by

0.2 percent to 81.5 million barrels a day, while consumption had risen by 1.1 percent to 85.22 million barrels a day. The gap was filled by oil produced from unconventional sources, including tar sands. Nevertheless, the balance of supply and demand appeared to be tilting over a precipice. The critical cushion to protect consumers from unforeseen calamities was shrinking, although experts disagreed over whether the gap was 300,000 or two million barrels a day. Across America and Europe, fuel and food prices rose, stock exchanges fell, airlines feared bankruptcy and politicians fretted. As usual, the Democrats in Washington berated Big Oil, and the Republicans berated the Democrats for preventing drilling offshore and in Alaska. Traders and speculators noted that despite talk of an emergency, there were no reports of queues of cars at empty gas pumps, or faltering electricity supplies, and that tankers laden with sour crude were anchored in the Persian Gulf waiting for customers. At the end of May oil prices slipped back to $120, but after a week's lull the storm re-erupted on June 5. Israel threatened to bomb Iran, violence was reported in Nigeria, and the dollar fell further. Within 36 hours, oil jumped by $16.24 to $139.12 a barrel. For the first time, American motorists were paying over $4 a gallon for gasoline. Exploiting the hysteria while visiting France, Alexei Miller summoned the media to predict that oil would hit $250 in 2009, and natural gas prices would triple.

The speculators' run appeared to be uncontrollable. Everyone was focused on the gap between the world's actual production of about 85 million barrels of oil a day and the anticipated pressure from China, India and elsewhere to immediately deliver an additional two million barrels a day. Over the following years, the experts anticipated that the world would demand 100 million barrels a day, and a plateau could be reached between 2015 and 2020. Doom merchants were swaying the markets. Their arguments appeared to be irrefutable: supplies from Russia would not increase above 10 million barrels a day, refineries across the globe were working at full capacity, and non-OPEC countries were producing less oil than expected. Although 250 companies had spent over $400 billion searching for oil during 2006, a remarkable

45 percent increase in one year, it would take at least eight years to produce an additional 20 million barrels a day, and in the meantime the
certainty of permanent shortages was the expression of a potent and
radical philosophy. The security of oil supplies, it appeared, could no
longer be guaranteed by the fundamental economic commandment of
supply and demand. Rather, prices were subservient to the decisions of
the producing countries that controlled the oil. By withholding supplies, they could dictate the world's fate. The advocates of peak oil were
celebrating.

The roots of the new mood had evolved after President Bush's State
of the Union address on January 31, 2006. "America," admitted the
president, "is addicted to oil [from unstable parts of the world]." "This
country," he urged, "[should] dramatically improve our environment,
move beyond a petroleum-based economy and make our dependence
on Middle Eastern oil a thing of the past." Some economists had predicted before 2006 that once prices passed $50 a barrel, a recession was
inevitable. During July and August 2006, oil hit $75. In Washing-
ton, OPEC's control over 77 percent of the world's remaining proven
reserves—922 billion out of 1,200 billion barrels—was reason for
apprehension. OPEC had abandoned the $22–$28 price range set in
March 2000, and by allowing prices to triple, its members had, by
some calculations, earned an additional $720 billion by 2006. Saudi
Arabia's motive was regarded in Washington as greed. The Kingdom
was spending billions of dollars to accommodate the huge growth of
its population, probably over 6 percent annually. Naturally, the new
buildings and infrastructure required extra revenue, but the Saudis'
conduct aroused suspicions after prices fell below $60 in October. "Oil
prices can't rise forever," commented *Oil & Gas*, the industry's journal.
BP's review for 2006 would show that the world's demand for oil during that year had grown by just 0.7 percent, half the average for the past
decade, and oil consumption in Western countries had declined.

Just as the journal was published, prices fell on October 20, 2006, to $55, and OPEC cut its output by 1.2 million barrels a day. There was little doubt in Washington and London that OPEC was manipulating prices. "The cartel," declared Congress's Joint Economic Committee in early November, "is the single greatest cause of market instability as it fans market fears with intermittent quota and output cuts to extend the price surge." Saudi Arabia was no longer producing oil to finance luxury foreign trips for the extended ruling family or to balance its own finances, but was endangering its customers. Few questioned America's decision to exclude Iran's huge reserves from the market. Rather, Washington blamed the producers, and alighted on a new culprit.

President Bush, briefed by the collective wisdom of the oil industry, accused China of tilting the balance. OPEC's ruse, American politicians were convinced, was intended to profit from China's surging demands, although the country's consumption was unquantified. China's state agencies had aggressively bought oil in Sudan, Iran and Burma, all countries excluded from America's orbit, and had invested in 27 oil projects in 14 African countries, including $2.3 billion for a 45 percent stake in Nigeria's deepwater Akpo field. African governments were reassured by China's noninterference in their internal affairs, in marked contrast to America's moralizing, although China's $8 billion pipeline to transport Sudan's oil to the Red Sea would be protected from rebellious tribesmen by Chinese soldiers. More threatening, in Bush's opinion, was China's response to America's threat to oil supplies in the event of a conflict between China and Taiwan or Japan. Fearing that America's warships could blockade the Straits of Hormuz and Malacca to prevent 80 percent of China's oil imports, China planned a "Strings of Pearls" strategy, building naval bases for a new fleet in Pakistan, Burma and Thailand.

To head off the Chinese challenge, Bush warned President Hu Jintao during his visit to the White House in April 2006 against attempting to "lock up" global oil supplies. His admonition was ignored. In November 2006, about 40 heads of African states, including those of Angola, Sudan, Congo Brazzaville and Equatorial Guinea, were invited to visit Beijing for a China–Africa forum. After contracting to buy Saudi oil, China began building a refinery dedicated to sour crude. That same month, the oil price rose from $58 to $60 a barrel, and Bush openly accused China of provoking a price surge. Traders found many other reasons to explain the increase: oil stocks were being sold, Big Oil had failed to invest, oil nationalism, Russia's untrustworthiness after the Ukraine crisis, and the unreliability of LNG, nuclear power and tar sands as alternative sources of energy; but Andy Hall highlighted China's increasing demand and OPEC's cut in production.

In Westport, Connecticut, insiders said that Andy Hall calculated his bets had earned Phibro about $1 billion since 2003. On a winning streak, he was certain that rising demand would push prices up further. The market, he believed, was going in his direction; this was not the time to separate himself from the herd. "The trick," he reminded his staff, "is to run in the same direction when the herd's right." In Congress and Westminster, some politicians were convinced the market was being manipulated. The uncertainty was whether it was traders, the national oil companies or shady officials employed by the governments of the oil producers who were secretly speculating against the market. Was there a genuine oil shortage, or was it contrived? Was Hall, like other speculators, driving or following the market? However much the politicians demanded answers, Hall could not answer that question. The spotlight for explanations switched to the regulator, principally the CFTC monitoring contracts for the future physical delivery of oil.

The unknown in 1996, as oil prices seesawed, was whether speculators, contrary to the CFTC's founding wisdom, were inflating prices beyond the normal boundaries of supply and demand. Prices and profits had soared, prompting politicians' criticism of the oil majors and demands for windfall taxes. The CFTC could not answer President Clinton's questions about the oil majors or the speculators. To reinforce the regulator, Clinton appointed Brooksley Born, an accomplished derivatives lawyer, to chair the CFTC. Unlike her tame predecessors, Born was a suspicious outsider, uncertain whether traders and the major producers were manipulating prices by inventing "dangers" and rumors to exaggerate fears and increase their profits. Finding the proof, she knew, would require Herculean efforts.

A deliberate smokescreen concealed the over-the-counter (OTC) trade, the intense sale and purchase, or "spot trade," of physical oil for immediate or future delivery. Known also as "swaps," the OTC market had been deregulated in 1993 by Wendy Gramm, a laissez-faire economist and the wife of Senator Phil Gramm. Appointed chairman of the CFTC by President Reagan, who called her "my favorite economist," Wendy Gramm, like her husband, believed that capitalism blossomed with as little registration, record-keeping and regulatory supervision as possible. After resigning from the CFTC in 1992, Gramm became a director of Enron, and the company contributed to Phil Gramm's election campaigns. In sharp contrast, Brooksley Born favored speculation to stimulate the markets with cash, but wanted powers to prevent abuse. Too many major oil companies were buying "look-alike" OTCs from the banks so as to avoid the regulators. In 1996 the deregulatory atmosphere had encouraged one trader employed by the Japanese bank Sumitomo to mount a fraud trading copper futures on the London Metal Exchange and OTC copper "swaps," while in America, Bankers Trust had caused the multinational corporation Procter & Gamble to suffer huge losses through the sale of interest rate "swaps," and Orange County in California had become insolvent as a result of speculating on OTCs. By 1998, oil had become intrinsic to the OTC explosion. Many transactions were finalized by telephone conversations, with little committed to paper or e-mails. Vitol, Glencore and other traders spent millions of dollars gathering intelligence, but finding evidence even of normal trading was hard. Even if records existed, it was very difficult for the CFTC to obtain access to their computers and paperwork in Switzerland to scrutinize their "swaps" trade.

Challenging the oil magnates was beyond the CFTC's expertise and power. Fearing a crash, Born had deterred Enron's lobbyists, who were seeking approval to start trading energy derivatives on the OTC market. But in the remaining markets, although speculators were borrowing increasingly huge sums to trade swaps on the OTC market—much more than permitted by Nymex—this was beyond the regulator's supervision except in cases of fraud and manipulation. "It's so opaque,"

Born told her staff as they watched helplessly from the sidelines. "No one knows anything about this market, and we don't have any tools to police it." Multibillion-dollar investment funds like Long Term Capital Management in New York were speculating with unknown amounts in OTCs. Born wanted to reform the CFTC into a savvy, independent regulator, but her task was complicated. To reverse Wendy Gramm's liberalization required congressional approval for new regulations. Although the CFTC was only answerable to Congress and not the Treasury, Born circulated a "concept release" in April 1998 on OTC derivatives to several Washington politicians, including Bob Rubin, the Treasury secretary, and Alan Greenspan, the chairman of the Federal Reserve. Getting their support to extend the CFTC's authority over the "swaps loophole" was politically critical.

At a meeting with Rubin and Greenspan, Born was surprised when Rubin opposed her proposal. Under pressure from Wall Street, the traders and energy companies, especially Enron, Rubin insisted that the OTC market only existed because of the 1993 relaxation, and the CFTC was rightly denied any legal authority. Born disagreed about the legal nuances. When the two sides parted, Rubin believed he had triumphed. He was wrong. In May, Born issued her "concept release," recommending control. Rubin's assistant telephoned her and warned, "Brooksley, the big banks won't like this." Alan Greenspan, ideologically committed to free markets, successfully urged Rubin to initiate legislation to prevent Born's proposed reintroduction of controls. Greenspan's influence was marginal compared to the intense lobbying of Congress and the administration by John Damgard on behalf of the Futures Industry Association. In September, Born's premonition of disaster unfolded. Long Term Capital Management, a $4.7 billion fund, famous for employing mathematical geniuses, began crashing, notionally owing $1.5 trillion after reckless speculation on OTC financial derivatives. Ignoring the threat this posed to financial stability, Congress passed Rubin's legislation in October 1998. In June 1999 Born retired. "She just asked questions," observed one of her critics, "and she was crushed." William Rainer, her successor and a financier friend of Clinton's, was appointed to further reduce the CFTC's efficacy by removing all jurisdiction over OTCs and to sponsor the Commodity Futures Modernization Act 2000, liberating commodity traders from most meaningful controls. By 2000, the worldwide OTC derivatives market was notionally worth $105 trillion. The agency's staff was reduced to 500, and its budget cut. Effectively, traders and speculators were handed control of the market.

The new Bush administration's approval of unregulated OTC trading encouraged the major traders' desire to trade "swaps" electronically 24 hours a day, a facility Nymex could not provide. Their solution in 2000 was the Intercontinental Exchange (ICE), invented to enable them to trade on their own terms using a computer housed in Atlanta. Authorized by the Commodity Exchange Act, ICE's traders were required to keep records, but were exempt from the CFTC's supervision. To guarantee payment and cover any losses, traders deposited funds in a clearinghouse. Every day the clearinghouse collected the "margin" or losses on each contract from the exposed traders. If necessary, the "losing" trader was then required to deposit more money, or to increase the margin call to cover any future loss, protecting both parties from default. Haphazardly, the unregulated OTC market using ICE and the futures trade on Nymex converged, and, unseen, was exploited by Enron's traders, harming the integrity of the whole market.

The rising price of oil and the ease of investing in it through swaps and ICE attracted the managers of savings and hedge funds, who after 2000 became disenchanted with shares. Oil and other commodities were pounced on as a profitable safe haven against inflation and the weakening dollar. The number of energy hedge funds grew from 180 in 2005 to over 500 in 2006. By 2006, America's trade in commodities on overseas markets had grown sixfold. "Growth in our industry," admitted Jeff Sprecher in 2006, "is certainly exceeding the ability of the regulators to get their heads around it." Since 2000, the volume of commodities trading had risen fivefold, to three billion contracts a year, worth about $145 billion in 2006. The explosion had spawned a hybrid of over 1,000 different types of contracts. The CFTC, unable to supervise the market except notionally to prevent fraud, had ineptly lost a battle to prevent the reduction of its own budget. ICE's 13 founding members—the banks, oil companies and major traders—were the protectors of the entire OTC market's integrity. Despite that self-interest, after pushing aside smaller shareholders and buying London's International Petroleum Exchange, Sprecher lobbied the CFTC in January 2006 to allow ICE to trade oil futures in America from London, although the British FSA and not the CFTC would be the only regulator. The combination of poorly regulated trading in London and the rise of oil prices in April 2006 from $66 to $70 aroused an outcry among some politicians.

Motorists were complaining that prices of gas at the pump had soared—in America again passing $3 a gallon—while the oil majors simultaneously announced record profits and ran a multimillion-dollar advertising campaign in spring 2006 opposing a windfall tax. Even Bush's administration was uneasy about such contradictions. The previous year Congress had passed the Energy Policy Act, granting oil companies $2 billion in tax breaks to encourage exploration, yet Lee Raymond had received $398 million as a retirement payoff. The latest energy bill, badly drafted and unscrutinized by the politicians, appeared to allow the oil companies to pocket between $7 billion and $28 billion over the following five years by avoiding royalty payments on oil and gas produced in the Gulf of Mexico. Amid renewed suspicions of Big Oil, price gouging and rumors of an impending scandal, Bush urged Congress to revoke the $2 billion allowance. The oil majors were disdainful. Only the independents, anonymous spokesmen sniffed, would need the subsidy. To appease public irritation, on April 25, 2006, Bush announced at the Renewable Fuels Association an instruction to the Justice Department and the CFTC to investigate "illegal price manipulation" by the oil companies. Before either could report, the Senate Permanent Subcommittee on Investigations began examining whether speculators were influencing the steep price increases.

There had been previous well-known conspiracies to control markets. Secret manipulation of vast paper positions by traders had devastated

Sumitomo in 1996 and Enron in 2001, while reckless trading had destroyed Metallgesellschaft, Barings bank in 1995 and Long Term Capital in 1998. After each collapse, politicians and regulators spoke of lessons being learned and laws being tightened to prevent a recurrence of the abuses. Experience showed, however, that no regulations could ever prevent questionable speculation, especially if at the time it seemed to be reasonable. Nevertheless, basic facts about oil prices provoked questions. In 1998, oil was $15 a barrel. Since 2000, prices had risen from between $25 and $30 to between $60 and $75. The reason was unclear. Increasing demand had been matched by more production, and 347 million barrels of oil were in storage, an eight-year high. Yet prices continued to rise. The only reason, the committee surmised, was speculation. Over the previous three years, between $100 billion and $120 billion had been invested in energy funds. Oil had become an investment, and like gold, which does not rise in price because jewelers need more of it, oil prices were rising because of hype, political instability and speculators stoking misconceptions, not least about shortages. Philip Verleger, a veteran oil-market expert, compared the 19,624 very-long-term contracts held at the end of July 2001 with the 125,546 very-long-term contracts held at the end of July 2005. The proportion of long-term futures contracts favored by speculators had risen from 4.5 percent to 15 percent of all contracts, which, said Verleger, had added $25 to the price of a barrel. That speculation, he concluded, created additional demand for oil, which the speculators, rather than selling, had placed in storage. Faced with that vicious circle, the Senate committee concluded there was "a need to put the cop back on the beat." The "cop" was needed because indisputable evidence of manipulation was missing. The committee's investigators had failed however to prove how speculators had doubled prices. The shortcomings of their report were highlighted in September 2006 by the collapse of Amaranth, a gas trader, after losing $2 billion on mistaken speculation. Throughout their inquiries, the congressional committee's staff had no inkling that between 2002 and 2005, traders and utility corporations had manipulated the natural gas market by providing false prices to Platts and the National Gas Intelligence; the worst offender was Amaranth.

Amaranth had been founded in 2000 as a routine hedge fund. Investors had received annual returns of up to 29 percent, but in 2004 the fund's performance dipped to minus 1 percent. In response the managers hired Brian Hunter and other former Enron traders. Overnight the fund became an energy trader, and its fortunes were transformed. Hunter, lauded as "a one-man industry" in a "lopsided game," delivered 21 percent profits in his first year, and personally earned about $75 million.

In 2005, in the aftermath of Hurricanes Katrina and Rita, Amaranth traders were betting on continuing gas shortages and price rises. By early 2006 those bets appeared doubtful. January was the warmest on record, and the restoration of oil supplies from the Gulf of Mexico guaranteed ample reserves for the following winter. Amaranth's traders, using their $8 billion fund, nevertheless continued to bet on a shortage. Tilting the market amid huge volatility, the future price of natural gas rose. Asked for an explanation, officials at Nymex and CFTC denied that speculators were influencing prices or stability. Although five years earlier, in 2000 and 2001, Enron's traders had created California's energy crisis by transmitting power out of the state and selling it back to it at higher prices while celebrating the high life in Houston's steak houses and nightclubs, the regulators assumed that hedge funds made the market transparent and honest. Their complacency was surprising. Usually, any trader dealing over 12,000 natural gas contracts in one month would be subject to investigation; but during 2006 Amaranth held 100,000 short contracts in one month, representing 5 percent of all the natural gas used by the US in a year. The Amaranth fund, owning 75 percent of the outstanding futures contracts for November 2006, dominated the futures market, causing prices to rise. Each variation of 1 cent tipped the company's profits or losses by $10 million.

In August 2006, the regulators reexamined Amaranth's activities. Purchasers of natural gas for delivery in winter 2007 were paying high prices although there were ample supplies. Amaranth's control over the natural gas due to be consumed by American homes in January 2007 had clearly influenced prices. However, rival traders, judging the price to be too high, refused to buy Amaranth's contracts. After spotting the distortions, Nymex's officials ordered Amaranth to reduce its stake. The fund obeyed but, unknown to the regulator, after withdrawing from Nymex the fund managers sold 100,000 contracts short on ICE, assuming that the squeeze would push prices higher. By moving from the regulated to the unregulated market beyond Nymex's supervision, Amaranth's managers secretly owned about 70 percent of the speculative positions on the market, making them confident that their manipulation would be successful.

In September 2006, although the fundamentals of supply and demand had still not changed, gas prices soared. "Boy, I bet you see some CFTC inquiries," e-mailed one Amaranth trader to another. "Unless they monitor swaps," replied the other, "no big deal." ICE, he knew, was not regulated by the CFTC. Then suddenly the volatility collapsed. The price on Nymex of natural gas for delivery in October had been falling from $8.45 in late July, but the 50 percent crash in September, to about $4.80, the lowest "spot" price in two and a half years, was unexpected. Simultaneously, prices for 2007 fell by 75 percent. Brian Hunter and his traders were staggered. Rival traders, they realized, including many from the Enron diaspora, had held back in order to remanipulate the market against Amaranth. Hunter lacked the finance to cover the margin calls, and Amaranth's portfolio was liquidated. Hunter was again unemployed. American consumers, all of whom had paid too much for natural gas, had been the victims of a squeeze. The culpable included the CFTC's officials. The report assessing the agency's antiquated technology, shrinking staff and flat budget condemned "a broken regulatory system that has left our energy markets vulnerable to any trader with sufficient resources to alter energy prices for all market participants." Four years later, the Federal Energy Regulatory Commission (FERC) castigated Hunter, and the CFTC investigated him and others involved, including Coral Energy Resources, a Houston subsidiary of

Shell, for manipulating natural gas prices in 2001–02 by passing false information and dozens of false trade reports to Platts about prices in the western USA. Five other traders were fined, and one was convicted of manipulation of prices in Houston in November 2007.

The regulators' fumbling was exacerbated by the discovery during those same months that officials in the federal Department of the Interior had concealed their own mistaken payment of billions of dollars to the oil companies. The roots of the duplicity had grown since 1993. To encourage exploration in the Gulf of Mexico, the Clinton administration had granted a tax holiday on the exploration of about 1,000 offshore leases on federal land, especially in the Gulf. By mistake, government officials had omitted the right to levy royalties on oil extracted from those areas. Oil companies, silently noting the error, were not incentivized by Clinton's tax holiday, but relished pumping oil and natural gas from federal land without paying royalties. In 2000, federal officials spotted their blunder but made no public announcement. Instead, they quietly inserted royalty payments into future leases, to be paid when oil hit $34 per barrel. In 2004 oil prices passed that mark, but the department's officials again failed to charge royalties, and the historic oversight remained unrevealed. Johnnie Burton, the female director of the Minerals Management Service, was told about the lapse, but, it was later revealed in an official Department of the Interior report, "did not remember putting a great deal of thought into the matter." By 2006, $65 billion of oil and natural gas had been produced without any royalties being paid. Inquiries into her department revealed that unqualified officials, using inaccurate data supplied by the oil companies, had tolerated cronyism and cover-ups of major blunders. Those investigating what Bobby Maxwell, the Department of the Interior's auditor, called the "underpayments and outright cheating by companies that drill on property owned by the American public," estimated in 2006 that the corrupt relationship between the department and the oil companies had cost the US government as much as $10 billion over five years. While BP and Shell volunteered to repay the royalties, Exxon insisted on abiding by the law, and refused.

Exxon's defiance was followed by that of Anadarko, an independent oil company that had bought Kerr-McGee, the initiator of the action against the government. In successive trials and appeals in 2008 and 2009, judges in Louisiana decided that, despite the politicians' claims, the law did not authorize the Department of the Interior to collect royalties on the oil.

Bobby Maxwell alleged that the law had been undermined by corruption among government officials. Senior officials, he said, had blocked his attempts to recover hundreds of millions of dollars, ordering, "Don't bother the oil companies." Those officials, said Maxwell, enjoyed consultancy deals with the oil companies or romantic relationships with their employees. Politicians searching for other evidence of price manipulation in late 2006 were frustrated by dishonest government officials and by ineffective CFTC staff, still hampered by the restrictions imposed by Clinton's legislation in 2000. Their anger was somewhat assuaged by the CFTC accusing Marathon Oil of manipulating WTI prices at Cushing on November 26, 2003. Traders were always subject to scrutiny as to whether they had influenced the prices on Nymex by controlling the amount of oil being sold from the storage tanks in Cushing. On that occasion Marathon's directors were baffled by the CFTC's accusations. Marathon Oil, like BP and Shell, had lost money in the Platts "window" — the thirty minutes of trading to set prices — on that day. Careful scrutiny of the trades, confirmed by Platts, suggested that Marathon was innocent. But sensing that the CFTC needed a scalp, Marathon paid $1 million to get rid of the issue, exciting a residue of anger and "paranoia" about the government. Marathon was a minnow compared to BP, America's most powerful oil trader. In that year, the annual value of global oil transactions — physical, futures and OTC — was estimated to be $40 trillion. The key benchmark was the price of WTI in Cushing, which still appeared to be dominated by BP. There was a hunger among lawmakers and their agencies in New York and Washington to find a culprit.

Vulnerable after the disasters in Texas City and Alaska, BP's managers did not welcome the congressional reexamination of an incident that had taken place in 2002. The allegation was that prices had risen in New York Harbor because BP refused to supply additional oil down the pipeline from Cushing—a classic squeeze. In its defense, BP explained that in a tight market, the corporation was obliged to supply existing customers despite others suffering expensive pain, and dismissed any notion of squeezing the market. In 2003, after investigating the OTC trades on that day and BP's management of the oil in its storage tanks, Nymex's regulators had concluded that BP had probably manipulated prices in so-called "roundtrip trading" or "wash" trades and swaps for WTI futures contracts on crude on 10 occasions. BP was fined $2.5 million without admitting liability. After further investigation, the CFTC discovered that on six occasions between April and June 2000 a BP trader in Houston had executed prearranged trades at identical prices on an electronic exchange and then neutralized the sales by executing identical trades. The purpose of these so-called "wash sales" had been to fix a recorded price in the market to influence the payment of a real contract. BP was fined $100,000. In November 2006 the CFTC reopened the New York Harbor case, and warned BP it was taking action about the trade of gasoline futures contracts in 2002. The scope of the investigation increased after congressmen discovered that the CFTC had completed an investigation of BP's manipulation of the market in propane, a heating gas widely used in rural American homes.

Based in Houston, BP's propane traders under Cody Claborn operated independently of the oil traders in Chicago. Claborn and Mark Radley, his senior dealer, had plotted during 2003 to corner the market by buying more propane than would be produced for delivery in February 2004. On January 8, 2004, Radley told his subordinates that propane was "vulnerable to a squeeze." On January 13 he repeated that the market was "tight enough that if someone wanted to play games with it, potentially they could." Under Radley's direction, BP bought excessive amounts of propane, but to his misfortune, at the beginning of February the market turned. Supplies of propane increased, and Radley noted that "prices have been dropping like a stone." Potentially, BP was losing money. To rescue the company's position, Radley ordered BP's traders to buy as much propane as possible. His intention was to create a shortage on the market, keep the price high and force other traders (who were "short") to pay excessive rates to BP for propane to fulfill their own contracts. In the event, as fast as his traders bought propane, warm weather was reducing prices. Concerned that prices would fall further, Radley ordered his traders to buy yet more gas. Across the industry, BP's bid to squeeze the market became common gossip, and was reported in trade newspapers. By February 24, the squeeze was working: despite rising prices, rival dealers started buying. On March 1, the buying stopped. Prices slumped, and did not recover. BP's losses were not revealed.

Over the following months, Radley and other BP employees relaxed, convinced that there would be no recrimination. Despite the widespread knowledge of their ruse, the CFTC's officials appeared to have ignored the coverage in the trade newspapers. Seemingly convinced of his immunity, Radley used a PowerPoint presentation to explain BP's tactics and losses to his superiors at a conference in Britain. Only a year later, after a victim did complain, was a CFTC investigation launched. BP's position was hopeless. All the traders' conversations had been recorded, including an exchange in 2003 between Cody Claborn and Dennis Abbott, his superior.

ABBOTT: How does it feel taking on the whole market, man?

CLABORN: Whew, it's pretty big, man.

ABBOTT: Dude, you're the entire [expletive deleted] propane market.

BP argued that the recorded conversations should not be taken at face value, as testosterone-fueled traders habitually speak with bravado. No trader ever loses money on his own — it's always someone else's fault. The CFTC's investigators were unpersuaded. Besides the tape recordings, they had seized incriminating notes and obtained signed confessions. But proving manipulation as opposed to speculation is difficult. The prosecutor needed to prove that the price was artificial; that BP intended to create that artificial price; that BP had the power to dominate the market; and that there was evidence of intention to manipulate. Unusually, in this case all those points were covered by the evidence handed over by BP to the CFTC. Patently, BP's traders knew their squeeze was illegal. On June 28, 2006, the regulator charged BP with violating the Commodity Exchange Act. The corporation had already agreed to pay $303 million in fines and restitution and submit itself to three years' supervision. Many propane traders left BP, complaining that the compliance regulations imposed by the CFTC would limit their earnings.

BP was wounded and embarrassed, but only briefly. Its oil traders remained unaffected. Shortly after the propane settlement, they received advance warning that two major American refineries processing WTI crude—BP's at Whiting in Indiana and Valero's at McKee in Texas—were malfunctioning. Once the market heard the news, the demand for WTI from Cushing would fall, bringing down prices. According to Platts, using that unique intelligence BP's traders had within minutes sold huge stocks of WTI and bought Mars, the sour crude from the Gulf of Mexico that would be refined to compensate for the lack of WTI. "BP made a big killing," noticed a Platts reporter. BP would deny that scenario. Under its new code, the corporation resiled from profiting from its own misfortunes. An unannounced change occurred. Under permanent US government supervision, BP's traders had become markedly less aggressive, while Shell's traders, Platts and Argus reporters noticed, capitalizing on their company's refineries, were reincarnated as belligerent hawks. Reprimands were swift. Unusually, a Shell trader in London was warned to stop simultaneously trading for Shell and himself on the Platts window. The admonitions had limited effect. Undeterred, Shell's and BP's traders and managers hoped to earn a bounty from rising oil prices.

During 2007, shortages rather than speculation were blamed for the rising prices. In his State of the Union message in January 2007, President Bush urged that Americans should cut gasoline consumption by 20 percent over the next 10 years. His "20/10" message, combined with subsidies to build an additional 70 ethanol factories to promote the switch from gasoline to biofuels, gave credibility to *A Crude Awakening: The Oil Crash*, an apocalyptic peak-oil documentary film predicting the end of civilization because mankind had used half of the world's oil reserves.

To Colin Campbell, the leading "oil peakist," Bush's fears were confirmation of his beliefs. His latest prediction for the peak was 2011. Thereafter production, he foretold, would decline steeply, leading to a "world without oil." Optimists, said Campbell, were being deliberately misled by the oil majors. "Truth is not their game," he said, referring to BP's reassurances about sufficient oil for 40 years' consumption at the current rate. His warnings were supported by recent events. Global warming had previously been disparaged by oil executives, and there had been silence for two years after North Sea oil peaked in 1999. Now, said Campbell, no one admitted the probability of a global recession if oil remained above $100 a barrel. Campbell's credibility had been reinforced over the previous 18 months by the conversion of some senior oil executives to the oil peakists' camp.

Leading that group were the senior executives of Total. French mathematical geologists had become adamant adherents of peak oil. With the state providing generous subsidies for biofuels, the country's oil executives, tainted by suggestions of chicanery, understood the financial advantages of embracing peak oil. Thierry Desmarest, Total's chairman, had said in 2006, "The opinion of our geologists is we can go a bit beyond 100 million but not to 120 million." One year later, he sharply retreated. Oil production, he forecast, would peak in 2020: "This is the great secret everyone knows about and governments are too terrified to discuss because they don't know what to do and oil companies don't want to frighten their shareholders." Christophe de Margerie, Total's chief executive, spread similar anxiety by preaching that among those who "like to speak clearly, honestly and not...just to please people," even 100 million barrels a day was falsely "optimistic about geology. Not in terms of reserves, but in terms of how to develop those reserves: how much time it takes, how much realistically do you need." His opinion was shared by James Mulva, head of Conoco-Phillips. "I don't think we're going to see the supply going over 100 million barrels a day...where is all that oil going to come from?" Mulva committed his corporation to far higher investment than other oil companies. All three executives, contradicting the IEA's forecast that production would reach 116 million barrels a day by 2030, dismissed technology's capacity to produce even an additional 20 million barrels every day. The national oil companies, they believed, were unwilling and unable to allow greater extraction. Rex Tillerson's criticism of President Nazarbayev for causing delays at Kashagan endorsed their argument. "We've invested $17 billion," said Tillerson, "and haven't got a drop of oil yet." He blamed the Kazakh government for having—like the governments of Venezuela, Mexico and Russia—a lack of interest in reinvesting its income in oil production, preferring to spend billions building Astana, a new capital on the barren steppe, employing the British architect Norman Foster to design edifices glorifying Nazarbayev for winning elections with never less than 95 percent of the vote. The oil peakists' pessimism was welcomed by the anti-American oil producers. In the month of October 2007 oil had passed first $80 and then $90 a barrel. Convinced that the rising prices would be permanent, Colonel Gaddafi of Libya threatened to reduce production, while Hugo Chávez, with Iran's support, declared open warfare on the oil majors.

One year earlier, after government agents raided ExxonMobil's offices in San Ignacio Towers, Caracas, Chávez had ordered the corporation to submit its assets to Venezuelan control by May 1, 2007. ExxonMobil's development of the Orinoco's heavy oil was halted. The confiscation was not entirely unexpected. Venezuela's oil exports to America had been dramatically reduced, not least because the country's oil production had fallen since Chávez took power in 1999. To aggravate ExxonMobil, Chávez pledged to sell Venezuela's oil to China, even though the Chinese, having pledged $450 million to Venezuela, would need to build refineries to process Venezuela's high-sulphur crude. Nevertheless, the combination of the seizure and the threat to switch sales to China, Rex Tillerson knew, was a defining moment. If Chávez was successful, national oil companies in Kazakhstan, Nigeria and other South American countries would also consider partial nationalization. ExxonMobil and ConocoPhillips refused to hand over control of their assets to the confiscatory socialist. "If Chávez wants it," said an ExxonMobil executive in Caracas, "he'll have to pay market value and in cash." ExxonMobil's assets were worth about $2 billion, and ConocoPhillips's about $7 billion.

Disunity among the oil majors weakened Tillerson's position. Shell and Chevron, despite the confiscation, agreed to cooperate with Chávez, while the Italian corporation ENI, another victim of confiscation, agreed to develop the Orinoco and other projects confiscated from ExxonMobil, despite lacking the necessary expertise. ExxonMobil appeared wounded after Total agreed to consider taking over its investments in Venezuela. Considering his corporation's refusal to help others, Tillerson could not complain about disloyalty. Isolated but defiant, ExxonMobil's lawyers sought court orders to freeze Venezuela's assets in America, Britain, Holland and the Dutch Antilles. In early 2008, a judge in New York maintained the order over $315 million of Venezuela's assets, but in London a court ruled that ExxonMobil had failed to overcome legal hurdles, and unfroze Venezuela's $12 billion worth of assets in Britain.

"Petro-nationalism" was not new, nor was Chávez's grandstanding populism in summoning South America's leaders to an energy summit, but the prospect of a global oil crisis aggravated the dispute. Relishing their power to make more money by producing less, the national oil companies celebrated each cent added to the oil price, calculating the additional billions of dollars deposited in their treasuries. For the Gulf states, where production of a barrel of crude cost $2 in Saudi Arabia and at most $5 elsewhere, the daily windfall was enormous. Since oil had risen above $20, at least an additional $3 trillion had passed from consumers to the producers. The enhancement of the sovereign wealth funds was matched by the decline of the oil majors, although Tillerson's gesture at the company's staff Christmas party disguised any hint of shrinkage. The ground floor of the headquarters in Irving was covered with artificial snow, imported reindeer were tethered along the drive and the hospitality was uncommonly generous, reflecting record profits from high prices. Although the corporation's oil production in the first quarter of 2008 would fall by 5.6 percent, it accumulated $40.9 billion in cash during 2007, after spending $28 billion to buy back shares and distributing $1.9 billion in dividends.

In Tillerson's opinion, investment would be more profitable when oil prices inevitably fell. To a lesser degree, his rivals agreed. The oil majors were divided in their response to dictators, but had been united since 1998 about cutting investment. BP had spent $27 billion in 2005 and 2006 buying back shares, Chevron allocated $15 billion in 2007 to buy back shares over the following three years, and Shell, whose production fell by 6 percent in 2007 and was predicted by the company to continue to fall until 2011, spent $8.2 billion buying back shares in 2006. Only ConocoPhillips invested heavily in buying companies to replenish its reserves in compensation for poor discoveries, making itself vulnerable to a takeover if oil prices fell, a scenario Mulva dismissed. The end of 2007 was a brief calm before the storm.

During the last months of the year, banks in New York, London, Zürich and Frankfurt were reporting difficulties. Terms like "subprime" and "credit crunch" were entering the popular lexicon. Most assumed that after the boom, minor banking problems would restrict economic growth, but daily events suggested something worse. Increasing fuel prices, the national producers were warned by the governments in Washington and London, could tip the world into a recession. Since similar warnings had been issued by economists as oil neared the $50 and $100 benchmarks, the message was ignored. A greater fear was that the world would be endangered by disappearing oil supplies. The news that Abu Dhabi, Dubai and Oman were building coal-fired power stations because local natural gas was too expensive and coal was cheaper than crude confirmed the crisis.

The pessimistic scenario offered by Thierry Desmarest and

Christophe de Margerie had been contradicted in September 2007 by Shell's chairman Jeroen van der Veer. "There's a lot of psychology in the price," said van der Veer, finding no real reason for the high prices. Peter Davies, BP's economist, agreed: "An imminent peak in production has been repeatedly and wrongly predicted...It's not a resource issue, it's an investment issue." Globally in 2007, demand had increased by one million barrels a day, and was predicted to grow by a further 1.4 million barrels a day in 2008. Rex Tillerson, joining the chorus, urged calm: "Sufficient hydrocarbon resources exist to play their role in meeting this growing global demand, if industry is allowed access to them." At the end of 2007, the oil price fell back to $87, suggesting that the threat of a recession in America had ended the speculative boom.

In that fractious atmosphere, the Saudi oil minister Ali al-Naimi sought to rebut the predictions of gloom uttered by Total's executives. "Their skepticism is doing a lot of damage to the stability of the market," he said. "Only the speculators are being helped." Saudi Arabia, he insisted, could produce 11.3 million barrels a day, but was only producing nine million. "The demand is not there," he explained. Over four billion barrels of oil were in storage, more than was exported by the Middle East in six months. Nevertheless, to prove OPEC's sincerity, the Kingdom had invested $80 billion to increase output by 2009 to 12.5 million barrels a day. Other OPEC countries had also committed $40 billion to expand production. The unmentioned impediment was the bottlenecks among the world's refineries, constrained by new environmental regulations, which impeded their ability to process Saudi sour crude, especially to manufacture diesel. Underlying al-Naimi's exhortation was the truism that in the long term, OPEC and the oil producers were more dependent on consumers than the opposite. If OPEC restricted oil production, the West would find alternative sources of energy—nuclear, renewables and coal—and use it more efficiently. History had shown that the oil weapon was a blunt instrument with a limited life. Political manipulation of oil markets by producers and consumers had endured only for a short term.

Ali al-Naimi was not believed. Although Saudi Arabia refused to join Chávez in sabotaging America's economy, the IEA warned about an imminent "supply crunch." In the marketplace, anxieties about Saudi Arabia's restricted production were translated into an inevitable crisis. OPEC's supplies were stagnating, and output in Iraq and Venezuela was falling. Suspicions caused by OPEC's lack of transparency further excited volatility. In Congress, politicians urged the US administration to prosecute the OPEC countries for membership in a price-fixing cartel; legislation had been passed to remove those countries' sovereign immunity. Bush had opposed that law, a toothless piece of showboating, but could not ignore the new warning of Jeroen van der Veer. At the beginning of 2008, Shell's chairman somersaulted. Nations, he conceded to his staff, would "scramble" to secure oil and gas once demand outstripped supply by 2015. Predictions of shortages inspired Bush to telephone King Abdullah, asking for more supplies and price cuts. He was rebuffed. The king told him there was no shortage of sour crude. "We are idling at around nine million barrels," said al-Naimi. There would be no increase once capacity hit 12.5 million barrels in 2009. By May, prices had reached $130 a barrel. "We're running out of oil," claimed the oil peakists. The chorus of disaster delighted the traders and refiners. Expecting prices to rise further, the traders bet long, while refiners like Valero, with inventories at a nine-year high, bought extra storage capacity, betting on uncertainty and the churn of contracts in the markets. Few believed a BP spokesman's comment that speculators were not dictating price movements but had "just jumped on for the ride."

The mood ignored the facts. As the oil price rose, consumption in the Western world was falling. The fundamentals of supply and demand could not permanently support record high prices, and if the price increases were remorseless, the cost of processing "unconventional" tar sands into gasoline became realistic, undermining the peak oil case. High crude prices, however, were hitting profits. Refineries were earning less from their products, the cost of rigs and personnel was rising, and the taxes demanded by the producer nations soared. Rex Tillerson tried to talk prices down. "One trillion barrels of oil have been produced," he said, "and over 15 trillion barrels of energy remains." More conservatively, the energy consulting group CERA estimated that four trillion barrels of proven reserves existed; by including oil shale, tar sands and bitumen, that figure was increased to at least seven trillion barrels. A new CERA study by Peter Jackson of 811 oilfields, producing 19 billion barrels a year out of the world's total of 32 billion, showed that the 4.5 percent depletion was just half the rate previously assumed by the industry. "There is no technical evidence that Ghawar is about to decline," Jackson asserted. By 2017, global production could easily reach 100 million barrels a day, agreed Peter Davies of BP. Demand rather than supply would peak.

New technologies had revolutionized extraction rates. Old oil and natural gas fields were being rejuvenated. Shell had just reported that the proposal to import vast quantities of natural gas as LNG into America would be reconsidered. Using new technology, gas fields in Wyoming, previously regarded as redundant, had increased production by 9 percent in two years. New "tight gas technology" could release an extra 500 trillion cubic feet of natural gas, enough to satisfy American consumption for 20 years. The revolution affected every oil and natural gas field in the world, but was ignored in the frenzied reports that Qatar could not fulfill plans to double exports by 2011 to 77 million tons (20 percent of the world's LNG) or meet the increasing demand for natural gas, which was destined by 2030 to overtake oil.

The hysteria drowned out those who maintained during the early summer of 2008 that there were ample supplies. Bowing to the mantra of irreversible shortages, speculators appeared to have grabbed control of oil from traditional traders. The number of contracts held by Nymex traders rose from 850,000 in 2003 to 2,700,000 in 2008. Like a herd, pension funds, index funds, hedge funds and investment bankers introduced an estimated additional $80 billion into oil trading. Even that statistic fueled further speculation. Everyone knew that the growth of the swaps was huge, but beyond that part of the OTC market regulated by Nymex, no one knew its size. The mystery in New York and London was the identity of the participants. The *Washington Post* would report that the CFTC had noted that Vitol, which boasted an annual $147 billion turnover, had traded contracts for 57.7 million barrels by June 2008, three times the USA's daily needs, and at one point in July held a huge 11 percent of the futures market. Vitol stated that it was "not in the business of taking large positions speculating on the rise or fall of market prices," but others were unsure. Many in Congress regarded Vitol's activity as speculation rather than trading. "It's dirty hedge money and even dirtier sovereign wealth funds," suspected Ed Morse, the senior oil analyst at Lehman's. Russians and Arabs were assumed to be playing the markets by taking a position and then releasing price-sensitive rumors to move the market in a profitable direction.

Commodity markets throughout history had experienced periods of boom and bust, but no oil trader had ever witnessed the sheer weight of money now being bet on oil. No one could predict the effect of hundreds of hedge funds promising 25 percent profits in oil against 3 percent from bank deposits, or accurately balance the dollar's fall against rising oil prices. BP's experts were particularly bewildered by the argument advanced by Sharon Brown-Hruska, previously a CFTC commissioner, that hedge funds were good for liquidity. There was no relationship, said the former regulator, between speculation and prices. Nymex agreed: "Hedge funds have been unfairly maligned," said a spokesman, and "Simple answers are not supported by the available evidence." Some traders believed that Brown-Hruska's misunderstanding was profound. Hedge funds, insiders knew, were a one-way bet. Pouring money into the market — technically adding "liquidity" by betting long or short — meant that volatility had exploded. The hedge funds, BP's traders believed, had pushed prices up by 20 percent, or at least $10, but, unable to pinpoint a guilty hedge fund, the traders and speculators blamed the national oil companies for forcing prices to rise by holding back oil.

Few across Washington could disentangle the contradictory opinions. Walter Lukken, the CFTC's chairman, rejected any dishonesty among traders; Hank Paulson, now the US Treasury secretary, acknowledged that speculators might have a transitory influence, but said that in general they were just following the trend. Other finance ministers disagreed, but lacked any evidence. Their ignorance was shared by the EIA's director Guy Caruso, who admitted having no reliable data about trading or noncommercial speculation, or about supply and demand. "It's gone out of whack," he confessed. "We're not in a new paradigm where speculators rule, but it was a new situation when oil went over $90." Ten percent of the price rise was due to speculation, he told a Senate hearing. Pinpointing 2006 as the moment predictions became unreliable, he explained, "We were too optimistic and high about the NOCs' [national oil companies'] production. We thought more would be produced." The optimistic predictions were influenced by the "difficult" political environment, in particular that of his employer, the US government. No one, he realized, had grasped the influence of the dollar's decline, which had led speculators to "rush out of currencies into oil." On Nymex, speculators were paying $100 for crude to be delivered in 2012.

Caruso's confession exposed the administration's impotence. He was accused of misunderstanding markets. His worship of supply and demand and denials about speculators' influence had guided congressional policies. Overnight, his status was washed away. "The EIA is behind the eight ball," complained a congressman. The agency was following, not leading. "The EIA is always lowballing." Blinkeredly, the EIA and the administration had allowed speculators to run up prices toward the same peak as 1973, and became powerless to exert any influence. The confusion sparked heated accusations in Congress on May 21, 2008. Carrying out the familiar routine of lashing oil-company bosses to prove to constituents how their pain was shared, Dianne Feinstein, the Democrat senator for California, snapped at Stephen Simon, a senior vice president of Exxon, during a heated hearing, "You rack up record profits and apparently have no ethical compass about the price of gasoline." Her repeated demands for "energy independence," "security" and more oil at lower prices willfully ignored the fact that oil was priced globally, far beyond America's control. Even ExxonMobil, the mightiest corporation, had been exposed as a pygmy in the face of the hurricane. The big six oil majors controlled just over 5 percent of oil reserves, and access to another 30 percent through partnerships with the national oil companies. The remaining two thirds was beyond the West's access. The existence of ExxonMobil's $40 billion bank deposit was evidence of Tillerson's doubts. If oil prices could move $10 within one day, how could he justify committing billions to a project over 20 years? The crisis exposed the oil majors as minors. No one in America or Europe could apparently exercise any control over oil. Fearful of bankruptcy, the airlines' knee-jerk reaction was to hedge their fuel bills for the future at over $100. They risked being among the biggest losers if oil prices tanked.

By June, with billions of dollars ceaselessly churning through the markets, the dispute over whether the oil producers or the speculators dominated the markets had become irreconcilable. Even the traditional players—Morgan Stanley, BP and OPEC—were only marginally influencing the market. As prices rose, there was anger that anonymous fund managers and traders were the dogs setting prices, while Saudi Arabia appeared as the reactive tail. Governments were immobilized, unable to plan. A chorus screamed abuse of the free market and appealed to governments to intervene. "Competition is crazy," John D. Rockefeller used to moan. "It's ruining us." Now the free movement of prices was creating a clamor for controls and a windfall tax. "Outside of satanic cults," said Republican senator John McCain, "these people have the worst PR of anybody in the world." Manipulation, he suspected, could have been effortlessly crushed if BP and other stakeholders in Cushing had unleashed unlimited crude into all the pipelines. Prices would have fallen, but the owners had decided against it. Some commentators were now stating that they had lost faith in capitalism and markets. Others took comfort in the fact that while the financial players could introduce volatility, their influence in the markets was limited to less than 90 days. In BP's opinion, it was laughable to suggest that oil prices could be manipulated from Cushing.

A notable voice of sanity was Lehman's Ed Morse. Under pressure for contradicting Murti by forecasting that prices would fall "perhaps violently" in early 2009 to $80, he said, "The mania is a self-fulfilling momentum driving a misconception of a shortage. The 130-year cycle of oil has not disappeared." High prices, he predicted, would depress American consumption, and Saudi Arabia's increased production would restore sanity. "The best cure for high prices," Neal Shear at Morgan Stanley told a worried investor, "is high prices." The speculators, financier George Soros told a Senate commerce committee, were inflating a bubble into a "super bubble" superimposed on a naturally rising market. Data produced by Barclays bank in June suggested that sentiment rather than facts was influencing prices: only 2 percent of the "long" positions, reported Nymex, was owned by speculators without seeking ownership, so no physical oil was being held back from the market. Speculative bets, agreed the CFTC, had fallen by 48.5 percent. The regulators did not realize that index fund managers were compelled by internal rules to sell their future contracts as prices rose. Contrary to the politicians' jargon, some speculators were actually dumping oil. Beyond the politicians' understanding, more critical events were unfolding.

Eager to lead the public outcry, British Prime Minister Gordon Brown joined Bush in accusing Saudi Arabia of manipulating prices by cutting production between March 2006 and April 2007 by about a million barrels a day. OPEC's control over oil supplies, declared Brown, was "a scandal" and he appealed on May 19 to European governments to break OPEC's "stranglehold" on world economic growth. When Saudi Arabia summoned an emergency OPEC meeting in Jeddah on June 22, Brown decided to fly to the Kingdom for the day to publicly berate the producers for holding the world for ransom and demand that they pump more oil. Chakib Khelil, head of OPEC, was perplexed. Brown, he told his aides, was "irrational and illogical," preferring to ignore the high taxes he himself had imposed on oil, and his request for more oil was "simplistic." Khelil was convinced that "The demand for oil is exceeding the supply of oil, not just now but in the medium- to long-term future." Others felt that Khelil's explanation, while possibly true, was equally too simplistic. A complicated matrix of events, partly related to diesel, was also influencing the price increases.

Demand for diesel, manufactured from light sweet crude, had escalated during 2008. The Chinese government ordered diesel to be stockpiled to prevent any shortages during the Beijing Olympics, and European motorists, responding to tax incentives, were switching from gasoline. Because of limited refining capacity in Europe and Asia, the extra supplies were bought in America. The extra demand coincided with a fall in the supply of light sweet crude. New violence in Nigeria had reduced production, and the US Department of Energy chose that period to buy 30,000 barrels of light sweet every day to refill the Strategic Petroleum Reserve. Compounding the shortage, at the same time America's refineries were obliged to implement the regulations introduced by President Clinton in January 2000 to reduce sulphur in diesel. Overnight, the number of refineries able to produce diesel under the new regulations was reduced, and the production of diesel from sour crude fell. The unexpected bottleneck in the refineries would have been avoided if America's refineries had all been owned by the major oil companies, which could have pooled their knowledge through their trade association. Instead, scattered among diverse independents, there was no nerve center to warn the government about the consequences of Clinton's laws. All those unrelated circumstances overturned normality.

Customarily, in summer diesel was cheaper than gasoline, but now for the first time tight supplies not only pushed diesel prices up, but also increased the price of light sweet crude. As the price of WTI rose on Nymex, the Saudis automatically increased the price of their sour crude. Since few refineries had the capacity to process sour into low-sulphur diesel as required by American and European environmental regulations, OPEC countries were producing about four million barrels of excess sour crude every day. A complication was also distorting prices, especially between WTI and Brent, because the amount of crude flowing through Cushing was fast declining, artificially increasing the price of Brent oil—a distortion that the Saudis would remove in October 2009 by abandoning Nymex's prices as a benchmark for their crude oil and use instead a new index developed by Argus. Gordon Brown was oblivious, and his welcome in Jeddah on June 22—he was the only non-OPEC leader to attend the meeting—was cool.

Brown's high taxes had hastened a sharp fall of investment in the North Sea, an 8 percent annual fall of production and declining discovery of new fields. Although about 20 billion barrels of oil remained there, the cost of extraction had risen from $5 to $16 a barrel, making further prospecting unattractive, dramatically cutting the number of new wells and halving to 15 years the fields' lifespan. Unwilling to acknowledge his own responsibility for the fall in production, Brown had theatrically summoned an international oil summit in London, inviting President Bush, Hugo Chávez and Muammar Gaddafi. He had also convened all the North Sea operators to meet in Aberdeen. In front of the television cameras, he urged them to increase production and find new oil immediately. How, wondered al-Naimi, did he imagine that new North Sea oil, if it existed, could be produced in less than four years? Brown, he decided, was not a serious player but a dupe who believed the warning of Shokri Ghanem, the chairman of Libya's national oil company, that "Prices have nowhere to go but higher. OPEC is reaching its peak production and there is no more that countries can produce. We are squeezing the rocks as hard as we can." Al-Naimi knew there was a surplus of Saudi sour crude in storage in the Gulf and no buyers. Brown returned to London claiming credit for the Saudi agreement to increase production by 500,000 barrels a day, but still no wiser about why, despite increased supply, prices were heading toward $140 a barrel. Ten years earlier, crude had sold for $10.

Uncertainty and politicians' gestures encouraged more speculation. On June 26, 2008, the oil price shot above $140. Shares fell, gold hit a record $915 an ounce and reports mentioned "fear in Wall Street and Washington." The world, sages warned apocalyptically, was on the edge of a new oil shock. Shell, Marathon, ConocoPhillips and Lukoil locked themselves into trading short at prices between $120 and $140. Lord Meghnad Desai, a British economist, noting that $260 billion was invested in commodity markets, mostly in oil, compared with $13 billion five years earlier, wrote that not only speculation was driving up prices. Influenced by the testimony to the Senate on May 20, 2008, of Michael Masters, a former Goldman Sachs fund manager, Desai was convinced the problem was caused by a shortage of crude oil, which would continue unchanged for another 12 months. Looking for culprits, some observers in Congress of the market frenzy, ritually suspicious about manipulation, repeated their conviction of Big Oil's conspiracy to "talk up peak oil" and "talk down supplies" in order to push prices up. Those considering themselves more sophisticated seized upon ICE as the menace. For them the unregulated exchange in London, perniciously undermining Nymex, a properly policed market, was the obvious culprit.

The absence of American supervision over ICE in London irked regulators in New York and Washington. No one trusted the low-caliber officials employed by the British Financial Services Authority (FSA). Oil traders, with similar disdain for the CFTC's staff, were accused of distorting the market by trading one position on ICE and the opposite one on Nymex in New York. Denied information, the CFTC could not confirm that speculators in London were squeezing prices in America. "To date," it reported, "there is no statistically significant evidence" that speculation had influenced prices. Unknown to the CFTC, on one day in July Vitol owned a massive 11 percent of all the oil contracts traded on Nymex. Contrary to its official position as a trader, much of that huge position was purely speculative, and placed through ICE. When this was discovered, it contradicted Vitol's statement in May denying any speculation. The "London loophole," thundered Democratic senator Joseph Lieberman, was a blank check for the manipulators. Convinced that traders were arbitraging between Nymex and ICE to manipulate prices and even commit fraud, the politicians demanded that the CFTC supervise American citizens trading on foreign markets. Yielding to the outcry, the CFTC limited Americans speculating on WTI prices in London.

Any satisfaction that move caused was brief. Only 15 percent of WTI contracts were traded out of New York. Frustrated by the global nature of the oil trade, senators thrashed around seeking powers to compel traders to stop speculating. None could explain the distinction between a hedge fund and an airline protecting its commercial interests. Confused that prices on ICE were identical to Nymex's, and unable to name a single speculator, the senators, led by Carl Levin, demanded power to order American traders to limit trading in London or to ban speculation altogether. James Newsome, Nymex's president, was worried—especially because his rival, ICE's founder Jeff Sprecher, appeared relaxed. His criticism of ICE risked backfiring on Nymex. The politicians' threats, Newsome realized, were exaggerated, and none understood one likely consequence. Nymex's profitable trade would simply move from New York to London, the natural trading place for the Middle East because of the time zone, the laws and the financial environment. Only ICE would benefit, and Nymex would lose its lucrative income. The facts were ignored, and a flurry of bills was announced to ban speculators.

Immune to the bewilderment in Washington, Andy Hall was counting his profits. Going in and out of the market since his original bet in 2003 had notionally earned Phibro over $2 billion in profits. The losers were mostly banks, although even they had made healthy profits, albeit by taking greater risks. Hall had enjoyed the market of a lifetime. His envious rivals spoke of him "shooting the lights out over the profits." Of course, Hall had not anticipated in 2003 that high demand for diesel—but never a shortage—would spook the market, but he knew the value of trading on ignorance. In his small way, he had squeezed the world. Quietly, he decided to call it a day. Oil had hit $140 a barrel. Recession, he believed, was looming. "The whole thing is in danger of collapsing," he realized. "If the fundamentals suck, it's no use being 'long.'" The time had come to part with the herd. "There's no point in standing in front of an oncoming train." Over just two days Phibro anonymously offloaded contracts worth more than $1 billion. Hall's personal profit was at least $125 million, but the credit crunch, triggered in part by oil speculation, would sting Phibro. Citibank would need government funds to survive, and in summer 2009 Hall's $100-million pay supplement would be frozen as a result of Washington's edict banning bank employees from receiving bonuses. Blissfully unaware of the position to come, in summer 2008 Hall noted, "In liquid markets, it's easy to sell off." The buyers, he believed, would be burned; but even he did not anticipate the speed of the turn.

During those early days of July 2008, Walter Lukken, the CFTC's director, sought to disprove his agency's critics. Despite his limited resources, and still convinced that prices were rising because of shortages of all crude oils rather than speculators, he became involved in the blame game, looking for scapegoats. Convinced by a tip-off, Lukken believed that malicious traders were spreading false rumors that Cushing was empty, reporting false data about oil stocks and holding back tankers to push prices higher. His belief in the possibility of a conspiracy revealed his agency's limitations. Experienced traders knew that the stocks in tanks and pipelines were constantly changing, data was always history, forecasts were frivolous, and holding stocks in tankers was hugely expensive. Psychology was critical, but rumors could only influence prices for hours, perhaps a few days, but no longer. Understanding the oil market still seemed to be beyond the regulator's grasp.

On July 11, US light sweet crude oil hit $147.50 a barrel, a record high. How long, bystanders wondered, would it be before the price soared through $150? On the trading floors, uncertainty emerged. The spike coincided with banking collapses and talk of recession. Traders and speculators paused. By the end of the day, the price had fallen to $144. Sharp movements had become normal, but nevertheless a change of nearly $4 within five minutes was nerve-racking. "Volatility is only your friend if your positions are right," an experienced New York trader murmured. "All traders think they know how to make money, but not all know when to switch from a rising to a falling market." His aggressive rivals at Morgan Stanley, he suspected, were "floundering in oil." In fact, their model of balancing their positions was withstanding the strain, and they would earn good profits despite some inevitably bad bets. Sensing imminent catastrophe, a handful of sharp investors placed an apparently losing bet. For a mere $1.80 they bought options to sell oil to banks and other traders in November at prices of around $100 a barrel. If the actual price in November was below $100, the difference would be pure profit. On July 14, few traders at the major institutions imagined oil falling anywhere close to $120. Over the following week, prices seesawed around $145, and in the next week they began to slide to $130. As some spoke about a "temporary correction," the signs of a profound crash emerged. The SemGroup in New York filed for bankruptcy protection on July 22, citing $2.4 billion in losses from trading energy futures and OTC derivatives. At the beginning of August, the price was down to $120.

Many did not believe it was the end of the boom. During August, hedge funds were buying back in on the dip. At Morgan Stanley, losses were mounting in the infrastructure—the oil tankers and oil storage complexes. As the market swung into backwardation—oil prices being lower in the future—it made no sense to store oil. Morgan Stanley's tankers were empty, and the leasing costs were cash down the drain. On August 20, oil was down to $114. Michael Witnner, global head of oil research at Société Générale in London, predicted that the bottom in 2008 would be $105, and oil would rise in 2009 to $120 because "supplies are tight." Unseen by the traders and analysts was a change in demand for refined products that was about to cause havoc.

One source of healthy profits for traders, especially Vitol's and Glencore's, was playing a technical arbitrage called the "distillate-to-petrol spread." Traders would bet that diesel prices—contrary to the usual summer season, when refineries produced more gasoline for motorists—would remain high. But unexpectedly, in August demand for diesel dived and prices fell. Only later would all the reasons become apparent: in the post-Olympic weeks, China stopped importing diesel along with other fuels, and began consuming its stockpile; the American government stopped filling the Strategic Petroleum Reserve; Russia invaded Georgia; and general consumption slid as the recession grew. Suddenly, diesel and light sweet crude were abundant. "The market's changed," said one trader. "With swings like this, someone's lost a lot of money." The hedge funds that had bought into the market on the dip took a hit. Vitol was rumored to be among the losers on jet fuel, dropping $500 million. "They're so toxic, they eat their own," observed a rival. The EIA, which had predicted in June that gas prices in August would be $4.15 a gallon, was speechless as they fell to $3.69. Although consumption in the West had been falling for over a year, the IEA, based in Paris, still predicted that overall oil consumption in 2008 would rise by 1 percent to 89 million barrels a day. The average in fact would be 85.8 million barrels. Tantalizingly, subsequent research would suggest that speculators had not unduly influenced markets. As the hedge funds were buying, the more passive pension funds were selling. The markets were operating perfectly. Yet misjudgments about the panic had sparked a lust for regulation, especially for "forward" contracts. Suspiciously, while the CFTC, acknowledging the research, would abandon regulating "forward" deals, European regulators appeared determined to impose costly rules despite the threat of trade moving to Switzerland, to New York or to Dubai. Politics threatened to corrupt reasonable protection.

In early September, the oil producers were apprehensive. Igor Sechin was seeking explanations for why oil had fallen 30 percent to $106, and whether the Saudi announcement that it should be $80 was reliable. In Venezuela and Iran, the governments were dismayed, but consoled themselves that $80 was still a good price. For months they had pleaded their helplessness to stop rising prices, but between November 2006 and January 2007 they had cut production by 1.5 million barrels a day to push prices higher. Now they resolved to stop prices falling. Meeting on September 9 in Vienna, the 11 OPEC ministers welcomed Sechin to their discussions, and agreed to cut production to keep prices at $100. On the same day, stock markets across the world plunged, and the banking crisis appeared uncontrollable. The specter of the oil ministers debating how to control prices at the beginning of a global recession suggested exceptional artlessness.

On October 28, Mohamed bin Dhaen al Hamli, the UAE's minister for energy, addressed about 1,000 oil experts at the annual Oil and Money conference in London. "There are no smiles this year," he admitted. "We are in an unhealthy position." He glanced at Abdullah bin Hamad Al-Attiyah, the Qatari minister of energy, who agreed, "We have lost control of the market." Al Hamli continued, "I am confused. I am seeing collapse everywhere. Low oil prices are dangerous for the world economy." Oil was at $62, well below the $80 that OPEC had agreed four days earlier in Vienna would sustain their economies. Saudi Arabia and the Gulf states could survive if prices fell lower, but Venezuela, Iran and Algeria were suffering, and everyone suspected prices would continue to fall despite OPEC's cuts in production. Wistfully, al Hamli recalled the mood four months earlier: "Before, everyone made money and it was very exciting for all of us." That was not a sentiment shared by most of his audience, especially those relieved that gasoline prices in America had fallen to $2.59 a gallon. Oil markets, they knew, always assume different forms, making manipulation by producers and consumers impossible beyond the short term. "We're producing oil," continued al Hamli, "and no one is buying it. We are in a crisis period for new projects." Demand for oil had in fact fallen faster than anyone could have imagined, to just 81.8 million barrels a day, the fastest decline since 1982, and was unlikely ever to return to its former levels in the Western economies.

Graphs produced by Philip Verleger showed that the 2008 price spike was unlikely to be repeated for another 20 years. Al Hamli voiced the universal regret of all the oil producers at having refused to accept deals offered by the oil majors when the price was $80. Their revenge on the West had been short-lived. Hugo Chávez was caught between exulting in the crisis of capitalism and resolving its repercussions on Venezuela's economy as production fell to the lowest level in 20 years and he confiscated exploration equipment worth billions of dollars. Financial meltdown hit Dubai and Russia first. Dubai could rely on neighboring states to underwrite its debts, and Putin could draw on Russia's wealth fund to prevent the country's immediate slide toward bankruptcy. In the longer term, Putin's narrow nationalism concerning Russia's oil and natural gas was determining the world's financial stability. In common with other presidents of oil-producing countries,

Putin had resisted investing past profits to improve the industry. Leonid Fedun, the vice president of Lukoil, estimated that Russia would need to invest $300 billion over the following eight years just to keep production at its current level. Instead, Lukoil, bolstered by appropriating Yukos's assets and relying on Khodorkovsky's diminishing legacy, was mirroring the country's general decline. Russia's output, Putin knew, could have grown by 12 percent every year to 14 million barrels a day by 2008, but because Western investment had been deterred, it was hovering around nine million barrels a day.

In the midst of plummeting oil prices, Russia's falling production was not uppermost on Putin's agenda. But he could not ignore Gazprom's financial crisis. Alexei Miller, who only months earlier had swaggered across the globe, confessed that Gazprom was heading toward a huge deficit, and was unable to fund new exploration or the construction of pipelines. For a start, the dividend would need to be cut and the contracts for natural gas with the Caspian states renegotiated. Gazprom's plight was replicated across the world. Everywhere big projects were being canceled. In Canada, the tar sands were no longer economic; the North Sea was unattractive for new drilling; and Petrobras in Brazil calculated that offshore production would be loss-making if oil fell under $40. Even Beijing's technocrats were scrabbling to unwind long-term, self-destructively high contracts to supply oil at inflated prices.

In late October 2008, few imagined the oil price dipping below $60, but on November 14 WTI slipped beneath $55. OPEC announced a cut of another four million barrels a day, and traders spoke about chartering tankers for peppercorn amounts to store oil and profit when prices inevitably returned to $60 in March 2009. Just before Christmas, oil fell below $34. That was good news for the handful of astute investors who had bought options for $1.80 in July, betting that oil would fall below $100. Most of the difference between $34 and $100 was pure profit. In New York, *schadenfreude* greeted King Abdullah of Saudi Arabia's pronouncement that he favored an oil price of $75. Since July, Saudi Arabia had cut production by 20 percent. The king resembled King Canute as he lamented the market's unprecedented volatility.

On January 7, 2009, oil rose to $50. Two days later, the low was $39. Natural gas followed the dive. With some self-congratulation, Shell had started pumping LNG into tankers at Sakhalin 2, but supplies from Indonesia and Qatar meant the world was unexpectedly awash with LNG, and prices crashed from $13 per British thermal unit to $4. As Richard Fuld, the former chairman of Lehman's, had once observed, "Markets rise arithmetically but then collapse geometrically."

In these unprecedented conditions, only those with unusual courage would make predictions, but a principal purpose of the EIA in Washington and the IEA in Paris was to forecast the future. The divergence between the two undermined their credibility. The EIA prophesied that oil's average price until 2015 would be $57; the IEA said it would be $100. "The era of cheap oil is over," announced the OECD agency, warning that underinvestment between 2008 and 2015 would cause an energy crunch in 2030, with oil at $200. In 2007, the IEA had suggested that oil would be $108 in 2030. Within one year it had doubled its prediction on the grounds that the world was heading for a shortage even if oil rose to $70. "The current global trends in energy supply are patently unsustainable," said the IEA's experts, predicting that output would fall by 6 percent if current investment was maintained, but more probably by 9 percent. The scenario favored by the IEA and oil peakists pictured the recession ending, the oil "shortage" returning and prices soaring. The crash had exposed an alternative scenario: there was no shortage of oil, but there were restrictions to the development of existing reserves, limited finance to find new fields and bottlenecks in refineries. Leadership could solve all these problems, but competing interests prevented the major oil companies from thinking beyond their own profits.

On February 3, 2009, after touching $37 and lower in some markets, oil was priced at $40, and most of the majors admitted they were in trouble. ConocoPhillips revealed that it had paid $34 billion too much for companies with oil reserves over the previous three years. Assets worth $10 billion would need to be sold immediately and expenditure cut, albeit that its investment in exploration was low. BP acknowledged that unless oil rose to between $50 and $60 most of its investments would be unprofitable, albeit its 2009 cash costs would be reduced from $32 billion to $28 billion, and its investment to $20 billion. Shell, like BP, would acknowledge that its profits would fall from the 2008 peak, but it would try to sustain its $32 billion investment program in 2009 (Shell's profits for 2008 were $31.3 billion, 14 percent higher than 2007; BP's were $25.6 billion, up 39 percent). Only ExxonMobil, with record profits of $45.2 billion in 2008, pledged that there would be no change to its five-year $125 billion investment program, and unlike the other majors it even bought back $7 billion worth of shares. But, in the broader picture of 85 million barrels of oil being supplied every day across the globe, ExxonMobil was insignificant. In 2008, the corporation invested $26.1 billion to produce oil and spent $35.7 billion to buy back shares. Compared to Chevron's $22.8 billion investment, ExxonMobil paraded the virtue of conservatism just as the crash reconfirmed that the oil majors were minnows. Vulnerable to blackmail, stricken by falling prices, unable to influence markets and controlling just 5 percent of the world's oil, they were squeezed by the national oil producers, and forced to survive by their wits. Strong only in marketing gas from the pumps, their fate depended on acquisition, and each merger constricted the investment to find more oil. Shell and BP planned to squeeze costs by more staff cuts. Uncertainty, instability and speculation about prices had been a bonanza for the traders but a disaster for economies. Their profits in 2009 would slide by about 70 percent. The "crisis" was about not diminishing oil supplies, but a potentially destructive fall in demand.

A short-term cure had been successfully applied by Ali al-Naimi, the Saudi oil minister, who had carefully delivered soundbites before the 153rd OPEC meeting in Vienna in May, with the result that oil had risen that week to $68 a barrel and was expected to rise further. Speculators, switching away from equities and the weakening dollar and into commodities, were driving prices up. Last year's villains had become al-Naimi's Good Samaritans. At $70 a barrel, the oil companies could afford to continue investing. Steady production to maximize resources was the obvious solution. The obstacle was the oil majors' inability to fashion a sensible relationship with the national oil companies.

Corrupt and inefficient, the national oil companies were allowing production to diminish. Impatient with the lengthy timescale involved in producing new oil, their countries' leaders contemplated the pleasures of permanent high prices. They ignored the lesson of the 130-year cycle, that oil producers are more dependent on consumers than the opposite. While Saudi Arabia learned the advantage of stability, states like Iran, Venezuela and Algeria dismissed history. Lee Raymond, John Browne and Phil Watts had failed to seduce the leaders of most of the oil-producing states into understanding the advantage of stability, and their successors have done no better. Only President Putin appeared to reflect on the recent lessons and on Gazprom's falling income, squeezed by disadvantageous contracts. In June 2009 the Russian government asked Shell to develop Sakhalin 3, a coup for the beleaguered company, and a parallel offer on another site was made to Total of France. Exxon and the other American oil companies were excluded from Russian deals, albeit that in September, ExxonMobil was included among the oil companies invited to meet Putin in Salekhard, a remote Siberian oil town, to hear the president offer new terms for Western investment. But ExxonMobil would struggle to benefit from the president's promise of a new era. The backwash of the Yukos negotiations still rankled.

In common with other oil producers, the Russians were bewildered by the statistics. Over the previous months economists employed by Goldman Sachs had issued a succession of contradictory forecasts, suggesting first that oil would slump to $30 a barrel, then that it would stay at $55, then in June 2009 that it would surge to $70, and then toward $95. The bank was reflecting the confusion at the IEA. On June 11, 2009, the agency had announced, as oil hit $73 a barrel, that prices would rise further although demand had fallen since 2008. Less than three weeks later, on June 29, IEA forecast that in the future demand would drop sharply. In 2014 the world would need 84.9 million barrels a day rather than, as previously predicted, 89 million. Prices inevitably fell toward $60 a barrel. The volatility reflected incompatible pressures:

fear of peak oil, uncertainty about whether new reserves could be developed, the weakness of the dollar and the development of renewable energy. OPEC was an immediate casualty, followed by the forecasters.

The solution to the deadlock between the disputing interest groups could be the Arctic. Just as North Sea oil undermined OPEC in the 1980s, oil from the Arctic could trigger an era of surplus oil and animate cooperation. At least 100 billion tons of hydrocarbons can be extracted from the Arctic, although the technological obstacles are considerable and the starting cost will be at least $20 billion. Only four Western companies possess the skills — ExxonMobil is not among them. So far, two suitable drill ships have been built, both under contract to Shell. Alaska took three years to master. The Gulf of Mexico has so far taken 40 years, and its exploration is still handicapped by a myriad of obstacles. But extracting oil off Brazil's coast has confirmed that no technical obstacle is insuperable, if the finance is available. Drawing on that new technology, the Arctic riches will eventually be released, despite Russia's attempts to establish territorial ambitions. In that unusual battleground, Russia has neither the financial nor the technical ability to execute the task. The fate of Arctic oil will depend on the oil majors repudiating their former cowardice. As the pressure of the price crash has eased during 2009, the chairmen have had time to conduct an autopsy on the past hectic months and consider the shrinkage of their corporations. Their salvation could be brought about by renegotiating their relationship with the governments of the oil-producing countries. Weakened by the collapse of prices, those governments could be encouraged to sign more reasonable contracts in order to expand production, reducing the risk of instability over the remainder of the century.

Andy Hall disagrees. The scarcity of oil, he believes, is indisputable, and that will enrich the oil majors, whose unique expertise will be needed to produce an increasingly valuable commodity. Oil, throughout Hall's career, has been pitched in a crisis between glut and famine. He will bet both ways, on the oil majors' success and on their failure. In 2008, Phibro declared profits of $700 million, and Hall's personal bonus was $125 million. His success had become an embarrassment for

Citigroup, beleaguered and under the US government's control. The bank's solution to Hall's bonus entitlement was Phibro's sale to Occidental Petroleum for a net value of about $250 million, a giveaway if Hall maintained his genius. He possesses a financial acumen that has eluded others in the trade. In the course of 35 years, he has encountered all the traits of oil's personalities—self-glorification, greed, hubris, nemesis, deception and self-destruction; but also genius, integrity and bravery. Each individual, whether explorer, geologist, executive or trader, seeks to establish a profitable balance between governments, regulators, markets and nature. Over the last 20 years that balance has been elusive. Navigating around the mystery, Hall says, recalls the Red Queen's words in *Through the Looking Glass*. To survive in her country, the queen explains to Alice, depends on a single rule: "Now, here, you see, it takes all the running you can do, to keep in the same place." The last 20 years confirm that paradox, uniting both sides of an irreconcilable argument.

Acknowledgments

No previous book has attempted to tell the story of oil over the past 20 years. Compressing in effect several books into one narrative is a challenge. *Oil* follows the stories of BP, Shell, ExxonMobil, Chevron, the traders, the Russian oligarchs and the environmentalists. Fortunately, I received generous cooperation from nearly all those involved. I interviewed about 250 people who over the years have been employed in diverse aspects of the oil industry. Besides their recollections, the principal sources of information for this book were government regulators' and company reports, specialist magazines, the major newspapers in America and Britain and the best of the huge library of previous books.

While I am not identifying or thanking my interview sources here, they know who they are, and I want each of them to know how grateful I am for their support and assistance. There are, however, many people whose invaluable help can be acknowledged. I could never have started this book without the guidance of Patrick Heren and the subsequent introductions to a wide range of people. I am also grateful to Shamil Yenikeyeff for his help in Russia. In Moscow, Lyuba Vinogradova was a terrific researcher. Several people helped in research, but in particular I am grateful to Olly Figg, a trusted friend. Many individuals read parts of the book, and in particular I owe thanks to Peter Gignoux. In the USA, Mike Lance and Bob Royer were good friends. Others whose contributions were invaluable were the directors and staff at BP, Shell,

ExxonMobil, Chevron, Cairn and Tullow. All wanted their help to remain anonymous.

Among many others, I am particularly grateful to Terry Adams, Meg Anesley, Faucon Benoit, Mike Bowlin, Mike Bradshaw, Axel Busch and Paul Sampson at Energy Intelligence; Carl Calabro, Daniel Carr, Guy Caruso, Guy Chazan, Judith Chomsky and Jennifer Green at the Center for Constitutional Rights; Brent Coon, Bill Cran, the producer and director of the BBC series *The Prize*; Mark Crandall, Ed Crooks of the *Financial Times*; John D'Ancona, Alan Detheridge, Kenneth Dickerson, Paul Domjan, Ray Drafter, Mary Dwyer, Chris Fay, John Fitzgibbon, Doug Ford, Herman Franssen, Lars Garrison, Fadel Gheit, Chuck Hamel, Tom Hamilton, Robert Healy, Albert Helmig, John Hoffmeister, Dennis Holden at the CFTC, Alex Kemp, Lutz Kleveman, Steve Levine, Laney Littlejohn, Richard Mabley of Reuters, Helen Manning, Jorge Montepeque, Ed Morse, Carl Mortishead, Colin Moynihan, John O'Connor, Teo Oerlemans, Willy Olsen, Tim Osborne, Stephen O'Sullivan, Ron Oxburgh, Greg Pytel, Michael Ritchie, Laurent Ruseckas, Professor Richard Sakwa, Adam Shrier, Adam Sieminski, Chris Skrebowski, Nick Starritt, Mark Stephens, Jonathan Stern, Doug Terreson, Charlie Tuke, Phil Verleger, Mike Watts, Chris Weafer, Julian West, Mike Wiley and Ian Wybrew-Bond.

The picture research was executed by Anna Phillips. The legal chores were undertaken by Godwin Busuttil. Jonathan Lloyd of Curtis Brown was as ever remarkably steadfast in his support. I am particularly grateful to Mitch Hoffman, Grand Central's astute editor, and to Kim Hoffman, for their advice and help. At HarperCollins, I owe a great debt to Robert Lacey, whose editing as always transformed the manuscript, Martin Redfern and Richard Johnson.

As always, my principal gratitude is to my family—Veronica, Nicholas, Oliver, Sophie and Alexander—who with my mother, Sylvia, always provide stability, humor and love.

Notes

————◇————

43 Exxon invested in: Ibid. p. 665
44 In the early 1980s these restrictions: Nigel Lawson pp. 163ff
54 the 600 staff struggled: Senate Ag Com 6/5/96

CHAPTER FOUR: THE CASUALTY

58 Shell's engineers had considered: *Financial Times* 8/23/97
61 In 1990 they fell by: *Wall Street Journal* 2/3/92
63 after Fay telephoned: *New York Times* 6/21/95
65 I arrived in this job: Harvard Business Study
66 The new dictator repressed: *New York Times* 9/6/94
68 in helping to preserve: *Mail on Sunday* 4/4/04
71 become inward-looking: *The Guardian* 5/14/97
73 stewardship over: Harvard Business Study
73 begin to take: *Financial Times* 10/14/97
74 There will be a coming crisis: *Financial Times* 3/3/98
75 while contributing to the well-being: Harvard Business Study
75 Beethoven's "Ode to Joy": *New York Times* 3/12/04

CHAPTER FIVE: THE STAR

77 D'Arcy's team ignored: Yergin p. 146
78 The problem was: James Bamberg, *British Petroleum and Global Oil, 1950–1975*
     p. 209
81 This is a long-term business: *New York Times* 11/13/90
82 the fundamental realities: Statement to shareholders, AGM, April 1992
88 Oil prices, David Simon predicted: *New York Times* 8/9/94
89 Browne's admirers spoke of: *Wall Street Journal* 11/5/93
89 After substantial criticism: *Financial Times* 2/8/97
90 This is the classic way: *New York Times* 3/20/94

CHAPTER SIX: THE BOOTY HUNTERS

93 Gorbachev appealed to Germany: Marshall Goldman, *Oilopoly: Putin, Power and
     the Rise of the New Russia* p. 52
94 Oil prices had in fact fallen: Ibid.
95 the Yamal Peninsula: *New York Times* 11/27/94
95 Conoco excitedly signed deals: *New York Times* 4/12/94
97 cataclysmic event: *Wall Street Journal* 5/17/99
97 its oil reserves would slump: *New York Times* 6/12/04
98 The knowledge that production: *New York Times* 2/4/91
98 gave Russia's media: *New York Times* 8/16/91
99 Unlike the oil majors: *New York Times* 3/23/91
99 Gorbachev bowed to threats: *New York Times* 8/6/91
104 deal of the century: Steve Levine, *Oil and the Glory: The Pursuit of Empire and
     Fortune on the Caspian Sea*

106 Phibro would lose: *New York Times* 3/20/94, 3/31/94

108 increasingly tilted in Azerbaijan's favor: Levine p. 161

108 the demand by Marat Manafov: *Sunday Times* 3/26/00

108 Circumstances change: *Wall Street Journal* 4/25/94

109 In Washington, Bill White: Levine pp. 211ff

110 Amoco's ambassador to the NSC: Ibid. p. 232

110 an outbreak of violence: Ibid. pp. 203ff

112 Both countries enjoyed: vol. 93/19, 5/8/95

113 The fate of the $1.4 billion pipeline: Levine

113 he slashed investment: *Wall Street Journal* 2/13/95

CHAPTER SEVEN: THE OLIGARCHS

115 Western analysts' estimates: *BusinessWeek* 11/29/94

117 Yeltsin gave his formal approval: Freedland pp. 171ff

118 Arco became the first: *New York Times* 8/31/95

118 At every level: Freedland p. 274

119 One obstacle was: *New York Times* 11/8/98

120 Our long-term security interests: *New York Times* 5/9/97

121 Yeltsin restored tax: *New York Times* 12/9/97

123 On auction day: *Wall Street Journal* 7/8/98

123 In July, the White House: Freedland p. 297

125 Browne was mocked: *New York Times* 12/2/99

127 aggressive meddling: *Oil & Gas* 9/27/99

CHAPTER EIGHT: THE SUSPECT TRADERS

138 BP was fined £125,000: www.fsa.gov.uk/pubs/additional/sfa004-00.pdf

139 Being in the physical business: *Wall Street Journal* 3/3/05

139 Their expectation: *New York Times* 12/26/90

140 You know it's going to be: *Wall Street Journal* 6/30/89

141 Alaska's officials would assert: *Wall Street Journal* 2/7/90; *New York Times* 11/19/92

141 the corporation was "sloppy": *New York Times* 1/6/90

141 without any punishment whatsoever: *Wall Street Journal* 7/13/92

142 The settlement: *Wall Street Journal* 1/31/91

143 you mention oil: *New York Times* 8/19/90

144 In 1996 he would agree: *Wall Street Journal* 9/26/96

144 Unexpectedly, the public outcry: *New York Times* 6/28/90

145 Lawmakers are driven: *Oil & Gas* 3/4/91

151 paying the paramilitary leader: *The Observer* 7/1/01

CHAPTER NINE: THE CRISIS

153 Costs had fallen: *New York Times* 10/22/96

154 I hope this doesn't: *Wall Street Journal* 12/1/99

155   This is corporate: *Wall Street Journal* 5/2/94

156   I was shocked: *Wall Street Journal* 7/6/89

156   Raymond argued: *New York Times* 11/19/92

156   scientific studies: *Wall Street Journal* 3/25/93

156   400,000 birds: *New York Times* 4/30/93

156   The verdict is: *Wall Street Journal* 9/19/94

157   The damages could: *Financial Times* 6/25/08

158   the biggest criminals: Yergin p. 92

159   Uncertainty and gyrating prices: *New York Times* 11/13/90

162   We put them through: Iwst 4/8/05

163   the destructive power: *Oil & Gas* 11/9/02

164   It had successfully: *New York Times* 9/21/99, 9/24/99; *Wall Street Journal* 9/10/99

164   The people of: *Wall Street Journal* 5/11/92

164   Exxon did pay: *Wall Street Journal* 11/18/94

165   Daily rentals: *New York Times* 3/7/97

167   At a price of $10: *New York Times* 12/2/98

168   up to five million barrels: *New York Times* 2/19/00

169   I want to confirm: *New York Times* 11/2/90

169   Gasoline fumes had seeped: *New York Times* 11/27/97

169   His imprisonment: *New York Times* 4/27/92

170   The United States can trust: *New York Times* 10/8/90

171   It's back to a high degree: *New York Times* 1/26/97, 1/2/97

172   The oil majors, he believed: Daniel Yergin and Joseph Stanislaw, *The Commanding Heights: The Battle for the World Economy*

CHAPTER TEN: THE HUNTER

173   We want to overtake Shell: *Financial Times* 3/13/97

174   a $10 billion union: *Wall Street Journal* 10/7/96

177   he anticipated dismissing: *Oil & Gas* 3/1/99

181   It is clear: *Oil & Gas* 2/22/99; Institute of Petroleum

181   We're beginning to create: *Wall Street Journal* 4/8/98

183   a conspiracy or a common: Sampson p. 37

183   The discovery that: Ibid. p. 112

183   Noto allowed the conversation: *New York Times* 12/2/98

185   His comment about: *Wall Street Journal* 12/2/98

185   Both chairmen expected: *New York Times* 12/2/98

185   The merger will mean: House Energy and Power Subcommittee 3/22/99

186   their motive may be: *Wall Street Journal* 12/31/98

187   The company's profits fell: *Wall Street Journal* 5/6/98, 8/5/98

187   People say we are: *New York Times* 4/1/99

194   There's now a pronounced trend: *Wall Street Journal* 12/31/98

197   the new salvation against OPEC: *Oil & Gas* 12/13/99

200   At $3 a barrel: *Wall Street Journal* 3/15/00

200   BP had suspected: Sampson pp. 178–9

CHAPTER ELEVEN: THE AGGRESSORS

208 The most extreme: Report p. 66
208 The idea that: Ibid. pp. 58ff, 67

CHAPTER TWELVE: THE ANTAGONISTS

212 I came close to being: Shell history p. 296
212 Van Wachem's deadline: *Wall Street Journal* 12/15/98
212 Analysts do not value: *Financial Times* 11/2/00
218 after five years: *Wall Street Journal* 9/21/99
219 To justify the increase: Issue 17, 4/24/95
222 the world's reserves of oil: *Oil & Gas* 12/17/01
222 New discoveries and cuts: *Daily Telegraph* 6/19/01
222 enabling it to deposit: *Wall Street Journal* 7/30/01
225 Earning profits: *New York Times* 4/12/94
226 the world's biggest producer: *The Times* 4/3/02
226 They've bought flowerpots: *The Guardian* 4/12/02
226 Failure to meet: *The Times* 11/4/02

CHAPTER THIRTEEN: THE SHOOTING STAR

231 to recover £1.4 million: *Financial Times* 8/23/97
232 Burning the Planet: *The Independent on Sunday* 9/3/00
235 Early results confirmed: Robert Healy, *Public Relations Quarterly* winter 2004
236 he had withdrawn the company's: *The Independent* 28.2.02; *Wall Street Journal* 3/1/02
237 his preference for delivering speeches: *Fortune* magazine, "Global 500," 2005
237 felt more brutal: *Wall Street Journal* 3/31/98
242 the public had become: *The Guardian* 12/3/02

CHAPTER FOURTEEN: THE TWISTER

246 As Itera expanded: *Financial Times* 11/1/00, 10/25/00
246 Miller squeezed its directors: *New York Times* 5/5/02
247 stiffened Putin's inclination: *New York Times* 12/15/01
248 actually encouraged Russia: *New York Times* 11/20/01
248 Completion was set: *New York Times* 4/18/03
251 It won't be a problem: *New York Times* 10/28/04
254 a flagship for the Russian: *New York Times* 4/23/03
254 Putin was aggravated: *New York Times* 1/13/03
256 encouraged that Putin: *Financial Times* 9/19/03
260 Mulva had also mentioned: *Wall Street Journal* 9/24/03
261 We might see how: Goldman p. 114
265 We bring capital: *Wall Street Journal* 11/6/03

CHAPTER FIFTEEN: THE GAMBLE

270 Putin and his advisers: *New York Times* 1/3/04

271 It is now evident: http://www.peakoil.net/about-peak-oil

272 we conclude that: "The End of Cheap Oil," *Scientific American*, March 1998

272 close to its depletion: *Oil & Gas* 3/20/00

273 he had underestimated: CERA report 11/27/06

274 the rate of replacement: Centre for Global Energy Studies 11/6/06

274 after 50 percent of the oil: *Hubbert Revisited*

276 can be expected to be acute: wells.neftex@virgin.net 2/21/05

282 There are defects: Permanent Subcommittee on Investigations—US Strategic Petroleum Reserve 3/5/03

CHAPTER SIXTEEN: THE DOWNFALL

285 Prove it to us: 2005 CERA report p. 15, n. 33

286 too big to swallow: Polk Report p. 12

287 his own promotion: Ibid. p. 4

295 he had consolidated: *The Times* 10/13/03

295 He had even cut: *Wall Street Journal* 2/8/02

295 I did not know: *Wall Street Journal* 3/19/04

300 It has been a very: *The Times* 8/29/04

301 Our policy is the most: *Wall Street Journal* 1/11/04

301 ExxonMobil simply resisted: *Wall Street Journal* 11/22/04

CHAPTER SEVENTEEN: THE ALARM

309 from 8.6 billion barrels: *Financial Times* 3/30/04

310 BP had sold its shares: *New York Times* 1/13/04

310 The best is yet to come: *Daily Telegraph* 2/9/05

311 In 2004, Goldman Sachs: *New York Times* 1/15/06

CHAPTER EIGHTEEN: THE STRUGGLE

314 If I were dying: Sampson p. 196

315 What I was interested in: *Wall Street Journal* 6/3/99

315 O'Reilly agreed: *New York Times* 10/17/00

317 deposited $2.3 million: *Financial Times* 11/27/07

317 signature bonuses: *The Guardian* 3/24/04

318 Summoning the army: *Wall Street Journal* 5/28/99

319 Vitol would plead guilty: *Wall Street Journal* 11/15/07, 5/9/07, 10/11/04

321 a price O'Reilly said: *Wall Street Journal* 3/3/05, 4/11/05; *New York Times* 6/23/05

322 On the first day: *New York Times* 7/20/05, 8/11/05

323 That expansion: *Wall Street Journal* 11/22/02

323 It's about time: *New York Times* 11/16/02

325 The original plan: *Wall Street Journal* 9/5/03

CHAPTER NINETEEN: THE SURVIVOR

327   Even Exxon: *Financial Times* 11/12/07

329   legal irregularities: *Wall Street Journal* 8/28/97

330   had not deterred: *Financial Times* 11/12/07

332   You need to keep: Iwst 4/8/05

333   Some executives appeared: *New York Times* 11/17/05

333   Four weeks later: *Petroleum Review* January 2006

334   Most Americans think: Senate Commerce Committee Hearings 11/9/05, p. 45

334   Raymond was convinced: Ibid.

334   but the billions of dollars: *New York Times* 12/22/06; Department of the Interior report on Minerals Management Service

334   higher wholesale gasoline: GAO study, May 2004

335   Raymond had boasted: *New York Times* 3/30/06

336   the public's simmering distrust: *New York Times* 12/30/06

336   Nature, not man: 15th World Petroleum Congress, Beijing; *Wall Street Journal* 9/21/06

336   have yet to be proven: *Wall Street Journal* 8/29/01

336   I don't make moonshine: *Wall Street Journal* 10/14/97

338   he appeared to support: *New York Times* 1/7/03

338   the next he greeted: *New York Times* 7/24/04

339   Victory in a referendum: *New York Times* 8/17/04

339   It's absurd they were: *New York Times* 10/12/04

340   Simultaneously, he ordered: *New York Times* 5/9/05

341   In 2006, O'Reilly increased: *Wall Street Journal* 12/8/06

341   Raymond's public persona: "Hard Truths" p. 244

341   In 1997 he had urged: *Wall Street Journal* 10/14/97

CHAPTER TWENTY: THE BACKLASH

344   Only a fifth: *The Times* 6/24/02

346   An official investigation: *Mail on Sunday* 10/29/00

350   things not getting fixed: *Financial Times* 12/19/06

350   no passion, no curiosity: *Wall Street Journal* 12/12/06

352   Surprising and deeply: *Wall Street Journal* 5/18/05

352   in May 1988: *Wall Street Journal* 10/21/93

353   Single fatalities: *Wall Street Journal* 8/24/90, 4/29/91; *New York Times* 8/23/91

353   His announcement: *Wall Street Journal* 9/23/05

354   systemic lapses: *Financial Times* 8/18/05

354   he wanted to demolish: *Wall Street Journal* 10/31/06

355   I hate big corporations: *Wall Street Journal* 11/10/06

355   there was no evidence: *Wall Street Journal* 1/17/07

355   did not know which: *Wall Street Journal* 5/5/07

355   I wasn't aware: *The Guardian* 7/24/08

356   Texas City was: Speech to the American Society of Safety Engineers 11/26/07

357   and encouraged the: *Wall Street Journal* 4/6/06

358   the company paid Alaska: *Wall Street Journal* 11/18/94

359   The next hope: *The Times* 2/23/06

360   he did commission: *The Guardian* 7/26/01

361   We will never expose: BP Concerns/Comments on the Final Draft Coffman report, November 2001

362   without taking adequate: *Financial Times* 1/21/03

363   corporate greed: *Wall Street Journal* 12/20/05

365   Eight percent: *Daily Telegraph* 8/8/06

365   severe corrosion: *Wall Street Journal* 8/7/06

367   I am even more: *New York Times* 9/8/06

CHAPTER TWENTY-ONE: THE CONFESSION

374   The only certainty: *Wall Street Journal* 7/15/05

378   expressed distress: *New York Times* 10/6/06

379   international standards: cf Mike Bradshaw, "The Greening of Global Project Financing"

381   enjoyed reminding: *New York Times* 5/17/05

381   totally unfair: *Financial Times* 8/31/07

382   China had lent: *New York Times* 2/2/05

382   reaffirmed his commitment: *New York Times* 1/22/05

382   A new Russian: Prevnov06

382   Putin witnessed: *New York Times* 4/22/06, 3/27/07

383   but later agreed: *Wall Street Journal* 7/13/07

387   I'm pleased that: *New York Times* 12/12/06, 12/22/06

388   with sufficient reserves: IEA forecast, OECD, "Energy Technology Perspectives, 2008"

388   Shell was losing: *New York Times* 4/11/03

388   Shell paid ransoms: *New York Times* 2/19/06, 8/26/06

389   More than Shell's: *The Times* 2/1/08

390   Miller was on: *Financial Times* 1/5/08

390   quite aggressive: *Financial Times* 6/25/08, 7/10/08, 9/17/08

391   the volume of oil: *Financial Times* 2/1/08

391   self-confident image: *The Times* 2/1/08; *Financial Times* 3/18/08

392   The use of biofuels: *Financial Times* 1/21/08, 1/15/08

394   van der Veer questioned: *The Guardian* 8/31/07

394   actively involved if not: *Financial Times* 6/19/07

394   First, in December 2007: *The Times* 12/21/07; *Financial Times* 12/11/07

394   technology baths: *The Guardian* 3/18/09

395   sheer greed: *The Guardian* 5/2/08, 12/11/07, 2/21/08; *Financial Times* 12/6/07

395   In March 2008: *The Guardian* 3/18/08; Annual Strategy Review

CHAPTER TWENTY-TWO: THE OLIGARCH'S SQUEEZE

398   BP's reserves guaranteed: *Financial Times* 2/6/08

400   Browne's meeting with Putin: *Financial Times* 4/23/05

405 BP opened the bidding: Goldman p. 122

405 Hayward also agreed: *New York Times* 3/24/07

407 $70 million in fines: *Wall Street Journal* 10/26/07

407 Browne's penny-pinching: *Wall Street Journal* 10/11/07

407 I've had to eat: *Financial Times* 10/12/07

407 a "fundamental shift": *Financial Times* 10/12/07

414 are only starting the offensive: *Financial Times* 6/9/08

415 Russia is not yet ready: *Financial Times* 6/13/08

416 Their argument is commercial: *Financial Times* 7/7/08

418 up 56 percent: *Financial Times* 7/30/08

CHAPTER TWENTY-THREE: THE FRUSTRATED REGULATOR

419 Oil was still cheap: LehRes, March 2008

420 Over the previous five years: Michael Masters, Senate testimony 5/20/08

421 Matt Simmons predicted: *The Times* 6/30/08

422 Alexei Miller summoned the media: *The Times* 6/11/08

423 oil consumption in Western countries: OECD 6/18/07

424 Bush warned President Hu Jintao: *New York Times* 4/19/06

428 The number of energy hedge funds: Energy Hedge Fund Center

429 between $7 billion and $28 billion: *New York Times* 3/28/06

429 Only the independents: *New York Times* 4/27/06

429 illegal price manipulation: *Wall Street Journal* 5/10/06

430 The proportion of long-term: *Petroleum Economics Monthly*

431 the regulators assumed: *Wall Street Journal* 1/15/06

432 Four years later: US Wire 11/13/07

433 Five other traders were fined: CFTC press releases 5411-0, 5435-08

433 did not remember: *New York Times* 1/17/07

433 as much as $10 billion: *New York Times* 12/7/06, 12/30/06

433 Exxon insisted on: *New York Times* 1/13/07

434 Those officials, said Maxwell: *New York Times* 3/29/07

434 estimated to be $40 trillion: Barclays Cap

435 BP was fined $2.5 million: *Wall Street Journal* 8/29/06

435 warned BP it was taking action: *The Times* 8/11/07; *Daily Telegraph* 9/18/03

435 vulnerable to a squeeze: CFTC report p. 12

435 prices have been dropping: Ibid.

438 Now, said Campbell: *The Independent* 6/14/07

439 The national oil companies: *Financial Times* 11/1/07

439 We've invested $17 billion: *Financial Times* 7/2/08

441 it accumulated $40.9 billion: *Financial Times* 5/2/08

441 The news that Abu Dhabi: *The Times* 5/19/08

442 There's a lot of psychology: *Sunday Telegraph* 9/16/07

442 Other OPEC countries had also committed: *Financial Times* 11/16/07

443 In the marketplace: *Financial Times* 11/13/07

443   Suspicions caused by: *Financial Times* 1/23/08

443   would "scramble" to secure: *The Times* 1/25/08; RoyalDutchShellplc.com

443   He was rebuffed: *Financial Times* 2/1/08

443   We are idling: *The Times* 3/3/08; *Financial Times* 4/21/08

443   Refineries were earning less: *Wall Street Journal* 11/2/07

444   Demand rather than supply: *The Times* 1/18/08

444   frenzied reports: *Newsweek* special 2007; *Financial Times* 10/16/07

445   not in the business of: *Wall Street Journal* 8/20/08

445   had traded contracts: IHT 5/17/09, Oil Trading & Geneva

445   acknowledged that speculators: *Financial Times* 6/16/08

446   Ten percent of the price rise: Senate Energy Committee; *New York Times* 3/4/08

446   rush out of currencies: AP/IHT 3/4/08

446   speculators were paying $100: *The Times* 3/19/08

447   Outside of satanic cults: *New York Times* 5/3/06

448   Data produced by Barclays: *Financial Times* 6/5/08

448   A complicated matrix: Verleger, "The Anatomy of the Recent Oil Price Cycle," *Petroleum Economic Monthly*

449   OPEC countries were producing: Ibid.

450   Although about 20 billion barrels: *Financial Times* 2/26/08; *The Observer* 5/4/08

450   Prices have nowhere to go: *Financial Times* 6/13/08

451   Desai was convinced: *Financial Times* 7/4/08

451   much of that huge position: *Washington Post* 8/21/08

452   The facts were ignored: *The Economist* 10/5/08

453   rumors could only influence: *Wall Street Journal* 9/4/08

453   their model of balancing: *Financial Times* 7/4/08

454   supplies are tight: *New York Times* 8/23/08

458   The current global trends: IEA report 11/5/08

Index